Springer Series in Computational Mechanics

Edited by S. N. Atluri

W. B. Krätzig, E. Oñate (Eds.)

Computational Mechanics of Nonlinear Response of Shells

With 159 Figures

Springer-Verlag Berlin Heidelberg NewYork
London Paris Tokyo Hong Kong

Editor of the series:

Prof. S. N. Atluri
Georgia Institute of Technology
Center for the Advancement of Computational Mechanics
School of Civil Engineering
Atlanta, GA 30332
USA

Editors of this volume:

Prof. Dr.-Ing. Wilfried B. Krätzig
Institut für Statik und Dynamik
Ruhr-Universität Bochum
Universitätsstraße 150
4630 Bochum
Germany

Prof. Dr. Eugenio Oñate
Universidad Politècnica de Catalunya
Escola Tècnica Superior d'Enginyers de Caminos
Canales y Puertos
Jordi Girona Salgado, 31
08034 Barcelona
Spain

ISBN 3-540-52035-X Springer-Verlag Berlin Heidelberg New York
ISBN 0-387-52035-X Springer-Verlag New York Berlin Heidelberg

Offsetprinting: Color-Druck Dorfi GmbH, Berlin
Bookbinding: Lüderitz & Bauer, Berlin

2161/3020 543210

Preface

Shell structures and their components are utilized in a wide spectrum of engineering fields reaching from space and aircraft structures, pipes and pressure vessels over liquid storage tanks, off-shore installations, cooling towers and domes, to bodyworks of motor vehicles. Of continuously increasing importance is their nonlinear behavior, in which large deformations and large rotations are involved as well as nonlinear material properties.

Recently, theoretical and numerical research of nonlinear shell problems has accelerated considerably, with strong demands from industries. Concepts of continuum mechanics and the rising power of computing machines are striving to model the response behavior of these types of structures up to collapse, e.g. in the highly nonlinear range. This trend became very clear at the ICES'88 Conference in April 1988 in Atlanta, Georgia, where dozens of papers were aiming into this direction. The editors have the great pleasure to present in this volume a careful selection of highly distinguished contributions to this field, presented at the above mentioned conference.

The book starts with a survey about nonlinear shell theories from the rigorous point of view of continuum mechanics, this starting point being unavoidable for modern computational concepts. There follows a series of papers on nonlinear, especially unstable shell responses, which draw computational connections to well established tools in the field of static and dynamic stability of systems. Several papers are then concerned with new finite element derivations for nonlinear shell problems, and finally a series of authors contribute to specific applications opening a small window of the above mentioned wide spectrum.

As editors, we wish to express our gratitude to all contributors for their efforts in producing this volume in time, to the series-editor for his continuing scientific involvement and finally to the publisher for his friendly co-operation in numerous detailed questions during the preparation term. We sincerely hope this volume to be a stimulating reading to scientists as well as to practising engineers.

Wilfried B. Krätzig and Eugenio Oñate

Contents

Part 4: SPECIFIC APPLICATIONS OF NONLINEAR SHELL CONCEPTS

Part 1
Large Deformation and Large Rotation Theories

Introduction into Finite-Rotation Shell Theories and Their Operator Formulation

Y. BAŞAR, W.B. KRÄTZIG

Institut für Statik und Dynamik
Ruhr-Universität Bochum

Summary

In this paper a five-parametric shell theory valid for finite
rotations is derived and discussed under different aspects.
Its transformation into a KIRCHHOFF-LOVE type theory is demonstrated. The paper continues with the operator formulation of
this theory, highly suitable for numerical applications. This
formulation permits, in particular, a general discussion of
consistency properties. Finally, the discretization of the nonlinear operators by finite-elements is shown and examples are
given demonstrating the efficiency of finite-rotation shell
theories presented.

1. Introduction

Geometrically nonlinear shell theories permitting the analysis
of thin walled shell structures in the nonlinear range are be-
coming increasingly important because they can be transformed
into efficient numerical models, e.g. finite shell elements
(see e.g. [6,7,11]). Accordingly, the derivation of nonlinear
shell theories should be oriented to new aspects in close connection with numerical applications [3,4]. Firstly, we remind
the reader that nonlinear theories are, in general, accessible
to the numerical analysis in form of an incremental formulation. But the corresponding relations are always linear in the
unknown incremental variables, independently of the structure
of the initial nonlinear theory and its order of nonlinearity.
Thus, classical simplifications i.e. for the transformation of
nonrational terms into ordinary polynomials [16] are superfluous
and of no significant advantage for numerical applications. For
the development of numerical models applicable to arbitrary
nonlinear phenomena, theories for arbitrary large (**finite**) rotations, so-called **finite-rotation** theories, should be used per
preference.

The structure of a shell theory depends on the kinematic assumption used for the definition of two-dimensional shell variables. The KIRCHHOFF-LOVE hypothesis usually applied [5,11, 13,16] permits the avoidance of some numerical difficulties, such as shear locking, and an easy determination of the deformed unit normal vector, even in the deep nonlinear range [6,7,11]. On the other hand, boundary variables and rotational constraints to be satisfied are of a rather complicated structure and require a special care during the numerical analysis. In contrary, five-parametric theories [3,4,7,15] considering transversal shear deformations possess a more systematic structure and are of significant importance for the analysis of sandwich and composite shells. Thus, we prefer to start with a shear deformation theory and transform it at a later state into a KIRCHHOFF-LOVE-type theory.

Every shell theory has, of course, to satisfy the well-known requirements of consistency [1,2,3,10]. This can be fulfilled by a three-dimensional variational derivation leading, by virtue of GREEN's theorem, to formal **adjoint** kinematic and equilibrium operators.

The force variables introduced by the mentioned variational procedure are, in particular, for finite rotation problems not interpretable on two-dimensional shell element, in this sense they present **pseudo**-variables. For engineering applications, their transformation into geometrically interpretable ones seems unavoidable and this can be carried out if the theory in question contains the corresponding transformations [1,2,3]. Many existing formulations, however, take no notice of this important aspect.

The main quality requirements for the development of nonlinear shell theories now summarize as follows: Any theory should be applicable to finite-rotation problems, present a consistent formulation and permit the calculation of real force variables. The possibility to consider shear deformations is also suitable for some practical applications.

In this sense the aim of this paper is the presentation of finite-rotation shell theories satisfying the aspects discussed above. The paper shows also the operator formulation of the shear deformation theory and its transformation into finite-element displacement models.

2. Geometrical relations

The shell equations will be formulated in tensor notation which presents considerable advantages in the description of nonlinear phenomena as well as in the numerical modelling. Curvilinear convected coordinates Θ^α will be used to fix points of the middle surface. As usual, Greek indices represent the numbers 1, 2 while Latin ones stand for the numbers 1, 2, 3 . Let $\overset{\circ}{r} = \overset{\circ}{r}(\Theta^\alpha)$ be the position vector of a point $\overset{\circ}{P}$ of the undeformed middle surface $\overset{\circ}{F}$. Geometrical elements associated with $\overset{\circ}{P}$ will be denoted in usual manner [1] suffixed by $(\overset{\circ}{...})$, if referred to the undeformed state:

base vectors:
$$\overset{\circ}{a}_\alpha = \overset{\circ}{r}{,}_\alpha \ , \qquad \overset{\circ}{a}{}^\alpha = \overset{\circ}{a}{}^{\alpha\beta} \overset{\circ}{a}_\beta \ , \qquad (2.1)$$

metric tensor:
$$\overset{\circ}{a}_{\alpha\beta} = \overset{\circ}{a}_\alpha \cdot \overset{\circ}{a}_\beta \ , \qquad \overset{\circ}{a}{}^{\alpha\beta} = \overset{\circ}{a}{}^\alpha \cdot \overset{\circ}{a}{}^\beta \ , \qquad (2.2)$$

determinant:
$$\overset{\circ}{a} = \overset{\circ}{a}_{11} \overset{\circ}{a}_{22} - (\overset{\circ}{a}_{12})^2 \ , \qquad (2.3)$$

curvature tensor:
$$\overset{\circ}{b}_{\alpha\beta} = -\overset{\circ}{a}_\alpha \cdot \overset{\circ}{a}_{3,\beta} \ , \qquad \overset{\circ}{b}{}^\beta_\alpha = \overset{\circ}{b}_{\alpha\rho} \overset{\circ}{a}{}^{\rho\beta} \ . \qquad (2.4)$$

Herein, $\overset{\circ}{a}_3$ denotes the unit normal vector of $\overset{\circ}{F}$ and $(...),_\alpha$ partial derivatives with respect to $\overset{\circ}{F}$. Covariant derivatives with respect to $\overset{\circ}{F}$ will be denoted by $(...)|_\alpha$. In contrast to the geometrical elements of $\overset{\circ}{F}$, those associated with the deformed state F will be denoted without the suffix $(\overset{\circ}{...})$. Thus, $a_\alpha = r,_\alpha$ are the base vectors connected with the deformed position P of $\overset{\circ}{P}$.

Let Θ^3 be the distance of an arbitrary point $\overset{\circ}{P}{}^*$ of the undeformed shell continuum from $\overset{\circ}{F}$, measured in the direction of $\overset{\circ}{a}_3$. Thus, from the position vector of $\overset{\circ}{P}{}^*$ (Fig. 1)

$$\overset{\circ}{r}{}^* = \overset{\circ}{r} + \Theta^3 \overset{\circ}{a}_3 \qquad (2.5)$$

we find for the base vectors $\overset{\circ}{a}{}^{*}_{i}$ associated with $\overset{\circ}{P}{}^{*}$:

$$\overset{\circ}{a}{}^{*}_{\alpha} = \overset{\circ}{a}_{\alpha} + \overset{\circ}{\mu}{}^{\rho}_{\alpha}\,\overset{\circ}{a}_{\rho} \quad , \qquad \overset{\circ}{a}{}^{*}_{3} = \overset{\circ}{a}_{3} \quad , \tag{2.6}$$

where

$$\overset{\circ}{\mu}{}^{\rho}_{\alpha} = \delta^{\rho}_{\alpha} - \Theta^{3}\,\overset{\circ}{b}{}^{\rho}_{\alpha} \quad . \tag{2.7}$$

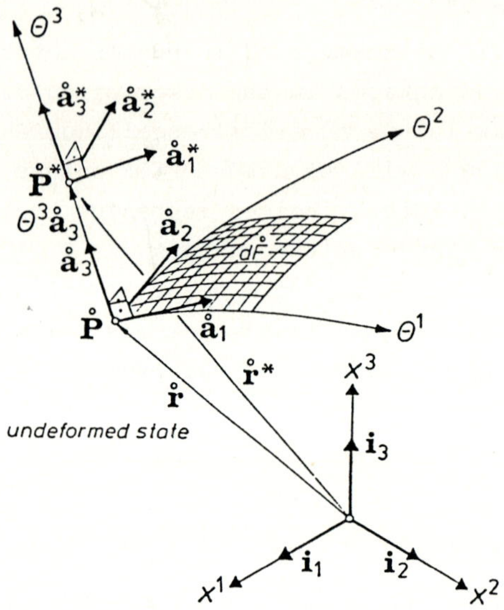

Fig. 1. The geometry of the shell continuum

In the following the notation $(...)^{*}$ characterizes variables related to the shell continuum, e.g. $\overset{\circ}{a}{}^{*}$ the determinant $|\overset{\circ}{a}{}^{*}_{\alpha\beta}|$ connected with $\overset{\circ}{P}{}^{*}$. Accordingly, the volume element of the shell is given by

$$d\overset{\circ}{V} = \sqrt{\overset{\circ}{a}{}^{*}}\,d\Theta^{1}d\Theta^{2}d\Theta^{3} \quad . \tag{2.8}$$

For later use we introduce along the boundary curve $\overset{\circ}{C}$ of $\overset{\circ}{F}$ the unit tangent vector $\overset{\circ}{t}$ and the unit outward normal $\overset{\circ}{u}$ by

$$\overset{\circ}{t} = \frac{d\overset{\circ}{r}}{d\overset{\circ}{s}} = \overset{\circ}{t}_{\alpha}\,\overset{\circ}{a}{}^{\alpha} \quad , \qquad \overset{\circ}{u} = \overset{\circ}{t} \times \overset{\circ}{a}_{3} = \overset{\circ}{u}_{\alpha}\,\overset{\circ}{a}{}^{\alpha} \quad . \tag{2.9}$$

These vectors satisfy the relations

$$\overset{\circ}{t}_{\alpha} = \overset{\circ}{\varepsilon}_{\beta\alpha}\,\overset{\circ}{u}{}^{\beta} \quad , \qquad \overset{\circ}{u}_{\beta} = \overset{\circ}{\varepsilon}_{\beta\alpha}\,\overset{\circ}{t}{}^{\alpha} \tag{2.10}$$

and form together with $\overset{\circ}{\mathbf{a}}_3$ a right-handed triad $(\overset{\circ}{\mathbf{u}}, \overset{\circ}{\mathbf{t}}, \overset{\circ}{\mathbf{a}}_3)$ to be used for the definition of physical boundary variables. In (2.10), $\overset{\circ}{\varepsilon}_{\alpha\beta}$ is the permutation tensor referred to $\overset{\circ}{F}$.

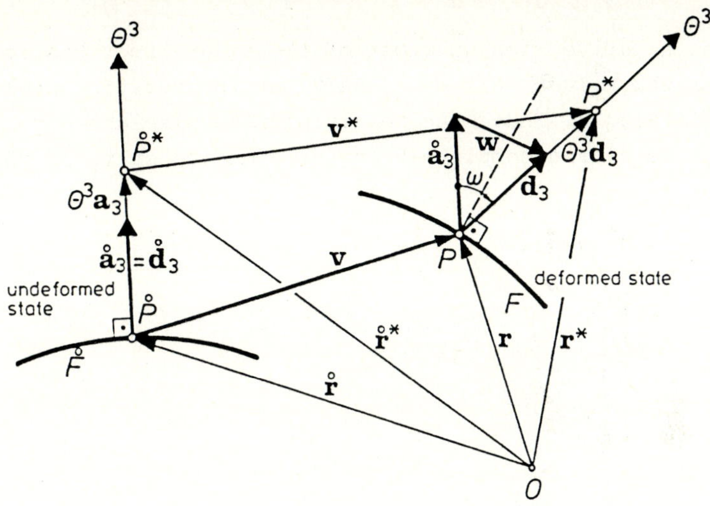

Fig. 2.   The deformation state

3.   The state of deformation

In order to introduce two-dimensional displacement variables we assume that the straight coordinate line Θ^3 remains straight during deformation and no change of length occurs in Θ^3-direction. Thus, the displacement vector $\mathbf{v}^*$ of the shell continuum can be expressed in terms of the displacement vector of the middle surface $\mathbf{v}$ and the difference vector $\mathbf{w}$ by (Fig. 2):

$$\mathbf{v}^* = \mathbf{r}^* - \overset{\circ}{\mathbf{r}}^* + \Theta^3 (\mathbf{d}_3 - \overset{\circ}{\mathbf{a}}_3) = \mathbf{v} + \Theta^3 \mathbf{w} \, , \qquad (3.1)$$

where $\mathbf{d}_3$ describes a unit vector in the direction of the deformed coordinate line Θ^3 . Both vectors $\mathbf{v}$ and $\mathbf{w}$ will be decomposed with respect to the base vectors $\overset{\circ}{\mathbf{a}}_i$ of the undeformed state:

$$\mathbf{v} = \mathbf{r} - \overset{\circ}{\mathbf{r}} = v_\alpha \overset{\circ}{\mathbf{a}}^\alpha + v_3 \overset{\circ}{\mathbf{a}}^3 \, , \qquad (3.2)$$

$$\mathbf{w} = \mathbf{d}_3 - \overset{\circ}{\mathbf{a}}_3 = w_\alpha \overset{\circ}{\mathbf{a}}^\alpha + w_3 \overset{\circ}{\mathbf{a}}^3 \, . \qquad (3.3)$$

In view of the definition $\mathbf{d}_3 \cdot \mathbf{d}_3 = 1$ the third component w_3 of $\mathbf{w}$ is a dependent variable subjected to the constraint

$$\mathbf{d}_3 \cdot \mathbf{d}_3 = 1 \;\rightarrow\; w_3(2+w_3) + w_\alpha w^\alpha = 0 \;\rightarrow\; w_3 = -1 \pm \sqrt{1-w_\alpha w^\alpha}\,, \qquad (3.4)$$

where the negative sign in front of the square root has to be taken for the values $\pi/2 \leq \omega \leq 3\pi/2$ of the rotation angle defined in Fig. 3. Thus, the remaining five displacements v_i and w_α determine completely the deformation state of the shell continuum.

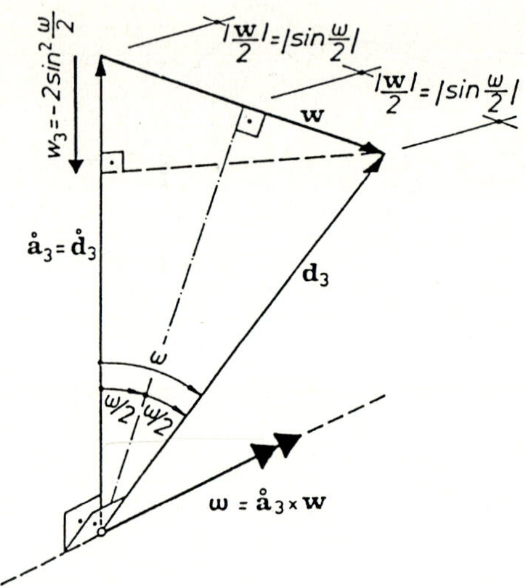

Fig. 3.   The rotation vector

For later use we introduce as an equivalent variable to the tangential part of $\mathbf{w}$ the rotation vector

$$\boldsymbol{\omega} = \omega_\alpha \, \mathring{\mathbf{a}}^\alpha = \mathring{\mathbf{a}}_3 \times \mathbf{w}\,, \qquad (3.5)$$

satisfying by virtue of (3.3) and (3.5) the transformations

$$\omega_\alpha = \mathring{\varepsilon}_{\beta\alpha}\, w^\beta\,, \qquad w_\alpha = \mathring{\varepsilon}_{\alpha\beta}\, \omega^\beta\,. \qquad (3.6)$$

From (3.5), we recognize the perpendicularity of $\boldsymbol{\omega}$ to the plane defined by the rotation of $\mathring{\mathbf{a}}_3$ into $\mathbf{d}_3$. Fig. 3 demonstrates furthermore that its magnitude is given in terms of the rotation angle ω by

$$|\omega| = |\overset{\circ}{a}_3||w||\sin(\tfrac{\pi}{2} - \tfrac{\omega}{2})| = 2|\sin\tfrac{\omega}{2}\cos\tfrac{\omega}{2}| = |\sin\,\omega| \ .$$

$$(3.7)$$

Again from Fig. 3 we deduce the transformation

$$w_3 = -2\sin^2\tfrac{\omega}{2} = -\tfrac{1}{2}\,w\cdot w = -\frac{\sin^2\omega}{2\cos^2\tfrac{\omega}{2}} = -\frac{\omega\,\cdot\,\omega}{\cos^2\tfrac{\omega}{2}}$$

$$(3.8)$$

and hence

$$1 + w_3 = \cos\,\omega.$$

By virtue of (3.5) and (3.8) the vector (3.3) can be expressed in terms of ω as

$$w = d_3 - \overset{\circ}{a}_3 = \omega\times\overset{\circ}{a}_3 - \frac{1}{2\cos^2\tfrac{\omega}{2}}\,(\omega\cdot\omega)\,\overset{\circ}{a}_3$$

$$(3.9)$$

$$= \omega\times\overset{\circ}{a}_3 + \frac{1}{2\cos^2\tfrac{\omega}{2}}\,\omega\times(\omega\times\overset{\circ}{a}_3) \ .$$

Hence, the vector ω is recognized as the rotation vector transforming the unit normal vector $\overset{\circ}{a}_3$ into its deformed position d_3 .

If we use the deformation gradients

$$\varphi_{\alpha\rho} = v_\rho|_\alpha - \overset{\circ}{b}_{\rho\alpha}\,v_3 \ , \quad \varphi_{\alpha 3} = v_{3,\alpha} + \overset{\circ}{b}{}^{\lambda}_{\alpha}\,v_\lambda \ , \quad (3.10)$$

$$\psi_{\alpha\rho} = w_\rho|_\alpha - \overset{\circ}{b}_{\rho\alpha}\,w_3 \ , \quad \psi_{\alpha 3} = w_{3,\alpha} + \overset{\circ}{b}{}^{\lambda}_{\alpha}\,w_\lambda \quad (3.11)$$

as abbreviations, we find from (3.2) and (3.3)

$$v_{,\alpha} = \varphi_{\alpha\rho}\,\overset{\circ}{a}{}^{\rho} + \varphi_{\alpha 3}\,\overset{\circ}{a}{}^{3} \ , \tag{3.12}$$

$$w_{,\alpha} = \psi_{\alpha\rho}\,\overset{\circ}{a}{}^{\rho} + \psi_{\alpha 3}\,\overset{\circ}{a}{}^{3} \ . \tag{3.13}$$

Thus, from (3.2) the base vectors a_α of the deformed middle surface F can be evaluated as

$$a_\alpha = \overset{\circ}{a}_\alpha + v_{,\alpha} = (\delta^\rho_\alpha + \varphi_{\alpha.}{}^{\rho})\overset{\circ}{a}_\rho + \varphi_{\alpha 3}\,\overset{\circ}{a}_3 \ , \tag{3.14}$$

while (3.3) leads to

$$d_3 = \overset{\circ}{a}_3 + w = w^\alpha\,\overset{\circ}{a}_\alpha + (1+w_3)\overset{\circ}{a}_3 \ . \tag{3.15}$$

In order to define two-dimensional strain measures we refer
to GREEN's strain tensor of the shell continuum defined by [1]:

$$\gamma_{ij} = \frac{1}{2} (\overset{\circ}{a}{}^*_i \cdot v^*_{,j} + \overset{\circ}{a}{}^*_j \cdot v^*_{,i} + v^*_{,i} \cdot v^*_{,j}) \ . \qquad (3.16)$$

Considering (2.6), (2.7), (3.1), (3.12) and (3.13) and neglecting quadratic terms in Θ^3 , we obtain for the tangential
components

$$\gamma_{\alpha\beta} = \alpha_{\alpha\beta} + \Theta^3 \beta_{\alpha\beta} \ , \qquad (3.17)$$

where the abbreviations

$$\alpha_{\alpha\beta} = \frac{1}{2} (\varphi_{\alpha\beta} + \varphi_{\beta\alpha} + \varphi_{\alpha\lambda} \varphi_{\beta.}^{\ \lambda} + \varphi_{\alpha 3} \varphi_{\beta 3}) \ , \qquad (3.18)$$

$$\beta_{\alpha\beta} = \frac{1}{2} (\psi_{\alpha\beta} + \psi_{\beta\alpha} - \overset{\circ}{b}{}^\lambda_\alpha \varphi_{\beta\lambda} - \overset{\circ}{b}{}^\lambda_\beta \varphi_{\alpha\lambda} + \varphi_{\beta.}^{\ \lambda} \psi_{\beta\lambda} +$$

$$+ \varphi_{\beta 3} \psi_{\alpha 3} + \varphi_{\alpha 3} \psi_{\beta 3}) \qquad (3.19)$$

denote the first and second strain tensor of the middle surface, respectively. Using in addition the identity

$$(\mathbf{d}_3 \cdot \mathbf{d}_3)_{,\alpha} = 0 \ \rightarrow \ \overset{\circ}{a}_3 \cdot \mathbf{w}_{,\alpha} - \overset{\circ}{b}{}^\rho_\alpha \overset{\circ}{a}_\rho \cdot \mathbf{w} + \mathbf{w}_{,\alpha} \cdot \mathbf{w} = 0 \ , \qquad (3.20)$$

we obtain similarly for $\gamma_{\alpha 3}$

$$\gamma_{\alpha 3} = \frac{1}{2} \gamma_\alpha = \frac{1}{2} \left[w_\rho (\delta^\rho_\alpha + \varphi_{\alpha.}^{\ \rho}) + \varphi_{\alpha 3}(1+w_3) \right] \ , \qquad (3.21)$$

where γ_α describes constant shear deformations throughout
the thickness of the shell. Finally, it can be proved by virtue of (3.4) that the component γ_{33} vanishes identically.
Thus, the deformation state is described entirely by the strain
measures (3.18), (3.19) and (3.21) as:

$$\gamma_{ij} = \left[\begin{array}{c|c} \gamma_{\alpha\beta} & \gamma_{\alpha 3} \\ \hline \gamma_{3\alpha} & \gamma_{33} \end{array} \right] = \left[\begin{array}{c|c} \alpha_{\alpha\beta} + \Theta^3 \beta_{\alpha\beta} & \frac{1}{2} \gamma_\alpha \\ \hline \frac{1}{2} \gamma_\alpha & 0 \end{array} \right] \ .$$

$$(3.22)$$

4. The principle of virtual work and the definition of consistent force variables

Two-dimensional internal force-variables will be called **consistent** if related to the **first** variation of the strain variables (3.18), (3.19) and (3.21) in the corresponding expression of internal virtual work. Their derivation will start from the three-dimensional expression of internal virtual work which is given in terms of the PIOLA-KIRCHHOFF stress tensor of second kind s^{ij} by

$$\delta^* A_i = - \iiint\limits_{\overset{\circ}{V}} \overset{\circ}{\mu}\, s^{ij}\, \delta \gamma_{ij}\, \sqrt{\overset{\circ}{a}}\; d\Theta^1 d\Theta^2 d\Theta^3 \quad , \tag{4.1}$$

where

$$\overset{\circ}{\mu} = \sqrt{\dfrac{\overset{\circ*}{a}}{\overset{\circ}{a}}} \quad . \tag{4.2}$$

Substituting (3.22) into (4.1) and abbreviating the surface element by $d\overset{\circ}{F} = \sqrt{\overset{\circ}{a}}\; d\Theta^1 d\Theta^2 d\Theta^3$ we obtain

$$\delta^* A_i = - \iint\limits_{\overset{\circ}{F}} (\tilde{N}^{(\alpha\beta)} \delta\alpha_{\alpha\beta} + \tilde{Q}^\alpha \delta\gamma_\alpha + M^{(\alpha\beta)} \delta\beta_{\alpha\beta})\, d\overset{\circ}{F} \quad , \tag{4.3}$$

defining the following **consistent** force variables:

$$\tilde{N}^{(\alpha\beta)} = \int\limits_{-h/2}^{h/2} \overset{\circ}{\mu}\, s^{\alpha\beta} d\Theta^3, \quad \tilde{Q}^\alpha = \int\limits_{-h/2}^{h/2} \overset{\circ}{\mu}\, s^{\alpha 3} d\Theta^3, \quad M^{(\alpha\beta)} = \int\limits_{-h/2}^{h/2} \overset{\circ}{\mu}\, s^{\alpha\beta} \Theta^3 d\Theta^3 \quad . \tag{4.4}$$

Herein, $\tilde{N}^{(\alpha\beta)}$ abbreviates the pseudo-stress resultant tensor, $\tilde{Q}^\alpha$ the pseudo-shear stress vector and $M^{(\alpha\beta)}$ the moment tensor, h denotes the shell thickness. Because of the well-known symmetry condition $s^{\alpha\beta} = s^{\beta\alpha}$ both variables $\tilde{N}^{(\alpha\beta)}$ and $M^{(\alpha\beta)}$ preserve symmetry indicated by round brackets.

In contrast to $M^{(\alpha\beta)}$, the force variables $\tilde{N}^{(\alpha\beta)}$ and $\tilde{Q}^\alpha$ are not physically interpretable on two-dimensional middle surface element and called therefore **pseudo-force** variables. For engineering applications, their transformation into geometrically interpretable variables $N^{\alpha\beta}$ and Q^α is necessary. This can be carried out by using nonlinear relations given i.e. in [3,4].

In contrast to the internal forces, load variables will be introduced by a two-dimensional procedure. Thus, we introduce the load vector $\mathbf{p}$ and the load moment vector $\mathbf{c}$, both referred to the unit area of the deformed middle surface F. The vector $\mathbf{c}$ is supposed to be perpendicular to $\mathbf{d}_3$. Using the factor $dF/d\overset{\circ}{F} = \sqrt{a/\overset{\circ}{a}}$ we decompose the load components as follows:

$$\sqrt{\frac{a}{\overset{\circ}{a}}}\,\mathbf{p} = p^\alpha \overset{\circ}{\mathbf{a}}_\alpha + p^3 \overset{\circ}{\mathbf{a}}_3 \quad, \quad \sqrt{\frac{a}{\overset{\circ}{a}}}\,\mathbf{c} = (1+w_3)c^\alpha \overset{\circ}{\mathbf{a}}_3 \times \overset{\circ}{\mathbf{a}}_\alpha + c^3 \overset{\circ}{\mathbf{a}}_3 \quad . \quad (4.5)$$

Herein, small letters indicate components p^i, c^i with respect to the undeformed base vectors $\overset{\circ}{\mathbf{a}}_i$. The virtual work of the loads (4.5) along the independent virtual displacements is, by virtue of (3.5) and (3.6), given by:

$$\delta^* A_{aP} = \iint\limits_{\overset{\circ}{F}} \left[\left(\sqrt{\frac{a}{\overset{\circ}{a}}}\,\mathbf{p}\right)\cdot\delta\mathbf{v} + \left(\sqrt{\frac{a}{\overset{\circ}{a}}}\,\mathbf{c}\right)\cdot\frac{\delta\boldsymbol{\omega}}{1+w_3} \right] d\overset{\circ}{F} = \iint\limits_{\overset{\circ}{F}} (p^i \delta v_i + c^\alpha \delta w_\alpha)\, d\overset{\circ}{F} \quad .$$
$$(4.6)$$

The virtual work expression for the load moment $\mathbf{c}$ can be confirmed by a three-dimensional derivation [3,4].

For convenience, we now restrict our attention to a boundary value problem, with given displacement variables along the whole boundary curve $\overset{\circ}{C}$ $(\delta v_i = \delta w_\alpha = 0)$. By virtue of (4.3) and (4.6), the principle of virtual work can, in this case, be formulated as follows:

$$\delta^* A = \delta^* A_a + \delta^* A_i = \iint\limits_{\overset{\circ}{F}} (p^i \delta v_i + c^\beta \delta w_\beta)\, d\overset{\circ}{F} - \iint\limits_{\overset{\circ}{F}} (\tilde{N}^{(\alpha\beta)} \delta\alpha_{\alpha\beta} + \tilde{Q}^\alpha \delta\gamma_\alpha$$
$$+ M^{(\alpha\beta)} \delta\beta_{\alpha\beta})\, d\overset{\circ}{F} = 0 \quad . \qquad (4.7)$$

5. The equilibrium equations and the boundary force variables

Equilibrium equations and force boundary conditions will now be derived from the principle of virtual work by application of standard rules of the calculus of variation. This permits to obtain relations which are in the same order of accuracy as the kinematic relations (3.18), (3.19) and (3.21) and which can be, therefore, regarded as a consistent formulation.

We start with the substitution of the kinematic relations
(3.18), (3.19) and (3.21) into (4.7) and express the deformation gradients φ_{ai} and ψ_{ai} according to (3.10) and (3.11).
We further eliminate the dependent displacement δw_3 from the
relation

$$\delta w_3(1+w_3) + w^\alpha \delta w_\alpha = 0 \quad \rightarrow \quad \delta w_3 = - \frac{w^\alpha}{1+w_3} \delta w_\alpha \ , \tag{5.1}$$

obtained from the constraint (3.4). Finally, we transform in
(4.7) all terms connected with covariant derivatives of the independent displacements δv_i and δw_α by means of GREEN's
theorem, for instance, in the form:

$$\iint\limits_{\overset{\circ}{F}} M^{(\alpha\beta)} \varphi_{\beta.}^{\ \lambda} \ \delta w_\lambda |_\alpha \ d\overset{\circ}{F} = \oint\limits_{\overset{\circ}{C}} M^{(\alpha\beta)} \varphi_{\beta.}^{\ \lambda} \ \delta w_\lambda \ \overset{\circ}{u}_\alpha \ d\overset{\circ}{s}$$

$$- \iint\limits_{\overset{\circ}{F}} (M^{(\alpha\beta)} \varphi_{\beta.}^{\ \lambda})|_\alpha \ \delta w_\lambda \ d\overset{\circ}{F} \ , \tag{5.2}$$

where $\mathbf{u} = \overset{\circ}{u}_\alpha \mathbf{a}^\alpha$ abbreviates the unit normal of $\overset{\circ}{C}$. This procedure leads to the following **consistent** equilibrium equations:

$$n^{\alpha\beta}|_\alpha - \overset{\circ}{b}^\beta_\alpha q^\alpha + p^\beta = 0 \ , \quad \overset{\circ}{b}_{\alpha\beta} n^{\alpha\beta} + q^\alpha|_\alpha + p^3 = 0 \ ,$$

$$m^{\alpha\beta}|_\alpha - q^{*\beta} + c^\beta = 0 \ , \tag{5.3}$$

where

$$n^{\alpha\beta} = \tilde{N}^{(\alpha\rho)} (\delta^\beta_\rho + \varphi_{\rho.}^{\ \beta}) - M^{(\alpha\rho)} (\overset{\circ}{b}^\beta_\rho - \psi_{\rho.}^{\ \beta}) + \tilde{Q}^\alpha w^\beta \ ,$$

$$q^\alpha = \tilde{Q}^\alpha (1+w_3) + \tilde{N}^{(\alpha\rho)} \varphi_{\rho 3} + M^{(\alpha\rho)} \psi_{\rho 3} \ ,$$

$$m^{\alpha\beta} = M^{(\alpha\rho)} l_{\rho.}^{\ \beta} \ , \tag{5.4}$$

$$q^{*\beta} = \tilde{Q}^\rho l_{\rho.}^{\ \beta} + M^{(\alpha\rho)} \left\{ \overset{\circ}{b}_{\alpha\lambda}(\delta^\lambda_\rho + \varphi_{\rho.}^{\ \lambda}) \frac{w^\beta}{1+w_3} + \left[\overset{\circ}{b}^\beta_\alpha - (\frac{w^\beta}{1+w_3})|_\alpha \right] \varphi_{\rho 3} \right\} \ ,$$

and

$$l_{\rho.}^{\ \beta} = (\delta^\beta_\rho + \varphi_{\rho.}^{\ \beta} - \frac{w^\beta}{1+w_3} \varphi_{\rho 3}) \ . \tag{5.5}$$

The LAGRANGIAN force variables $n^{\alpha\beta}$, q^α and $m^{\alpha\beta}$ in (5.3) are

14

defined with respect to the undeformed reference frame $\overset{\circ}{\mathbf{a}}_i$, and have been denoted, therefore, by small letters [1,3,4]. The variable $q^{*\beta}$ is, however, not interpretable in this sense. The above mentioned transformation leads, in addition, to the definition of the tensorial force variables

$$n^\beta = n^{\alpha\beta}\,\overset{\circ}{u}_\alpha \ , \quad n^3 = q^\alpha\,\overset{\circ}{u}_\alpha \ , \quad m^\beta = m^{\alpha\beta}\,\overset{\circ}{u}_\alpha \ , \tag{5.6}$$

prescribable along the boundary curve $\overset{\circ}{C}$. The corresponding virtual work is given by

$$\delta^* A_{aC} \;=\; \oint_{\overset{\circ}{C}} (n^\beta \delta v_\beta + n^3 \delta v_3 + m^\beta \delta w_\beta)\,d\overset{\circ}{s} \ . \tag{5.7}$$

Thus, the relation (4.7) can be extended as

$$\delta^* A = \iint_{\overset{\circ}{F}} (p^i \delta v_i + c^\beta \delta w_\beta)\,d\overset{\circ}{F} + \oint_{\overset{\circ}{C}} (n^\beta \delta v_\beta + n^3 \delta v_3 + m^\beta \delta w_\beta)\,d\overset{\circ}{s}$$

$$- \iint_{\overset{\circ}{F}} (\tilde{N}^{(\alpha\beta)} \delta \alpha_{\alpha\beta} + \tilde{Q}^\alpha \delta \gamma_\alpha + M^{(\alpha\beta)} \delta \beta_{\alpha\beta})\,d\overset{\circ}{F} \;=\; 0 \tag{5.8}$$

to shell problems with additionally prescribed force variables along $\overset{\circ}{C}$.

The tensorial boundary force variables (5.6) the definition of which is closely connected with the LAGRANGIAN variables (5.4) can be transformed further into physical force variables defined by

$$\frac{ds}{d\overset{\circ}{s}}\,\mathbf{n} = n_t\,\overset{\circ}{\mathbf{t}} + n_u\,\overset{\circ}{\mathbf{u}} + n_3\,\overset{\circ}{\mathbf{a}}_3 \ , \quad \frac{ds}{d\overset{\circ}{s}}\,\mathbf{m} = m_t\,\overset{\circ}{\mathbf{t}} + m_u\,\overset{\circ}{\mathbf{u}} + m_3\,\overset{\circ}{\mathbf{a}}_3 \ ,$$

$$\tag{5.9}$$

according to:

$$n_t = n^\beta\,\overset{\circ}{t}_\beta \ , \quad n_u = n^\beta\,\overset{\circ}{u}_\beta \ , \quad n_3 = n^3 \ ,$$

$$m_t = (1+w_3)m^\beta\overset{\circ}{u}_\beta \ , \quad m_u = -(1+w_3)m^\beta\overset{\circ}{t}_\beta \ . \tag{5.10}$$

In (5.9), $\mathbf{n}$ and $\mathbf{m}$ denote the stress resultant vector and the stress couple vector per unit length of the deformed boundary curve C , respectively. Due to the orthogonality of $\mathbf{m}$ to $\mathbf{d}_3$, the component m_3 is a dependent variable which can be expressed in terms of m_t and m_u as [4]:

$$m_3 = -\frac{w^\alpha}{1+w_3} \, (m_t \, \overset{\circ}{t}_\alpha + m_u \, \overset{\circ}{u}_\alpha) \; . \qquad (5.11)$$

It therefore can not be prescribed along the boundary curve $\overset{\circ}{C}$.

6. Constitutive equations

The relations established above, valid for arbitrary large (finite) displacements and rotations, shall now be completed by constitutive relations. For this purpose, we consider a HOOKEAN material with YOUNG's modulus E and POISSON's ratio ν . If we suppose sufficiently small strains (3.22) the corresponding constitutive equations read as [1,3,4]:

$$\tilde{N}^{(\alpha\beta)} = DH^{\alpha\beta\rho\lambda} \alpha_{\rho\lambda} \; , \quad \tilde{Q}^\alpha = Gh \, \overset{\circ}{a}^{\alpha\lambda} \gamma_\lambda , \quad M^{(\alpha\beta)} = BH^{\alpha\beta\rho\lambda} \beta_{\rho\lambda} ,$$

$$(6.1)$$

with the elastic coefficients

$$H^{\alpha\beta\rho\lambda} = \frac{1-\nu}{2} \, (\overset{\circ}{a}^{\alpha\rho} \, \overset{\circ}{a}^{\beta\lambda} + \overset{\circ}{a}^{\alpha\lambda} \, \overset{\circ}{a}^{\beta\rho} + \frac{2\nu}{1-\nu} \, \overset{\circ}{a}^{\alpha\beta} \, \overset{\circ}{a}^{\rho\lambda}), \quad (6.2)$$

and the stretching, bending and shear stiffness

$$D = \frac{Eh}{1-\nu^2} \; , \qquad B = \frac{Eh^3}{12(1-\nu^2)} \; , \qquad Gh = \frac{Eh}{2(1+\nu)} \; . \qquad (6.3)$$

7. Kirchhoff-Love-type theory of finite rotations

The shear deformation theory presented above with five independent displacements v_i and w_α can easily be transformed into a finite-rotation theory of KIRCHHOFF-LOVE type. For this purpose we postulate the vector d_3 (3.1) to be perpendicular to the deformed middle surface F . This leads by means of (3.3) and (3.14) to:

$$a_\alpha \cdot d_3 = 0 \quad \rightarrow \quad w_\rho \, (\delta^\rho_\alpha + \varphi_{\alpha.}{}^\rho) + (1+w_3) \, \varphi_{\alpha 3} = 0 \; , \quad (7.1)$$

while the condition (3.4) remains unchanged:

$$d_3 \cdot d_3 = 1 \quad \rightarrow \quad w_3 \, (2+w_3) + w_\alpha \, w^\alpha = 0 \; . \qquad (7.2)$$

These equations can be, however, replaced equivalently by the relations [1,5,7]

$$w_\beta = -\rho^{-\frac{1}{2}}(\varphi_{\beta 3} + \delta_{\alpha\beta}^{\lambda\rho} \varphi_{\lambda .}^{\ \ \alpha} \varphi_{\rho 3}) \, ,$$

$$w_3 = \rho^{-\frac{1}{2}} (1+\varphi_{\rho .}^{\ \ \rho} + \frac{1}{2} \delta_{\rho\lambda}^{\alpha\beta} \varphi_{\alpha .}^{\ \ \rho} \varphi_{\beta .}^{\ \ \lambda} - 1) \qquad (7.3)$$

$$\rho \equiv \frac{a}{\overset{\circ}{a}} = 1 + 2 (a_\lambda^\lambda + \delta_{\lambda\mu}^{\alpha\beta} a_\alpha^\lambda a_\beta^\mu) \, , \qquad (7.4)$$

which express the fact that d_3 is now equivalent to the unit normal vector a_3 of the deformed middle surface F . We emphasize that the simultaneous use of the conditions (7.1), (7.2) and (7.3), (7.4) presents siginificant advantages for finite-element formulations [6,7].

The kinematic relations (3.18), (3.19) preserve their validity while the shear deformations γ_α (3.21) vanish by virtue of (7.1). By means of the well-known identity

$$a_\alpha\big|_\beta \cdot a_3 = -a_\alpha \cdot a_3\big|_\beta \qquad (7.5)$$

the kinematic relation (3.19) may, however, be replaced by the following one [5]:

$$\omega_{\alpha\beta} = -\big[(\varphi_{\alpha 3}\big|_\beta + \overset{\circ}{b}_{\rho\beta}\varphi_{\alpha .}^{\ \ \rho})(1+w_3) + \overset{\circ}{b}_{\alpha\beta}w_3 + (\varphi_{\alpha .}^{\ \ \rho}\big|_\beta - \overset{\circ}{b}_\beta^\rho \varphi_{\alpha 3})w_\rho\big] \, ,$$

$$(7.6)$$

which is more advantageous for finite element formulations since the difference vector w_i does not occur as derivative [6,7]. In (7.6), the notation $\beta_{\alpha\beta}$ is changed into $\omega_{\alpha\beta}$ in order to emphasize the dependence of the second strain tensor solely on v_i within a KIRCHHOFF-LOVE type theory.

In contrast to the shear deformation theory in which five boundary conditions are required, the present theory permits the prescription of four boundary conditions. The definition of the corresponding rotational and force variables is detailed in [5]. The expression given in (5.7) for the virtual work of the boundary forces can, however, be used in the present theory too, provided the constraints (7.1), (7.2) are satisfied also along the boundary curve $\overset{\circ}{C}$. In this case the principle of virtual work (5.7) reads as follows:

$$\delta^* A = \iint\limits_{\overset{\circ}{F}} p^i \delta v_i \, d\overset{\circ}{F} + \oint\limits_{\overset{\circ}{C}} (n^\beta \delta v_\beta + n^3 \delta v_3 + m^\beta \delta w_\beta) d\overset{\circ}{s}$$

$$- \iint\limits_{\overset{\circ}{F}} (\tilde{N}^{(\alpha\beta)} \delta\alpha_{\alpha\beta} + M^{(\alpha\beta)} \delta\omega_{\alpha\beta}) d\overset{\circ}{F} = 0 \quad , \qquad (7.7)$$

where, in view of the dependent displacements w_α , the load moment vector c^α has been neglected.

8. Operator formulation and the characteristics of consistency

The operator formulation introduced in this section permits a short presentation of nonlinear shell theories and, in particular, the enlightenment of characteristic properties of consistency. Furthermore, the rules to be obtained on this basis are of great usefulness for any systematical derivation of the equilibrium equations and the force boundary conditions satisfying a priori the consistency requirements. Finally, operator formulations gain large benefits in the development of finite elements.

loads
$$\mathbf{p}^* = \begin{bmatrix} p^\lambda \\ \hline p^3 \\ \hline c^\lambda \end{bmatrix}$$

displacements
$$\mathbf{u} = \begin{bmatrix} v_\lambda \\ \hline v_3 \\ \hline w_\lambda \end{bmatrix}$$

internal forces
$$\boldsymbol{\sigma} = \begin{bmatrix} \tilde{N}^{(\alpha\beta)} \\ \hline \tilde{Q}^\alpha \\ \hline M^{(\alpha\beta)} \end{bmatrix}$$

strains
$$\boldsymbol{\varepsilon} = \begin{bmatrix} \alpha_{\alpha\beta} \\ \hline \gamma_\alpha \\ \hline \beta_{\alpha\beta} \end{bmatrix}$$

tensorial boundary forces
$$\mathbf{T}_r = \begin{bmatrix} n^\lambda \\ \hline n^3 \\ \hline m^\lambda \end{bmatrix}$$

tensorial boundary displacements
$$\mathbf{u} = \begin{bmatrix} v_\lambda \\ \hline v_3 \\ \hline w_\lambda \end{bmatrix}$$

Table 1. Definition of the mechanical variables as column vectors

For our purpose we introduce the following column vectors detailed in Table 1:

$\overset{*}{p}$: loads, u : displacements,

σ : internal forces, ε : strains

τ_r : tensorial boundary forces $u_r = u$ tensorial boundary
                                                         displacements

Each of these vectors collect variables of the same group. Components presented on the same line are dual in the sense that they define virtual work contribution (8.1). By means of Table 1 the principle of virtual work (5.8) can be transformed into the short form:

$$\overset{*}{\delta}A \;=\; \iint\limits_{\overset{\circ}{F}} \delta u^T \overset{*}{p}\, d\overset{\circ}{F} \;+\; \int\limits_{\overset{\circ}{C}} \delta u^T \tau_r\, d\overset{\circ}{s} \;-\; \iint\limits_{\overset{\circ}{F}} \delta \varepsilon^T \sigma\, d\overset{\circ}{F} \;=\; 0 \;. \tag{8.1}$$

We now deal with the question if the nonlinear field equations can be transformed into a suitable operator formulation. In contrast to linear theories [1], the definition of the kinematic operator D_k will be achieved not on the basis of the kinematic relations (3.18), (3.19) and (3.21) themselves, but on that of their first variation with respect to the independent displacements v_i and w_α . Eliminating the dependent variations $\delta\varphi_{\alpha i}$, $\delta\psi_{\alpha i}$ and δw_3 from (3.10), (3.11) and (5.1), this procedure leads to

$$\delta\varepsilon \;=\; D_k\,\delta u \;, \tag{8.2}$$

with the **kinematic operator** D_k presented in Table 2.

The notation $d_\alpha = (\ldots)|_\alpha$ in Table 2 denotes covariant derivatives with respect to the undeformed state $\overset{\circ}{F}$ and applies to the whole following expression so that e.g.:

$$d_\alpha\, w^\lambda\, \tilde{Q}^\alpha \;=\; (w^\lambda\, \tilde{Q}^\alpha)|_\alpha \;=\; w^\lambda|_\alpha\, \tilde{Q}^\alpha + w^\lambda\, \tilde{Q}^\alpha|_\alpha \;. \tag{8.3}$$

In the representation of the column vectors (Table 1) and the operator D_k (Table 2) the indices are chosen such that, by writing out the relation (8.2), tensorial equations are to be obtained with the usual summation convention. Similarly, the equlibrium equations (5.3) can be, by means of (5.4), (5.5), transformed into the short form

$$-\overset{*}{p} \;=\; D_e\,\sigma \;, \tag{8.4}$$

The first variation of the kinematic relations: $\delta\boldsymbol{\epsilon} = \mathbf{D}_k\,\delta\mathbf{u}$

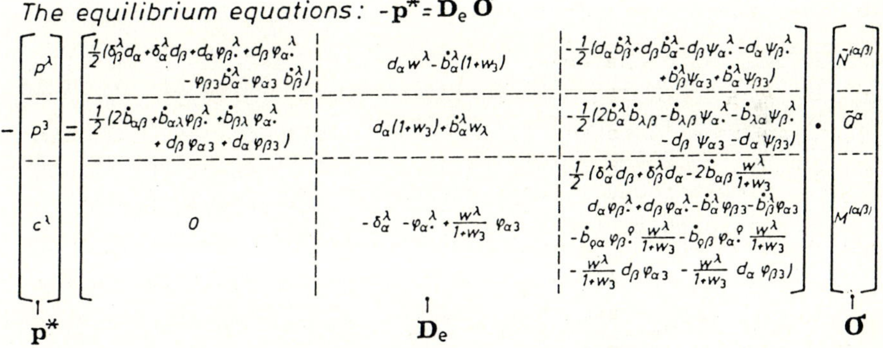

The equilibrium equations: $-\mathbf{p}^* = \mathbf{D}_e\,\boldsymbol{\sigma}$

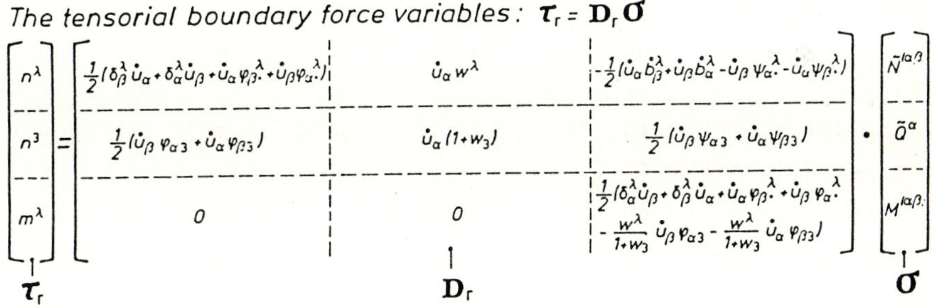

The tensorial boundary force variables: $\boldsymbol{\tau}_r = \mathbf{D}_r\,\boldsymbol{\sigma}$

$$
\begin{bmatrix} n^\lambda \\ n^3 \\ m^\lambda \end{bmatrix}
=
\begin{bmatrix}
\tfrac{1}{2}(\delta^\lambda_\beta \dot{u}_\alpha + \delta^\lambda_\alpha \dot{u}_\beta + \dot{u}_\alpha \varphi_{\beta\cdot}^{\ \lambda} + \dot{u}_\beta \varphi_{\alpha\cdot}^{\ \lambda}) & \dot{u}_\alpha w^\lambda & -\tfrac{1}{2}(\dot{u}_\alpha \dot{b}_\beta^\lambda + \dot{u}_\beta \dot{b}_\alpha^\lambda - \dot{u}_\beta \psi_\alpha^{\ \lambda} - \dot{u}_\alpha \psi_\beta^{\ \lambda}) \\
\tfrac{1}{2}(\dot{u}_\beta \varphi_{\alpha3} + \dot{u}_\alpha \varphi_{\beta3}) & \dot{u}_\alpha(1+w_3) & \tfrac{1}{2}(\dot{u}_\beta \psi_{\alpha3} + \dot{u}_\alpha \psi_{\beta3}) \\
0 & 0 & \tfrac{1}{2}(\delta^\lambda_\alpha \dot{u}_\beta + \delta^\lambda_\beta \dot{u}_\alpha + \dot{u}_\alpha \varphi_{\beta\cdot}^{\ \lambda} + \dot{u}_\beta \varphi_{\alpha\cdot}^{\ \lambda}) - \tfrac{w^\lambda}{1+w_3}\dot{u}_\beta \varphi_{\alpha3} - \tfrac{w^\lambda}{1+w_3}\dot{u}_\alpha \varphi_{\beta3})
\end{bmatrix}
\cdot
\begin{bmatrix} \bar{N}^{(\alpha\beta)} \\ \bar{Q}^\alpha \\ M^{(\alpha\beta)} \end{bmatrix}
$$

with $\boldsymbol{\tau}_r$, $\mathbf{D}_r$, $\boldsymbol{\sigma}$.

Table 2. Definition of the operators $\mathbf{D}_k$, $\mathbf{D}_e$ and $\mathbf{D}_r$

where D_e (Table 2) abbreviates the **equilibrium operator**. Considering again (5.4), (5.5) the tensorial boundary force variables (5.6) finally take the form

$$\tau_r = D_r \, \sigma \, ,$$ (8.5)

with the **boundary operator** D_r , also presented in Table 2. By means of GREEN's theorem (5.2), it can be shown that

$$\iint\limits_{\overset{\circ}{F}} \sigma^T \, D_k \, \delta u \, d\overset{\circ}{F} = - \iint\limits_{\overset{\circ}{F}} \delta u^T \, D_e \, \sigma \, d\overset{\circ}{F} + \oint\limits_{\overset{\circ}{C}} \delta u^T \, D_r \, \sigma \, d\overset{\circ}{s} \, .$$ (8.6)

This relation implies the operators D_k and D_e to be formal **adjoint**, the essential characteristic property of consistent formulations. By comparison of the operators from Table 2 we can easily deduce the simple transformation rules of D_k into D_e or vice versa. In order to obtain D_e , the operator D_k has to be transposed according to the appropriate rules of matrices. Simultaneously, each element of D_k has to be transformed according to the following rules [3]:

$$
\begin{array}{ccc}
F \, d_\alpha \, S & S \, d_\alpha \, F & S \, \overset{\circ}{u}_\alpha \, F \\[4pt]
F \, d_\alpha \longrightarrow & d_\alpha \, F \longrightarrow & \overset{\circ}{u}_\alpha \, F \\[4pt]
F & -F & 0 \\[4pt]
\underline{\qquad} & \underline{\qquad} & \underline{\qquad} \\[2pt]
D_k & D_e & D_r
\end{array}
$$ (8.7)

which are a consequence of GREEN's theorem (5.2) and the sign convention adopted for the definition of D_e (8.4). In (8.7), F and S are arbitrary tensorial functions presented for generality without indices. The first two rules are valid for terms involving the differential operator d_α and the last one for terms without d_α . Furthermore, the second rule can be regarded as a special case of the first one, where $S \to 1$. Obviously, all elements entering in D_k and D_e are of one of the given types. Since, in an actual expression, the order of d_α is of significance, all rules can be - apart from the change of sign in the last one - interpreted as transposition of matrix products. In this sense, the operator D_k (or D_e) consists solely of transposed submatrices of D_e (or D_k).

Finally, the boundary force operator D_r can be obtained from D_e if we replace in D_e all elements without d_α by zero and transform the other ones according to (8.7), where $\overset{\circ}{u}_\alpha$ is the unit outward normal of $\overset{\circ}{C}$ (2.9).

The operator formulation introduced here for a shear deformation theory is, of course, applicable to KIRCHHOFF-LOVE type theories. In this case the resulting operators are slightly more complicated than the present ones. This is due to the supplementary constraints (7.1) to be considered in the definition of the kinematic operator. The corresponding results are discussed in detail in [5].

We finally transform the constitutive relations (6.1) valid for a HOOKEAN material into an operator form to obtain:

$$\sigma = D\, \mathbf{E}\, \epsilon \ , \tag{8.8}$$

with the elastic matrix $\mathbf{E}$ defined as:

$$\underbrace{\begin{bmatrix} \tilde{N}^{(\alpha\beta)} \\[4pt] \tilde{Q}^{\alpha} \\[4pt] M^{(\alpha\beta)} \end{bmatrix}}_{\sigma} = D\ \underbrace{\begin{bmatrix} H^{\alpha\beta\rho\lambda} & 0 & 0 \\[6pt] 0 & \dfrac{1-\nu}{2}\,\overset{\circ}{a}{}^{\alpha\lambda} & 0 \\[6pt] 0 & 0 & \dfrac{h^2}{12}\,H^{\alpha\beta\rho\lambda} \end{bmatrix}}_{E} \cdot \underbrace{\begin{bmatrix} \alpha_{\rho\lambda} \\[4pt] \gamma_{\lambda} \\[4pt] \beta_{\rho\lambda} \end{bmatrix}}_{\epsilon} \tag{8.9}$$

9. Multiplicative decomposition of nonlinear operators

We now raise the question if the nonlinear operators D_k and D_e can be decomposed multiplicatively into a linear and nonlinear part. To this we introduce, according to Table 3, the following new column vectors:

φ : **deformation gradients**, τ : **Lagrangian force variables**,

which are obviously **dual** variables (9.7). The relations (3.10) permit the transformation of $\delta\varphi$ into the displacements δu , while τ can be, through the equilibrium equations (5.3), transformed into the loads p^* . This may be expressed by

$$\delta\varphi = T_k\, \delta u \ , \qquad -p^* = T_e\, \tau \ , \tag{9.1}$$

The first variation of the kinematic relations: $\delta\varepsilon = \mathbf{D}\,\delta\varphi = \mathbf{D}\mathbf{T}_k\,\delta\mathbf{u}$

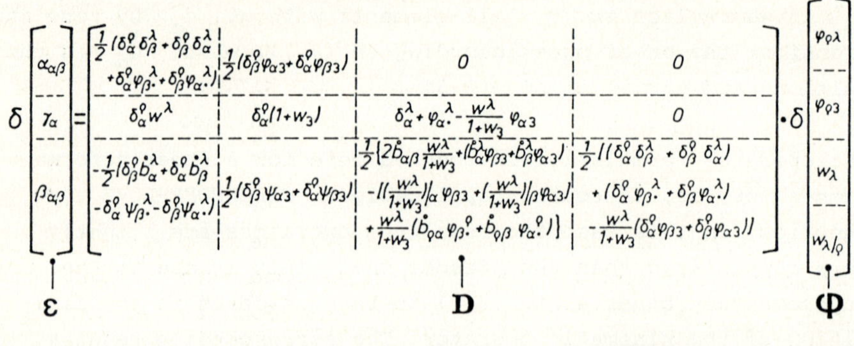

Equilibrium equations: $-\mathbf{p}^* = \mathbf{T}_e\,\tau = \mathbf{T}_e\,\mathbf{D}^T\sigma$

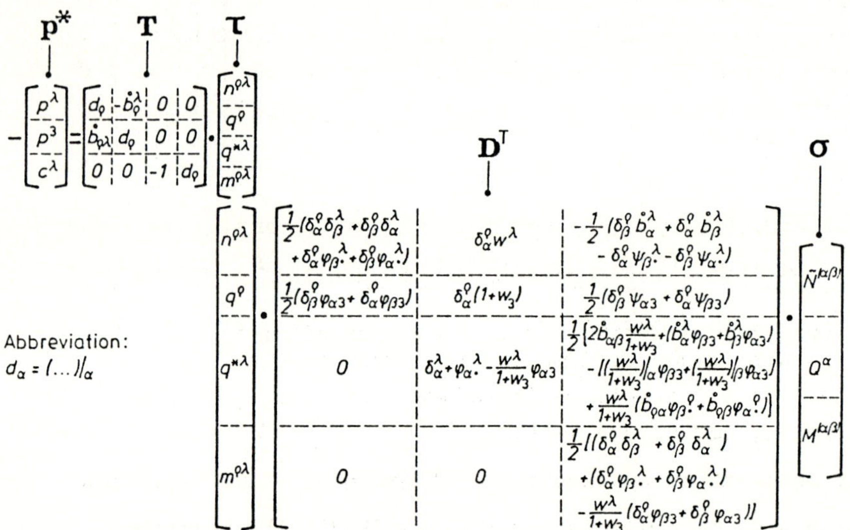

Table 3. Multiplicative decomposition of the operators $\mathbf{D}_k$
          and $\mathbf{D}_k$

in terms of purely linear operators T_k and $\overset{\circ}{T}_e$ depending on the operator d_α and the curvature tensor $b_{\alpha\beta}$ (Table 3). Similarly, LAGRANGIAN force variables (5.4) and, by virtue of (3.11), (5.1), the first variation of the kinematic relations (3.18), (3.19) and (3.21) can be transformed into

$$\delta\varepsilon \;=\; D\,\delta\varphi\,, \qquad \tau \;=\; D^T\,\sigma\,, \tag{9.2}$$

with a matrix D (Table 3) depending on the displacement u, but not on the operator d_α. Its transposed form is, as usual, denoted by D^T. Combining the equations (9.1) and (9.2), we obtain

$$\delta\varepsilon \;=\; D\,T_k\,\delta u\,, \qquad -p^* \;=\; T_e\,D^T\,\sigma\,, \tag{9.3}$$

and hence by comparison with (8.2) and (8.4)

$$D_k \;=\; D\,T_k\,, \qquad D_e \;=\; T_e\,D^T\,. \tag{9.4}$$

Each of the above decompositions is characterized by complete separation of the operator d_α and the displacements u occuring together in the inial operators D_k and D_e from each other. The displacements u appear only in the matrix D and the operator d_α solely in the linear operators T_k and T_e. A similar decomposition of the operator D_r

$$D_r \;=\; T_r\,D^T \tag{9.5}$$

defines finally the linear boundary operator T_r to be obtained from T_e by the same procedure as D_r from D_e.

Similar to D_k, D_e and D_r, the operators T_k, T_e and T_r are related to each other by

$$\iint\limits_{\overset{\circ}{F}} \tau^T\,T_k\,\delta u\,d\overset{\circ}{F} \;=\; -\iint\limits_{\overset{\circ}{F}} \delta u^T\,T_e\,\tau\,d\overset{\circ}{F} \;+\; \oint\limits_{\overset{\circ}{C}} \delta u^T\,T_r\,\tau\,d\overset{\circ}{s}\,, \tag{9.6}$$

which can be deduced from (8.5) using (9.2), (9.4) and (9.5). This relation defines the operators T_k and T_e as **adjoint** ones. Consequently, T_k can be transformed into T_e by the procedure $D_k \rightarrow D_e$ as can be easily confirmed from Table 3. It should be stressed that the multiplicative decompositions

24

(9.4) and (9.5) may help to reduce the computational efforts
in incremental-iterative treatments of nonlinear problems significantly.

In closing we present the operator relations established above
in form of a structure diagram (Fig. 4). The equations appearing on the right side of the diagram are the constraints to be
satisfied by the stain measures ε or φ in the principle of
virtual work. Their transformations by means of GREEN's theorem
(8.6) or (9.6) leads to the **natural** conditions presented on the
left side, namely, to equilibrium equations and dynamic boundary conditions.

According to (8.2) and (9.1), (9.2) the internal virtual work
(8.1) may be expressed alternatively by

$$\delta^* A_i \;=\; -\iint\limits_{\overset{\circ}{F}} \sigma^T \delta\varepsilon\, d\overset{\circ}{F} \;=\; -\iint\limits_{\overset{\circ}{F}} \sigma^T D_k \delta u\, d\overset{\circ}{F} \;=\; -\iint\limits_{\overset{\circ}{F}} \sigma^T D T_k \delta u\, d\overset{\circ}{F}$$

$$=\; -\iint\limits_{\overset{\circ}{F}} \tau^T \delta\varphi\, d\overset{\circ}{F} \;, \tag{9.7}$$

showing the duality of the LAGRANGIAN forces τ and the deformation gradients φ .

10. Finite-element-displacement models

Our final goal is to apply the relations established above to a
finite-element formulation of displacement shell models. Because incremental-iterative solution strategies are employed
for the nonlinear responses incremental models are desired.
Their derivation can be performed by a variational procedure
[2] according to which the nonlinear principle (8.1) takes the
form:

$$\delta^* \overset{+}{A}{}^+ = \iint\limits_{\overset{\circ}{F}} \overset{+}{p}{}^{*T}\, \delta\overset{+}{u}\, d\overset{\circ}{F} + \oint\limits_{\overset{\circ}{C}} \overset{+}{t}{}^T_r\, \delta\overset{+}{u}\, d\overset{\circ}{s} - \iint\limits_{\overset{\circ}{F}} (\overset{+}{\sigma}{}^T D_k + \sigma^T \overset{+}{D}_k)\, \delta\overset{+}{u} = \bar{\delta}(\delta^* A) \tag{10.1}$$

In (10.1), the notation $(.\overset{+}{.}.)$ characterizes all variables depending of the incremental displacements $\overset{+}{u} = \delta u$ defined as a

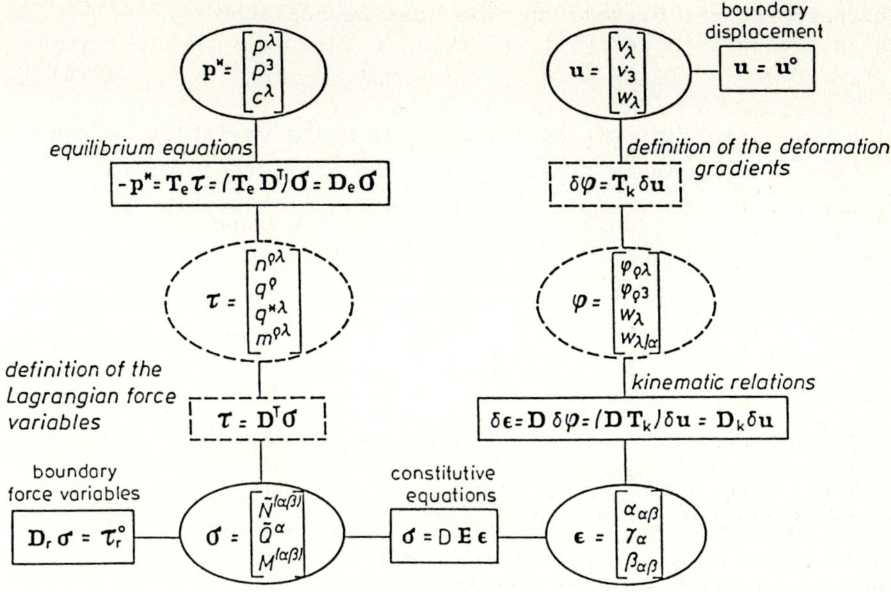

Fig. 4. Structure diagram of the nonlinear theory

special variation of u. The symbol $\bar{\delta}$ implies a variation with respect to $\overset{+}{u}$ and the operator $\overset{+}{D}_k = \delta D_k$ is the first variation of D_k with respect to the initial displacements $\overset{*}{u}$. Finally, $\bar{\delta}\,(\delta\,\overset{*}{}A)$ represents the corrector term to be iterated to zero during the iteration procedure.

For a suitable transformation of the internal virtual work $\bar{\delta}\,\overset{*+}{A}_i$ we refer to (8.2), (8.8) and (9.4) in order to write:

$$\overset{+T}{\sigma} D_k \overset{\bar{}+}{\delta u} = (D_k \overset{+}{u})^T D E D_k \overset{\bar{}+}{\delta u} = (T_k \overset{+}{u})^T D^T D E D T_k \overset{\bar{}+}{\delta u} \,,$$
$$\tag{10.2}$$

$$\sigma^T \overset{+}{D}_k \overset{\bar{}+}{\delta u} = (\overset{+T}{D} \sigma)^T T_k \overset{\bar{}+}{\delta u} \,,\tag{10.3}$$

where $\overset{+}{D}$ is the first variation of D with respect of u.

Furthermore, the expression $\overset{+T}{D}\sigma$ can be replaced by

$$\overset{+T}{D}\,\sigma \;=\; D_g^T\,T_k\,\overset{+}{u}\;, \tag{10.4}$$

with D_g depending of the fundamental state variables σ and u only. Its symmetry characterizes the consistency of the theory used. With the transformations (10.2) to (10.4) the internal virtual work in (10.1) takes the form:

$$\delta\overset{*++}{A_i} \;=\; -\iint\limits_{\overset{\circ}{F}} [(T_k\,\overset{+}{u})^T\,D^T\,D\,E\,D\,T_k + (T_k\,\overset{+}{u})^T\,D_g\,T_k]\,\delta\overset{+}{u}\;d\overset{\circ}{F}\;. \tag{10.5}$$

According to the standard displacement discretization, we approximate the independent displacements $\overset{+}{u}$ by the shape functions Ω and the nodal displacements $\overset{+}{v}$:

$$\overset{+}{u} \;=\; \Omega\,\overset{+}{v}\;. \tag{10.6}$$

Thus, the relation (10.5) becomes

$$\delta\overset{*++}{A_i} \;=\; \overset{+T}{v}\,(k_e + k_g)\,\delta\overset{+}{v}\;,$$

with the well-known **elastic** k_e and **geometric stiffness matrix** k_g

$$k_e \;=\; H^T\,D^T\,D\,E\,D\,H\;, \qquad k_g \;=\; H^T\,D_g\,H\;, \tag{10.7}$$

where

$$H \;=\; T_k\,\Omega\;. \tag{10.8}$$

The matrix H is a purely linear one depending on the linear operator T_k and the shape functions Ω adopted for the discretization. Thus, the matrix H is to be constructed once during the whole calculation. In contrary, the nonlinear matrix D must be regenerated at each new iteration step. This advantageous separation of the discretization procedure from the iterative procedure is due to the multiplicative decompositions (9.4).

11. Numerical Results

The finite-rotation shell theories presented above have been
transformed into finite-element models (finite-rotation elements) using a tensor-oriented procedure [6,7]. In the following, two examples are presented to demonstrate their applicability to strongly nonlinear phenomena.

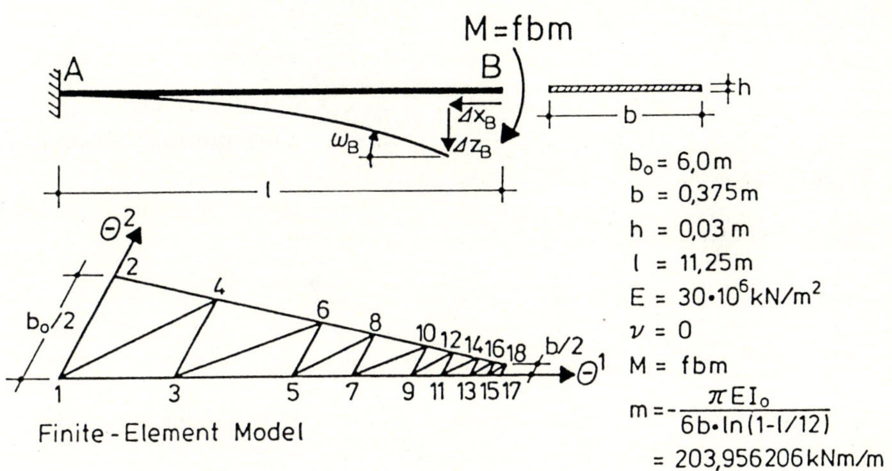

$$b_0 = 6,0\,\text{m}$$
$$b = 0,375\,\text{m}$$
$$h = 0,03\,\text{m}$$
$$l = 11,25\,\text{m}$$
$$E = 30\cdot10^6\,\text{kN/m}^2$$
$$\nu = 0$$
$$M = fbm$$
$$m = -\frac{\pi E I_o}{6b\cdot\ln(1-l/12)}$$
$$= 203,956206\,\text{kNm/m}$$

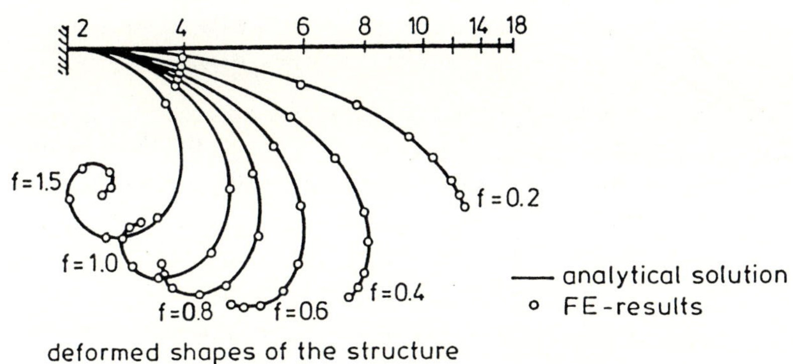

Fig. 5. Triangular cantilever plate subjected to a boundary
         moment

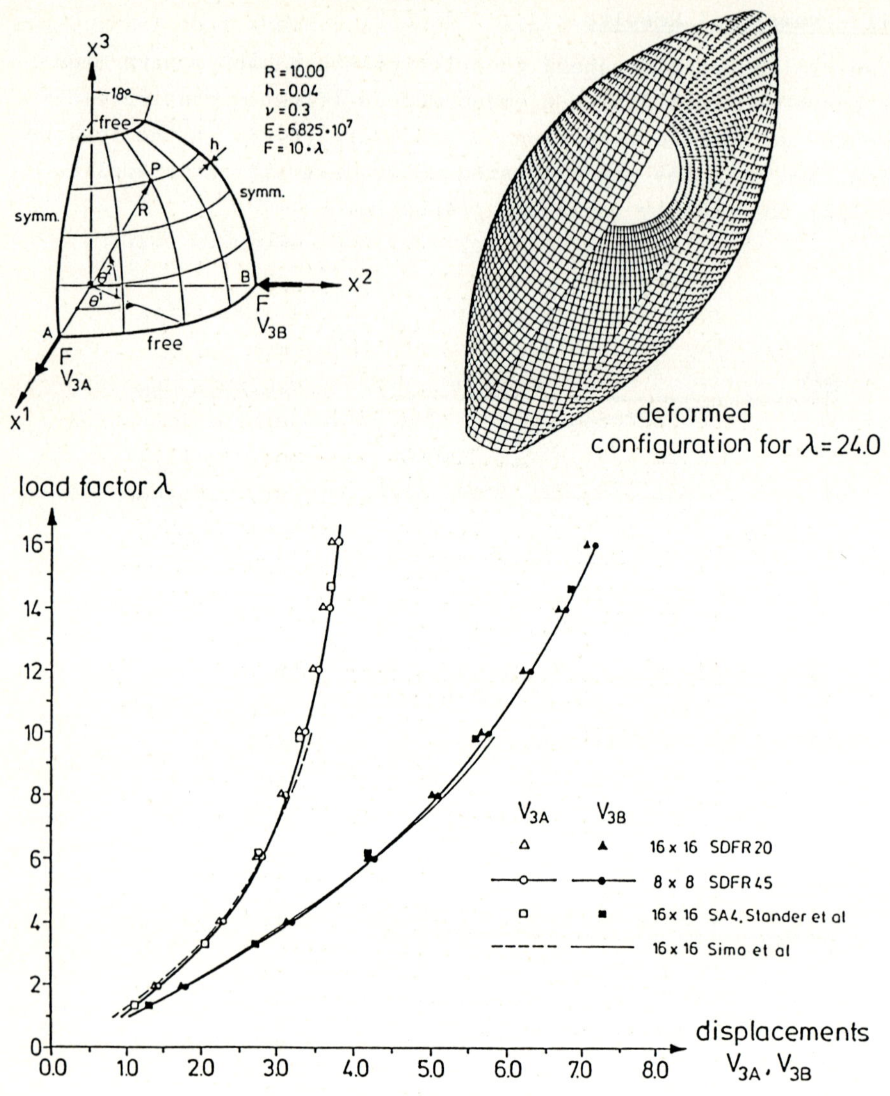

Fig. 6. Hemispherical shell under two point loads

The triangular cantilever plate under a concentrated load couple
(Fig. 5) has been analysed by finite-element displacement models
developed according to the KIRCHHOFF-LOVE type theory [6]. The
numerical results obtained are in a very good agreement with the
analytical solution constructed according to the classical beam
theory. Even for the load level with a rotational angle ω_B of
540°, the numerical errors are almost negligible (< 1%).

A second example (Fig. 6), the hemispherical shell under two
point loads, has been analysed by finite-element models SDFR20
and SDFR45 with 20 and 45 degrees of freedom. Their development
was accomplished on the basis of the shear deformation theory
using a mixed formulation. In Fig. 6 the numerical results obtained with an 8x8 and 16x16 mesh are compared with those given
in references [17,19] in form of load-displacement diagrams.
Fig. 6 presents also the deformed configuration of the structure
for $\lambda = 24,0$ as a three-dimensional plot.

Further strongly nonlinear examples, given in [6,7], demonstrate
that the finite-rotation theories discussed above present a
suitable basis for the development of high-precision finite-
element models applicable to arbitrary nonlinear phenomena.

REFERENCES

1. Başar, Y.; Krätzig, W.B.: Mechanik der Flächentragwerke.
   Vieweg Verlag, Braunschweig (1985).

2. Başar, Y.: Zur Struktur konsistenter inkrementeller Theo-
   rien für geometrisch nichtlineare Flächentragwerke und deren Operatordarstellung. Ingenieur-Archiv 56 (1986), pp.
   209-220.

3. Başar, Y.: Eine konsistente Theorie für Flächentragwerke
   endlicher Verformungen und deren Operatordarstellung auf
   variationstheoretischer Grundlage. ZAMM 66 (1986), pp. 297-
   308.

4. Başar, Y.: A consistent Theory of Geometrically Non-Linear
   Shells with an Independent Rotation Vector. Int. J. Solids
   Structures Vol. 23 (1987), pp. 1401-1415.

5. Başar, Y.; Krätzig, W.B.: A consistent Shell Theory for
   Finite Deformations. Acta Mechanica 76 (1989), pp. 73-87.

6. Başar, Y.; Ding, Y.: Finite-Rotation Elements for the Non-
    linear Analysis of Thin Shell Structures. Int. J. Solids &
    Structures (in preparation).

7. Ding, Y.: Finite-Rotations-Elemente zur geometrisch nicht-
    linearen Analyse allgemeiner Flächentragwerke. Dissertation
    der Ruhr-Universität Bochum, 1989.

8. Koiter, W.T.: On the Nonlinear Theory of Thin Elastic
    Shells. Proc. Kon. Ned. Ak. Wet., B 09 (1966), pp. 1-54.

9. Krätzig, W.B.: Allgemeine Schalentheorie beliebiger Werk-
    stoffe und Verformungen. Ingenieur-Archiv 40 (1971), pp.
    311-326.

10. Naghdi, P.M.: The Theory of Shells and Plates. Handbuch
    der Physik VI, A, (pp. 425-640). Springer-Verlag, New
    York, 1972, pp. 425-640.

11. Nolte, L.P.: Beitrag zur Herleitung und vergleichende Un-
    tersuchung geometrisch nichtlinearer Schalentheorien unter
    Berücksichtigung großer Rotationen. Mitteilungen aus dem
    Institut für Mechanik Nr. 39, Ruhr-Univerität, Bochum,
    1983.

12. Nolte, L.P.; Makowski, J.; Stumpf, H.: On the Derivation
    and Comparative Analysis of Large Rotation Shell Theories.
    Ingenieur-Archiv 56 (1986), pp. 145-160.

13. Pietraszkiewicz, W.: Finite Rotation and Lagrangean Des-
    cription in the Nonlinear Theory of Shells. Polish Scien-
    tific Publishers, Warschau 1979.

14. Pietraszkiewicz, W.: Lagrangean Description and Incremen-
    tal Formulation in the Nonlinear Theory of Thin Shells.
    Int. Journal Non-Linear Mechanics. Vol. 19 (1984), pp.
    115-139.

15. Reissner, E.: On the Theory of Tranverse Bending of Elastic
    Plates. Int. J. Solids Structures 12 (1976), pp. 545-554.

16. Schmidt, R.: A Current Trend in Shell Theory: Constrained
    Geometrically Nonlinear Kirchhoff-Love Type Theories Based
    on Polar Decomposition of Strains and Rotations. Computer
    & Structures, Vol. 20 (1985), pp. 265-275.

17. Simo, J.; Fox, D.; Rifai, M.S.: Formulation and Computati-
    onal Aspects of a Stress Resultant Geometrically Exact
    Shell Model. Proc. of the Int. Conf. on Computational En-
    gineering Science, April 10-14, 1988, Atlanta, GA, USA.

18. Simmonds, J.G.; Danielsen, D.A.: Non-linear Shell Theory
    with a Finite Rotation Vector. Proc. Kon. Ned. Ak. Wet.,
    Ser. B, 73 (1970), pp. 460-478.

19. Stander, N.; Matzenmiller, A.; Ramm, E.: An assessment
    of Assumed Strain Methods in Finite Rotation Shell Analy-
    sis. Journal of Eng. Computations (in preparation).

Formulation and Computational Aspects
of a Stress Resultant Geometrically Exact Shell Model

J. C. Simo, D. D. Fox and M. S. Rifai

Applied Mechanics Division
Stanford University
Stanford, CA 94305, U.S.A.

ABSTRACT

This paper considers the formulation and numerical implementation of a geometrically exact resultant based shell model for the analysis of large deformations of thin and moderately thick shells. The model is essentially a single extensible director Cosserat surface. Variable thickness and thickness stretch effects are properly modeled via the *extensibility* condition on the director field. A simple linear elastic constitutive model is given which possesses the correct asymptotic limits as the thickness tends to zero and recovers the plane stress constitutive relations in the thin shell limit. On the computational side, a configuration update procedure for the director field is presented which is *singularity free* and *exact* regardless of the magnitude of the director (rotation and thickness stretch) increment. The performance of the shell model is assessed through an extensive set of numerical examples.

1.   INTRODUCTION

The computational analysis of nonlinear shells has been dominated by the degenerated solid approach originally introduced by [1]; see e.g., [12,13], [5], [16] among many others. It is somewhat surprising that the classical and modern approaches to nonlinear shell theory, as discussed in [9], [17,18], [15] or [2,3], appear to have had little impact on the computational analysis of shells. In contrast with the established computational approach, the formulation considered in this paper finds its roots in the classical theory. The proposed geometrically exact shell model is formulated entirely in stress resultants and falls within the class of a single *extensible* director Cosserat surface. The present approach constitutes an extension of our earlier works in [22-25], in the sense that the restriction to an inextensible director field is now removed from the theory. In physical terms, the model discussed below allows for variable thickness shells and thickness stretch. Important aspects of this presentation are the following:

i. The parametrization of the weak form of the momentum balance equations avoids the explicit appearance of the Riemannian connection of the mid-surface. In particular, objects such as Christoffel symbols or the second fundamental form need not be computed.

ii. Thickness effects are modeled by an added degree of freedom which is coupled (interactively) to the membrane, shear and bending fields. This is at variance with previous attempts in the context of the degenerated solid approach, which treat thickness effects essentially as a post-processing procedure.

iii. A simple quadratic isotropic stored energy function is considered which possesses the proper asymptotic limits as the thickness of the shell tends to zero. For thin shells, this model reduces to the equations obtained by enforcing the plane stress assumption.

iv. A multiplicative decomposition of the director field into a unit vector (rotation part) and a magnitude parameter (thickness part) is introduced which circumvents numerical ill-conditioning of the formulation in the thin shell limit. Geometrically exact update procedures which are singularity free and exact for any magnitude of the incremental rotation of the director field are formulated in the context of this multiplicative decomposition.

From a computational standpoint, the present formulation, which incorporates thickness effects, is essentially identical to that presented in our previous work (which did not account for thickness effects) with the exception of two simple additional terms not present in the weak form of the momentum balance equations.

2. NONLINEAR RESULTANT SHELL THEORY WITH THROUGH-THE-THICKNESS STRETCH

2.1. Kinematic Description of the Shell

We will consider deformations of a shell which fall within the category of a single extensible director Cosserat surface. In contrast with the previous works on the stress resultant geometrically exact shell model ([22-25]), this formulation admits extension of the director, or more precisely, allows for variable thickness and through the thickness stretch.

The kinematic description of the shell is as follows. Points in the shell are parameterized by a coordinate system $(\xi^1, \xi^2, \xi) \in \mathcal{A} \times I$, where

$\xi \in I = [h_0^-, h_0^+]$ is the through the thickness parameter; $\mathcal{A}$ and I are fixed regions in $\mathbf{R}^2$ and $\mathbf{R}$, respectively. Any Euclidean placement of the shell $\mathcal{S} \subset \mathbf{R}^3$ is given by a mapping $\hat{\boldsymbol{\Phi}} : \mathcal{A} \times I \to \mathbf{R}^3$, where

$$\mathcal{S} := \left\{ \, \boldsymbol{x} \in \mathbf{R}^3 \,\middle|\, \boldsymbol{x} = \hat{\boldsymbol{\Phi}}(\xi^1, \xi^2, \xi) \text{ for } (\xi^1, \xi^2, \xi) \in \mathcal{A} \times I \, \right\}. \qquad (2.1)$$

Following the single director Cosserat surface kinematic assumption, any point in the shell $\boldsymbol{x} \in \mathcal{S}$ is identified as

$$\boxed{\hat{\boldsymbol{\Phi}}(\xi^1, \xi^2, \xi) := \boldsymbol{\varphi}(\xi^1, \xi^2) + \xi \boldsymbol{d}(\xi^1, \xi^2)}, \qquad (2.2)$$

where $\boldsymbol{\varphi} : \mathcal{A} \to \mathbf{R}^3$ is the mapping that defines the mid-surface of the shell and $\boldsymbol{d} : \mathcal{A} \to \mathbf{R}^3$ is the director field or fiber direction. Therefore, admissible configurations of the shell are given by pairs $(\boldsymbol{\varphi}, \boldsymbol{d}) \in \mathcal{C}$, where the abstract configuration manifold $\mathcal{C}$ is defined

$$\mathcal{C} = \left\{ \, \boldsymbol{\Phi} = (\boldsymbol{\varphi}, \boldsymbol{d}) : \mathcal{A} \to \mathbf{R}^3 \times \mathbf{R}^3 \,\middle|\, \det[\nabla\boldsymbol{\varphi}] > 0, \ \boldsymbol{d} \cdot \hat{\boldsymbol{n}} > 0 \, \right\}, \qquad (2.3)$$

with $\hat{\boldsymbol{n}} = \boldsymbol{\varphi}_{,1} \times \boldsymbol{\varphi}_{,2}$ being the normal to the mid-surface. Note that $\boldsymbol{d} \in \mathbf{R}^3$ is not restricted to have unit length. From a numerical standpoint, in order to avoid ill-conditioning in the thin shell limit, it proves essential to decompose the director field as follows

$$\boldsymbol{d}(\xi^1, \xi^2) = \lambda(\xi^1, \xi^2)\, \boldsymbol{t}(\xi^1, \xi^2), \text{ with } \|\boldsymbol{t}\| = 1 \text{ and } \lambda > 0. \qquad (2.4)$$

Accordingly, the abstract configuration manifold $\mathcal{C}$ can be re-phrased to incorporate the multiplicative decomposition (2.4) as

$$\mathcal{C} = \left\{ \, \boldsymbol{\Phi} = (\boldsymbol{\varphi}, \boldsymbol{t}, \lambda) : \mathcal{A} \to \mathbf{R}^3 \times \mathrm{S}^2 \times \mathbf{R}_+ \,\middle|\, \det[\nabla\boldsymbol{\varphi}] > 0, \ \boldsymbol{t} \cdot \hat{\boldsymbol{n}} > 0 \, \right\}. \qquad (2.5)$$

2.2. Stress and Stress Couple Resultants

Let $\boldsymbol{n}^\alpha$, $\boldsymbol{m}^\alpha$ and $\boldsymbol{l}$ represent the resultant stress, resultant stress couple and resultant through the thickness stress. See [22] for the definition of these resultants in terms of three-dimensional quantities. In components relative to the natural or convected basis

$$\{\boldsymbol{a}_1, \boldsymbol{a}_2, \boldsymbol{a}_3\} = \{\boldsymbol{\varphi}_{,1}, \boldsymbol{\varphi}_{,2}, \boldsymbol{d}\}, \qquad (2.6)$$

the resultants are given by

$$\begin{aligned}
\boldsymbol{n}^\alpha &= n^{\beta\alpha}\boldsymbol{\varphi}_{,\beta} + q^\alpha \boldsymbol{d} \\
\tilde{\boldsymbol{m}}^\alpha &= \tilde{m}^{\beta\alpha}\boldsymbol{\varphi}_{,\beta} + \tilde{m}^{3\alpha}\boldsymbol{d} \\
\boldsymbol{l} &= l^\alpha \boldsymbol{\varphi}_{,\alpha} + l^3 \boldsymbol{d}.
\end{aligned} \qquad (2.7)$$

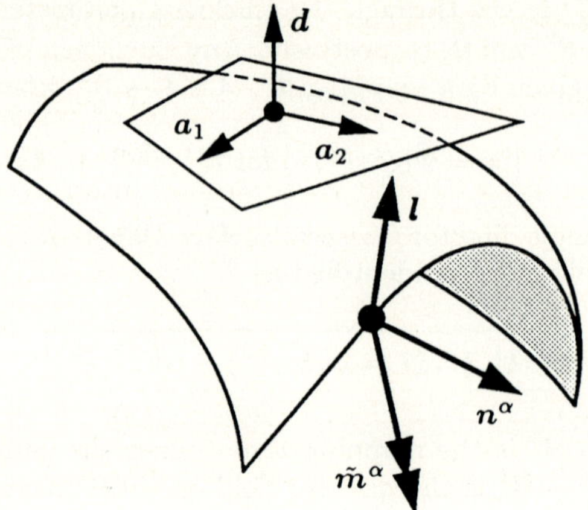

FIGURE 2.1. Convected (mid-surface) basis. Stress resultants, stress-couple resultants, and through the thickness stress resultant.

Figure 2.1 illustrates the resultants (2.7) and the basis (2.6).

2.3. Resultant Form of the Local Balance Equations

The local or strong form of the (static) momentum balance equations in terms of the resultants (2.7) are given by

$$\frac{1}{\bar{j}}(\bar{j}n^\alpha)_{,\alpha} + \bar{n} = 0\,,$$

$$\frac{1}{\bar{j}}(\bar{j}\tilde{m}^\alpha)_{,\alpha} - l + \bar{m} = 0\,. \tag{2.8}$$

A detailed derivation of these equations and the definition of the applied loads $\bar{n}$ and $\bar{m}$ is given in [22]. Equations (2.8) are the resultant balance of linear momentum and balance of director momentum (equivalent to balance of couples around the mid-surface), respectively. The *balance of angular momentum* equation (equivalent to $\boldsymbol{\sigma} = \boldsymbol{\sigma}^{\mathrm{T}}$ in the three-dimensional theory) is written as

$$n^\alpha \times \varphi_{,\alpha} + \tilde{m}^\alpha \times d_{,\alpha} + l \times d = 0. \tag{2.9}$$

2.4. Effective Stress Resultants and Balance of Angular Momentum

The balance of angular momentum equation (2.9) plays a crucial role in the definition of effective stress resultants conjugate to the desired strain measures. Define the *effective stress resultant* by the following rank two tensor

$$\tilde{n} := n^\alpha \otimes \varphi_{,\alpha} + l \otimes d - d_{,\alpha} \otimes \tilde{m}^\alpha\,. \tag{2.10}$$

It can be shown (see [26]) that the following equivalence holds

$$\text{Balance of angular momentum} \iff \boxed{\tilde{\mathbf{n}} = \tilde{\mathbf{n}}^{\mathrm{T}}}. \qquad (2.11)$$

The equivalence between balance of angular momentum and symmetry of the effective stress resultant admits a simple interpretation in component form. Consider component relations relative to the convected basis (2.6); $\boldsymbol{d}_{,\alpha}$ is written

$$\boldsymbol{d}_{,\alpha} = \lambda^{\gamma}_{\alpha}\boldsymbol{\varphi}_{,\gamma} + \lambda^{3}_{\alpha}\boldsymbol{d}. \qquad (2.12)$$

The effective stress components are then

$$\left.\begin{aligned}
\tilde{n}^{\alpha\beta} &= n^{\alpha\beta} - \lambda^{\alpha}_{\gamma}\tilde{m}^{\beta\gamma} && \text{(membrane)} \\
\tilde{n}^{3\alpha} &= q^{\alpha} - \lambda^{3}_{\gamma}\tilde{m}^{\alpha\gamma} =: \tilde{q}^{\alpha} && \text{(transverse shear)} \\
\tilde{n}^{\alpha3} &= l^{\alpha} - \lambda^{\alpha}_{\gamma}\tilde{m}^{3\gamma} =: \tilde{l}^{\alpha} && \text{(symmetric shear)} \\
\tilde{n}^{33} &= l^{3} - \lambda^{3}_{\gamma}\tilde{m}^{3\gamma} =: \tilde{l}^{3} && \text{(across-thickness)}
\end{aligned}\right\}. \qquad (2.13)$$

The equivalence (2.11), in components, becomes

$$\text{Balance of angular momentum} \iff \boxed{\tilde{n}^{\beta\alpha} = \tilde{n}^{\alpha\beta}} \text{ and } \boxed{\tilde{q}^{\alpha} = \tilde{l}^{\alpha}}. \qquad (2.14)$$

The effective stress components (2.13) are crucial to the development of the weak form of the momentum balance equations and the resultant hyper-elastic constitutive equations that follow.

2.5. The Weak Form of the Momentum Balance Equations

In what follows, we utilize the multiplicative decomposition of the director field given by (2.4). Thus, the tangent space to the abstract configuration manifold $\mathcal{C}$ (the space of kinematically admissible variations) is

$$T_{\boldsymbol{\Phi}}\mathcal{C} := \left\{ \, \delta\boldsymbol{\Phi} := (\delta\varphi, \delta t, \delta\lambda) : \mathcal{A} \to \mathbf{R}^{3} \times T_{t}\mathbf{S}^{2} \times \mathbf{R} \, \right|$$

$$\delta\varphi|_{\partial_{\varphi}\mathcal{A}} = \mathbf{0}, \; \delta t|_{\partial_{t}\mathcal{A}} = \mathbf{0}, \; \delta\lambda|_{\partial_{\lambda}\mathcal{A}} = 0 \, \Big\}. \qquad (2.15)$$

Here $\delta(\bullet)$ denotes the directional derivative of $(\bullet)$. In the notation of (2.15), the directional derivative of the director $\boldsymbol{d}$ is written as $\delta\boldsymbol{d} = \delta\lambda t + \lambda\delta t$.

The resultant linear and director momentum equations (2.8) are multiplied by variations in the mid-surface position and in the director, respectively. Integrating this result over the shell surface and using the divergence theorem, we obtain the (static) weak form of the momentum balance equations

$$\mathrm{G}(\boldsymbol{n}^{\alpha}, \tilde{\boldsymbol{m}}^{\alpha}, \boldsymbol{l}, \boldsymbol{\Phi}; \delta\boldsymbol{\Phi}) = \int_{\mathcal{A}} \left[\boldsymbol{n}^{\alpha} \cdot \delta\varphi_{,\alpha} + \tilde{\boldsymbol{m}}^{\alpha} \cdot \delta\boldsymbol{d}_{,\alpha} + \boldsymbol{l} \cdot \delta\boldsymbol{d} \right] \bar{\jmath} \, d\xi^{1} \, d\xi^{2}$$

$$- \mathrm{G}_{\mathrm{ext}}(\delta\boldsymbol{\Phi}) = 0, \qquad (2.16)$$

36

where $G_{ext}(\delta\boldsymbol{\Phi})$ is the external load term given by

$$G_{ext}(\delta\boldsymbol{\Phi}) := \int_{\mathcal{A}} \left[\bar{n}\cdot\delta\varphi + \bar{m}\cdot\delta d \right] \bar{j} d\xi^1 d\xi^2 + \int_{\partial\mathcal{A}} \left[n^\alpha\cdot\delta\varphi + \bar{m}^\alpha\cdot\delta d \right] \nu_\alpha \bar{j} d\Gamma.$$

$$(2.17)$$

In (2.17), ν_α are the components of the unit outward normal to the region $\mathcal{A}$ relative to the standard basis in $\mathbf{R}^2$.

2.6. Strain Measures and Component Expressions for the Weak Form

There are two alternative ways to express the shell strain measures and resulting component expressions for the weak form in terms of these strain measures. The two alternative representations correspond to the two possible representations of the abstract configuration manifold $\mathcal{C}$ given by (2.3) and (2.5).

First, we define the strain measures in terms of the director $\boldsymbol{d}$. These definitions correspond to those measures obtained from the Lagrangian strain tensor of the three-dimensional theory, $\boldsymbol{E} = \frac{1}{2}(\boldsymbol{F}^T\boldsymbol{F}-\boldsymbol{1})$, where $\boldsymbol{F}$ is the deformation gradient $\nabla\hat{\boldsymbol{\Phi}}\nabla\hat{\boldsymbol{\Phi}}_0^{-1}$. (The subscript o refers to any quantity particularized to the reference configuration). We have

$$\left.\begin{aligned}
\epsilon_{\alpha\beta} &= \tfrac{1}{2}(\varphi_{,\alpha}\cdot\varphi_{,\beta} - \varphi_{0,\alpha}\cdot\varphi_{0,\beta}) && \text{(membrane)} \\
\tilde{\rho}_{\alpha\beta} &= \varphi_{,\alpha}\cdot d_{,\beta} - \varphi_{0,\alpha}\cdot d_{0,\beta} && \text{(bending)} \\
\tilde{\delta}_\alpha &= \varphi_{,\alpha}\cdot d - \varphi_{0,\alpha}\cdot d_0 && \text{(transverse shear)} \\
\tilde{\chi}_\alpha &= d_{,\alpha}\cdot d - d_{0,\alpha}\cdot d_0 && \text{(symmetric shear)} \\
\tilde{\chi} &= \tfrac{1}{2}(d\cdot d - d_0\cdot d_0) && \text{(thickness stretch)}
\end{aligned}\right\} . \qquad (2.18)$$

We use a superposed $\tilde{}$ to differentiate the measures above from the strain measures used in the numerical implementation, defined subsequently, and denoted without the symbol $\tilde{}$. The weak form (2.16) is expressed in terms of the effective stress components (2.13) and the strain measures (2.18) as

$$G(\tilde{n}, \tilde{m}^\alpha, \boldsymbol{\Phi}; \delta\boldsymbol{\Phi}) = \int_{\mathcal{A}} \Big[\tilde{n}^{\alpha\beta}\delta\epsilon_{\alpha\beta} + \tilde{m}^{\alpha\beta}\delta\tilde{\rho}_{\alpha\beta} + \tilde{q}^\alpha\delta\tilde{\delta}_\alpha +$$
$$\tilde{m}^{3\alpha}\delta\tilde{\chi}_\alpha + \tilde{l}^3\delta\tilde{\chi} \Big] \bar{j} d\xi^1 d\xi^2 - G_{ext}(\delta\boldsymbol{\Phi}) = 0 . \quad (2.19)$$

Direct inspection of the expressions (2.18) reveals two noteworthy features. First, the through the thickness stretch is coupled with the bending and shear measures. Deformations leading to large thickness change can also result in large changes in the bending and shear strain measures $\tilde{\rho}_{\alpha\beta}$ and $\tilde{\delta}_\alpha$. Second, a quadratic stored energy function in the measures (2.18) (the usual model in classical shell theory) possesses an incorrect limit (i.e., finite energy) as the thickness of the shell decreases to zero. Taking the 3-dimensional theory as a guide, a more appropriate strain measure which produces the proper asymptotic behavior in a quadratic model is the logarithmic strain. With this motivation at hand, using the multiplicative decomposition of the director field, we define the following set of strain measures:

$$
\left.
\begin{aligned}
\epsilon_{\alpha\beta} &= \tfrac{1}{2}(\varphi_{,\alpha} \cdot \varphi_{,\beta} - \varphi_{0,\alpha} \cdot \varphi_{0,\beta}) && \text{(membrane)} \\
\rho_{\alpha\beta} &= \varphi_{,\alpha} \cdot t_{,\beta} - \varphi_{0,\alpha} \cdot t_{0,\beta} && \text{(bending)} \\
\delta_\alpha &= \varphi_{,\alpha} \cdot t - \varphi_{0,\alpha} \cdot t_0 && \text{(transverse shear)} \\
\chi_\alpha &= (\ln\lambda)_{,\alpha} - (\ln\lambda_0)_{,\alpha} = \ln(\lambda/\lambda_0)_{,\alpha} && \text{(symmetric shear)} \\
\chi &= \ln\lambda - \ln\lambda_0 = \ln(\lambda/\lambda_0) && \text{(thickness stretch)}
\end{aligned}
\right\} . \quad (2.20)
$$

The component expression (2.19) for the weak form, in terms of the strain measures (2.20), then becomes

$$
G(\tilde{n}, \tilde{m}^\alpha_i \Phi; \delta\Phi) = \int_{A} \left[\tilde{n}^{\alpha\beta}\delta\epsilon_{\alpha\beta} + \tilde{m}^{\alpha\beta}\delta\rho_{\alpha\beta} + \tilde{q}^\alpha\delta\delta_\alpha + \right.
$$
$$
\left. \tilde{m}^{3\alpha}\delta\chi_\alpha + \tilde{l}^3\delta\chi \right] \bar{j}\, d\xi^1 d\xi^2 - G_{\text{ext}}(\delta\Phi) = 0 \,, \quad (2.21)
$$

where the *modified* effective stress resultants in (2.21) are defined in terms of the effective stress resultants and the stress couple resultants as

$$
\left.
\begin{aligned}
\tilde{n}^{\alpha\beta} &= \tilde{n}^{\alpha\beta} \\
\tilde{m}^{\alpha\beta} &= \lambda\, \tilde{m}^{\alpha\beta} \\
\tilde{q}^\alpha &= \lambda\, \left[\tilde{q}^\alpha + (\ln\lambda)_{,\beta}\, \tilde{m}^{\alpha\beta} \right] \\
\tilde{m}^{3\alpha} &= \tilde{m}^\alpha \cdot d \\
\tilde{l}^3 &= l \cdot d + \tilde{m}^\alpha \cdot d_{,\alpha}
\end{aligned}
\right\} . \quad (2.22)
$$

The advantages of expressing the weak form in terms of the strain measures (2.20) are the following:

1. The through the thickness stretch is not coupled with the bending or shear strain measures.

2. A simple quadratic elastic constitutive model possesses the proper asymptotic behavior as the thickness of the shell tends to 0.

3. The numerical treatment of the equations is *identical* to that for the inextensible director formulation, with simply two additional terms in the weak form.

2.7. <u>Matrix Formulation and Constitutive Equations</u>

Next, we summarize the complete set of equations, in component matrix form, for the resultant based geometrically exact shell model with through the thickness effects. For completeness, we restrict our attention to a simple linear elastic isotropic constitutive model. (Note that $A_{\alpha\beta} = \varphi_{0,\alpha} \cdot \varphi_{0,\beta}$ are the components of the reference mid-surface metric tensor and the 3×2 transformation $\bar{\Lambda}$ is defined in section **3**.)

- Modified effective stress and stress couple vectors:

$$\check{n} = \bar{J} \left\{ \begin{matrix} \check{n}^{11} \\ \check{n}^{22} \\ \check{n}^{12} \end{matrix} \right\}, \quad \check{q} = \bar{J} \left\{ \begin{matrix} \check{q}^1 \\ \check{q}^2 \end{matrix} \right\}, \quad \check{m} = \bar{J} \left\{ \begin{matrix} \check{m}^{11} \\ \check{m}^{22} \\ \check{m}^{(12)} \end{matrix} \right\}, \quad \check{l} = \bar{J} \left\{ \begin{matrix} \check{m}^{31} \\ \check{m}^{32} \\ \check{l}^3 \end{matrix} \right\}$$

- Incremental displacement and rotation variations:

$$\delta\varphi = \left\{ \begin{matrix} \delta\varphi^1 \\ \delta\varphi^2 \\ \delta\varphi^3 \end{matrix} \right\}, \quad \delta T = \left\{ \begin{matrix} \delta T^1 \\ \delta T^2 \end{matrix} \right\}, \quad \delta t = \bar{\Lambda}\delta T, \quad \delta\Phi = \left\{ \begin{matrix} \delta\varphi \\ \delta T \\ \delta\mu \end{matrix} \right\}$$

- Matrix differential operators:

$$B_m = \begin{bmatrix} \varphi_{,1}^{\mathrm{T}} \frac{\partial}{\partial\xi^1} \\ \varphi_{,2}^{\mathrm{T}} \frac{\partial}{\partial\xi^2} \\ \varphi_{,1}^{\mathrm{T}} \frac{\partial}{\partial\xi^2} + \varphi_{,2}^{\mathrm{T}} \frac{\partial}{\partial\xi^1} \end{bmatrix}_{3\times3} \quad B_{sm} = \begin{bmatrix} t^{\mathrm{T}} \frac{\partial}{\partial\xi^1} \\ t^{\mathrm{T}} \frac{\partial}{\partial\xi^2} \end{bmatrix}_{2\times3} \quad B_{sb} = \begin{bmatrix} \varphi_{,1}^{\mathrm{T}} \\ \varphi_{,2}^{\mathrm{T}} \end{bmatrix}_{2\times3} \bar{\Lambda}_{3\times2}$$

$$B_{bm} = \begin{bmatrix} t_{,1}^{\mathrm{T}} \frac{\partial}{\partial\xi^1} \\ t_{,2}^{\mathrm{T}} \frac{\partial}{\partial\xi^2} \\ t_{,1}^{\mathrm{T}} \frac{\partial}{\partial\xi^2} + t_{,2}^{\mathrm{T}} \frac{\partial}{\partial\xi^1} \end{bmatrix}_{3\times3} \quad B_{bb} = B_m \bar{\Lambda}_{3\times2} \quad B_l = \begin{bmatrix} \frac{\partial}{\partial\xi^1} \\ \frac{\partial}{\partial\xi^2} \\ 1 \end{bmatrix}_{3\times1}$$

$$B_{\mathrm{total}} = \begin{bmatrix} B_m & 0 & 0 \\ B_{sm} & B_{sb} & 0 \\ B_{bm} & B_{bb} & 0 \\ 0 & 0 & B_l \end{bmatrix}_{11\times6} \qquad \check{r} = \left\{ \begin{matrix} \check{n} \\ \check{q} \\ \check{m} \\ \check{l} \end{matrix} \right\}$$

- Weak form $G : \mathcal{C} \times T_{\Phi}\mathcal{C} \to \mathbb{R}$: (Here, $\check{r} = \check{r}(\Phi)$)

$$G(\Phi; \delta\Phi) := \int_A \check{r}^T B_{\mathrm{total}} \, \delta\Phi \, \bar{j}_0 \, d\xi^1 d\xi^2 - G_{\mathrm{ext}}(\delta\Phi)$$

- Strain component vectors:

$$\epsilon = \left\{ \begin{matrix} \epsilon_{11} \\ \epsilon_{22} \\ 2\epsilon_{12} \end{matrix} \right\}, \quad \delta = \left\{ \begin{matrix} \delta_1 \\ \delta_2 \end{matrix} \right\}, \quad \rho = \left\{ \begin{matrix} \rho_{11} \\ \rho_{22} \\ 2\rho_{(12)} \end{matrix} \right\}, \quad \chi = \left\{ \begin{matrix} \chi_1 \\ \chi_2 \\ \chi \end{matrix} \right\}$$

- Elastic isotropic constitutive equations:

$$C_m = \begin{bmatrix} (1-\nu)A^{11}A^{11} & [\nu A^{11}A^{22} + (1-2\nu)A^{12}A^{12}] & (1-\nu)A^{11}A^{12} & \nu A^{11} \\ & (1-\nu)A^{22}A^{22} & (1-\nu)A^{22}A^{12} & \nu A^{22} \\ & & [\frac{1-2\nu}{2}A^{11}A^{22} + \frac{1}{2}A^{12}A^{12}] & \nu A^{12} \\ \text{Symmetric} & & & (1-\nu) \end{bmatrix}$$

$$
\mathbf{C}_b =
\begin{bmatrix}
A^{11}A^{11} & \begin{array}{c}[\nu A^{11}A^{22} + \\ (1-\nu)A^{12}A^{12}]\end{array} & A^{11}A^{12} \\
 & A^{22}A^{22} & A^{22}A^{12} \\
\text{Sym.} & & \begin{array}{c}[\frac{1-\nu}{2}A^{11}A^{22} + \\ \frac{1+\nu}{2}A^{12}A^{12}]\end{array}
\end{bmatrix}
$$

$$
\left\{ \begin{array}{c} \check{n} \\ \bar{J}\check{l}^3 \end{array} \right\} = \frac{\bar{\rho}_0 Eh}{(1+\nu)(1-2\nu)} \mathbf{C}_m \left\{ \begin{array}{c} \epsilon \\ \chi \end{array} \right\} ; \qquad
\check{q} = \kappa \bar{\rho}_0 Gh \begin{bmatrix} A^{11} & A^{12} \\ A^{12} & A^{22} \end{bmatrix} \delta
$$

$$
\check{m} = \frac{\bar{\rho}_0 Eh^3}{12(1-\nu^2)} \mathbf{C}_b \rho ; \qquad
\left\{ \begin{array}{c} \check{m}^{31} \\ \check{m}^{32} \end{array} \right\} = \frac{7\bar{\rho}_0 Eh^3}{240(1-\nu^2)} \begin{bmatrix} A^{11} & A^{12} \\ A^{12} & A^{22} \end{bmatrix} \left\{ \begin{array}{c} \chi_1 \\ \chi_2 \end{array} \right\}
$$

REMARK 2.1.

As previously alluded to, the simple linear elastic isotropic constitutive model presented above has several desirable features. First, the energy possess the proper asymptotic limit as the thickness goes to 0. Second, the plane stress assumption need not be enforced at the outset. Nevertheless, in the thin shell limit, plane stress constitutive equations are recovered *automatically.* ■

3. FINITE ELEMENT SOLUTION PROCEDURE

3.1. <u>Conceptual Iterative Solution Procedure</u>

The variational formulation of the (static) boundary value problem is posed: Find $\Phi = (\varphi, t, \lambda) \in \mathcal{C}$ such that

$$
G(\Phi; \delta\Phi) = 0, \quad \text{for all} \quad \delta\Phi = (\delta\varphi, \delta t, \delta\lambda) \in T_{\Phi}\mathcal{C}. \tag{3.1}
$$

Conceptually, equation (3.1) is solved through a Newton-Kantorovitch iteration scheme as follows: Given a configuration at iteration k, $\Phi^k = (\varphi^k, t^k, \lambda^k) \in \mathcal{C}$, solve the *linearized* problem

$$
G(\Phi^k; \delta\Phi) + D[G(\Phi^k; \delta\Phi)] \cdot \Delta\Phi^k = 0 \tag{3.2}
$$

for the incremental change in the configuration, an element $\Delta\Phi^k = (\Delta\varphi^k, \Delta t^k, \Delta\lambda^k) \in T_{\Phi_k}\mathcal{C}$. The incremental change $\Delta\Phi^k$ is then used to *update* the configuration $\Phi^k \mapsto \Phi^{k+1} \in \mathcal{C}$. Convergence of the solution is attained when $G(\Phi^{k+1}; \delta\Phi) = 0$. The *numerical implementation* of this solution procedure involves the following steps:

i. Approximation: Galerkin projection of the problem onto a finite dimensional manifold $\mathcal{C}^h \subset \mathcal{C}$. The interpolation of the "midsurface" part φ and $\Delta\varphi$ is accomplished in the standard fashion by means of the isoparametric concept. The interpolation of the "director part", however, is non-standard, and must be such as to ensure that t^h and Δt^h do remain in S^2 and $T_t S^2$, respectively. We refer to [24] for a detailed account.

ii. Linearization: Construction of the *consistent tangent operator* which is the bilinear form $\mathrm{B} : T_{\Phi^k} \mathcal{C}^h \times T_{\Phi^k} \mathcal{C}^h \to \mathbf{R}$ defined as

$$\mathrm{B} \equiv D\mathrm{G}(\Phi^k; \delta\Phi) \cdot \Delta\Phi^k := \left. \frac{\mathrm{d}}{\mathrm{d}\epsilon} \right|_{\epsilon=0} \mathrm{G}(\Phi^k_\epsilon; \delta\Phi) \ . \tag{3.3}$$

Here, $\Phi^k_\epsilon = (\varphi^k_\epsilon, t^k_\epsilon, \lambda^k_\epsilon) \in \mathcal{C}^h$ is a one-parameter family or curve of configurations whose "tangent" is in the direction of the element of the tangent space $(\Delta\varphi^k, \Delta t^k, \Delta\lambda^k) \in T_{\Phi^k}\mathcal{C}^h$, which is systematically constructed by means of the *exponential mapping*; see [24,26].

iii. Update: Given the configuration $\Phi^k \in \mathcal{C}^h$, and the incremental vector field $\Delta\Phi^k = (\Delta\varphi^k, \Delta t^k, \Delta\lambda^k) \in T_{\Phi^k}\mathcal{C}^h$, the configuration update procedure must ensure that the updated configuration Φ^{k+1} does in fact remain in $\mathcal{C}^h \subset \mathcal{C}$. (For convenience, the superscript h has been ommitted from quantities identified with the iteration parameter k.) A brief outline of the update procedure is given in the following section, and a detailed account is found in [22,24,26].

3.2. <u>Update Procedure: Geometrically Exact Director Field Update</u>

The director field update is composed of two parts: the unit or inextensible director update (t) and the thickness parameter update (λ). The update procedure for the unit director field which is *exact for any magnitude of the incremental director displacement* $\Delta t \in T_t S^2$, is developed by making use of the *exponential mapping* in S^2. We exploit the well-known fact that straight lines in the tangent plane $T_t S^2$ defined by $t + \epsilon\Delta t$, with $\Delta t \cdot t = 0$, are uniquely mapped onto geodesics in S^2 by the exponential mapping $\exp : T_t S^2 \to S^2$. Furthermore, one has the following simple explicit formula ([22])

$$t + \epsilon\Delta t \mapsto t_\epsilon = \cos\|\epsilon\Delta t\| t + \frac{\sin\|\epsilon\Delta t\|}{\|\Delta t\|} \Delta t \in S^2. \tag{3.4}$$

Thus, at the k-iteration of the Newton method, let $t^k_A \in S^2$ be the *known* director at node A, and let $\Lambda^k_A \in SO(3)$ be the *known* orthogonal transformation such that $t^k_A = \Lambda^k_A E$ (here $E = E_3$). Further, let $\Delta T^k_A \in T_E S^2$ be the incremental director variation in the material description. It follows that the spatial director variation is obtained as

$$\Delta t^k_A = \Lambda^k_A \Delta T^k_A. \tag{3.5}$$

Observe that $\Delta t_A^k \in T_{t_A^k} S^2$ since by definition $E = E_3$ and ΔT_A^k only has components relative to E_1 and E_2, then $t_A^k \cdot \Delta t_A^k = \Lambda_A^k E \cdot \Lambda_A^k \Delta T_A^k = E \cdot \Delta T_A^k = 0$. Direct application of (3.4) then yields

$$\boxed{t_A^{k+1} = \cos \|\Delta t_A^k\| t_A^k + \frac{\sin \|\Delta t_A^k\|}{\|\Delta t_A^k\|} \Delta t_A^k}. \tag{3.6}$$

It can be verified from (3.6) that in fact $\|t_A^{k+1}\| = 1$, as required. A geometric interpretation of the update formula (3.6) is contained in Figure 3.1.

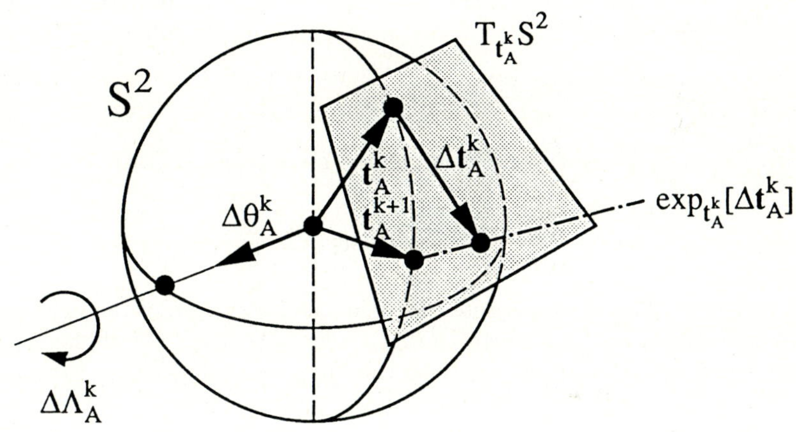

FIGURE 3.1. Geometric update procedure in S^2. Nodal directors, which are updated by means of the exponential map, rotate by the amount $\Delta\Lambda_A^k$.

Once the nodal director field is updated, the update of the orthogonal transformation $\Lambda_A^{k+1} : E \mapsto t_A^{k+1} \in S^2$ is accomplished as follows. We set

$$\Lambda_A^{k+1} = \Delta\Lambda_A^k \Lambda_A^k, \tag{3.7}$$

where $\Delta\Lambda_A^k : t_A^k \in S^2 \mapsto t_A^{k+1} \in S^2$ is the *unique orthogonal transformation* that rotates t_A^k to t_A^{k+1} *without drill* and is defined by

$$\boxed{\begin{aligned} \Delta\Lambda_A^k &= \cos \|\Delta t_A^k\| \mathbf{1} + \frac{\sin \|\Delta t_A^k\|}{\|\Delta t_A^k\|} \widehat{[t_A^k \times \Delta t_A^k]} \\ &+ \frac{(1 - \cos \|\Delta t_A^k\|)}{\|\Delta t_A^k\|^2} (t_A^k \times \Delta t_A^k) \otimes (t_A^k \times \Delta t_A^k). \end{aligned}} \tag{3.8}$$

where $\hat{\boldsymbol{w}}\boldsymbol{h} := \boldsymbol{w} \times \boldsymbol{h}$, for all $\boldsymbol{w}, \boldsymbol{h} \in \mathbf{R}^3$. Expression (3.8) is obtained by suitably restricting the exponential map in $SO(3)$ to the unit sphere S^2; see [22].

The construction of an update algorithm for the thickness stretch $\lambda : \mathcal{A} \to \mathbf{R}_+$ at the nodal points in the mid-surface of the shell is motivated by the following considerations:

i. Construct a one-parameter curve of admissible $\lambda > 0$ with the property that

$$\lambda_\epsilon|_{\epsilon=0} = \lambda$$
$$\frac{\mathrm{d}}{\mathrm{d}\epsilon}\bigg|_{\epsilon=0} \lambda_\epsilon = \Delta\lambda \, . \tag{3.9}$$

ii. Use the exponential map $\exp : \mathbf{R} \to \mathbf{R}_+$ to construct the one-parameter curve

$$\lambda = \exp[\mu]$$
$$\lambda_\epsilon = \exp[\mu + \epsilon\Delta\mu] \, . \tag{3.10}$$

iii. The variation of λ is defined as the directional derivative

$$\Delta\lambda = \frac{\mathrm{d}}{\mathrm{d}\epsilon}\bigg|_{\epsilon=0} = \lambda\Delta\mu \, . \tag{3.11}$$

Let $\lambda^k \in \mathbf{R}_+$ be the stretch value in the k-iteration of the iterative solution procedure at a given (nodal) point $\boldsymbol{x} = \varphi(\xi^1, \xi^2)$ of the mid-surface. Further, let $\Delta\mu^{k+1} \in \mathbf{R}$ be the incremental logarithmic stretch at the point $\boldsymbol{x} = \varphi(\xi^1, \xi^2)$ assumed to be *given*. The *updated* stretch at $\boldsymbol{x} = \varphi(\xi^1, \xi^2)$ consistent with the geometric structure of the problem is then obtained by the formula

$$\lambda^{k+1} = \lambda^k \exp[\Delta\mu^k] \, . \tag{3.12}$$

In the context of our formulation of shell theory, however, the preceding update takes a remarkably simple additive form which entirely avoids the use of the exponential map at each step of the iterative solution procedure. Recall that the weak form given by equation (2.21) and the matrix equations of section 2.7, are formulated entirely in terms of the *logarithmic stretch* $\mu := \ln[\lambda]$. Consequently, we take the logarithm of formula (3.12) and obtain the simple *additive* update procedure

$$\boxed{\mu^{k+1} = \mu^k + \Delta\mu^k} \, . \tag{3.13}$$

Thus, in terms of the logarithmic stretch, the update procedure is additive. The total finite stretch is then obtained once the calculation is completed by exponentiation as

$$\boxed{\lambda_{\text{final}} = \exp[\mu_{\text{final}}]} \, . \tag{3.14}$$

4. NUMERICAL SIMULATIONS

The behavior of the shell formulation presented above, and the performance of its numerical implementation are examined in this section. The results of the *mixed interpolations* proposed in [23,24] are presented for several problems selected from the literature to demonstrate the *accuracy* of the method for *coarse* and *skewed* meshes. Two examples are considered to illustrate the *large deformation* capabilities of the present formulation, and *singularity-free* nature of the rotational updates. The results of the present formulation are shown to be in complete agreement with *classical approximations* for the critical loads of the axial compression of cylinders, and two examples of *post-buckling* problems are studied using *continuation* methods, *extended* systems, and *mode superposition* methods. Sections 4.1-4.3 examine the *inextensible* formulation, and the effects of *thickness change* are shown in section 4.4 via a large deformation example.

4.1. Linear Benchmarks

The performance of the proposed shell element was evaluated with several discriminating problems selected from the literature in [23,24,26]. We present in this section the results of two selected linear problems.

4.1.1. *Cook's Membrane.* A trapezoidal plate is clamped on one end and subjected to a distributed in-plane bending load on the other end, as shown in Figure 4.1. This problem has a considerable amount of shear deformation, and is an excellent test of an element's ability to model membrane dominated situations with skewed meshes. A finite element converged solution of 23.91 is used to normalize the results, which are shown in Figure 4.1. The material properties are $E = 1.0$, $\nu = 0.33$, and $h = 1.0$. The results are compared to the 4-noded (4-ANS) and 9-noded (9-ANS) *Assumed Natural Strain* elements of [16], and the bi-linear quadrilateral element with uniform reduced integration (4-URI).

4.1.2. *Pinched Hemispherical Shell.* A pinched hemispherical shell with an 18° hole at the top, is subject to two inward and two outward forces 90° apart is modeled using symmetry boundary conditions on one quadrant. This problem is a good test of the inextensional bending behavior of an element, and an excellent test for the ability of an element to model rigid body modes. The material properties are $E = 6.825 \times 10^7$ and $\nu = 0.3$, the radius is $R = 10$, and the thickness is $h = 0.04$.

The displacement at the points of application of the forces is nor-

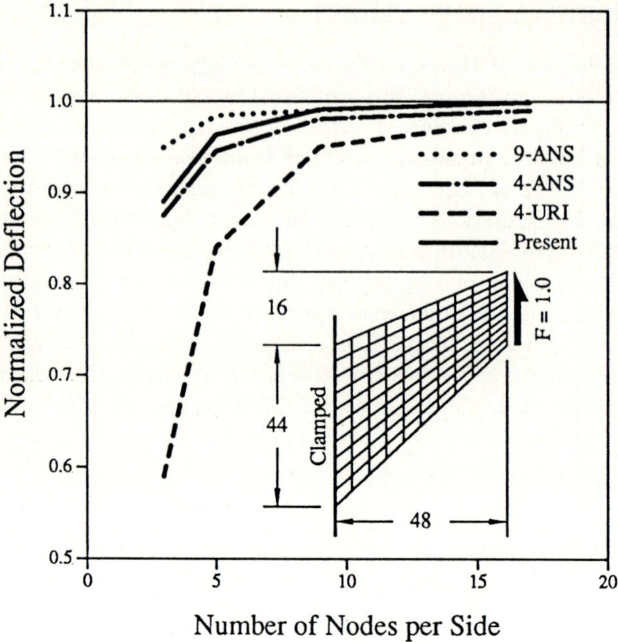

FIGURE 4.1. Description and Results of Cook's Membrane Problem.

malized by a value of 0.093 (obtained through analytical solutions based on asymptotic expansions). Figure 4.2 compares the proposed element's performance to the 4-noded Discrete Kirchoff element (4-DKQ) of reference [31], the bi-quadratic selective reduced integration (9-SRI) element, and the bi-quadratic uniform reduced integration element with γ-stabilization (9-GAMMA) of reference [6].

REMARK 4.1.

The results of the pinched hemisphere problem were obtained with both *inextensible* and *extensible* formulations, and the results were *identical* for both. This confirms that the present *extensible* formulation recovers the *inextensible* solution in the *thin-shell* limit without ill-conditioning.

4.2. Large Deformation Examples

We now consider two large deformation problems that demonstrate the ability of the formulation to capture large rotations and displacements and, at the same time, illustrate the robustness of the numerical implementation.

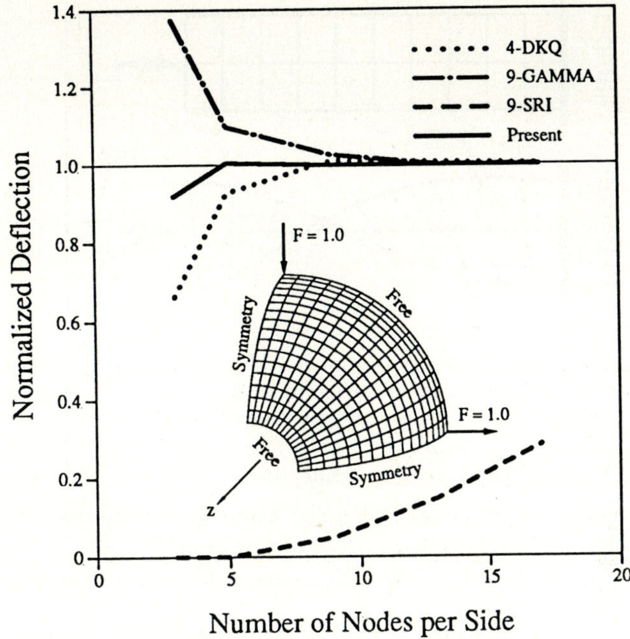

FIGURE 4.2. Description and Results of the Pinched Hemisphere with an 18° Hole. Symmetry is used and only one quadrant is modeled.

4.2.1. *Torsion of a Flat Plate Strip.* A torsional moment is applied to the end of an initially flat plate strip, leading to a relative torsional rotation of $\approx 180°$. The initial and deformed mesh configurations are shown in Figure 4.3. The material properties are $E = 12 \times 10^6$, $\nu = 0.3$, the length is $l = 1.0$, the width is $w = 0.25$, and the thickness is $t = 0.1$. Remarkably, the final configuration shown in Figure 4.3 is attained in *three load-steps*.

4.2.2. *Pinched Hemispherical Shell. Finite Deformation Version.* The pinched hemisphere problem, presented in section 4.1.2, is used to illustrate the large deformation capabilities of the formulation. The loads are increased by two orders of magnitude in order to obtain deflections of $\approx 60\%$ of the initial radius. In contrast with the linear case, the deflections under the loads, *are not equal.* A plot of the pinching load values versus the corner defections is shown in Figure 4.4 (b) for different mesh configurations. This calculation was carried out in *ten load steps.*

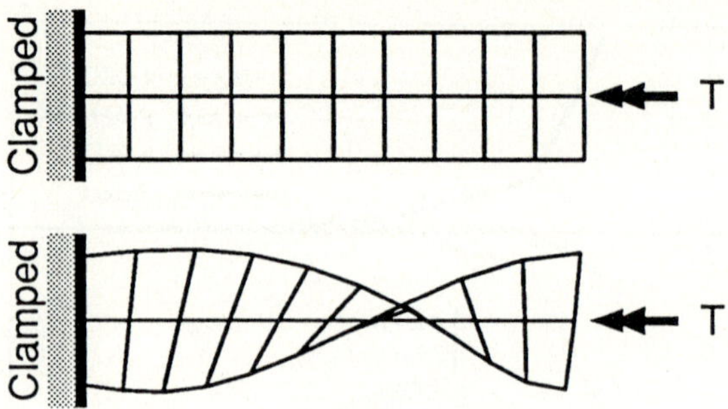

FIGURE 4.3. Initial and deformed mesh configurations for the torsion of a clamped plate strip. Final configuration is arrived at in *three* load steps. Deformations are plotted without magnification.

4.3. Buckling and Post-Buckling Analysis

This section is concerned with the accurate representation of the buckling and post-buckling/post-limit response of thin-shells. The linearized buckling analysis is simply the solution of the eigenvalue problem

$$\left[K_{\mathrm{mat}} - \phi_n K_{\mathrm{geo}}\right]\Psi_n = 0 \tag{4.1}$$

where K_{mat} and K_{geo} are the *material* and *geometric* stiffness matrices at the reference configuration, ϕ_n are the buckling load levels, and Ψ_n are the buckling mode shapes. Our solution procedure for tracing the non-linear load-deflection paths in the post-critical range makes essential use of standard *continuation techniques* as in [19,20], [14] and others. In particular, we employ an arc-length constraint in the form of *updated hyperplane arc-length*. Our implementation is essentially that described in [30], and [21].

4.3.1. Effect of L/R Ratio on the Buckling Load of an Axially Compressed Cylinder. Clamped, Simply-Supported, and Free End Boundary Conditions. The buckling of axially compressed cylinders is a classical problem extensively studied from an analytical, experimental and numerical perspective in the literature; see e.g., the recent review account of [8]. It is a classical result that, except for the case of free boundaries, the magnitude of the buckling load is highly insensitive to the edge boundary conditions provided that "the length of the cylinder is not small (say $L > 2R$)" ([32, pp. 464]). On the other hand, for extremely small values of the length to radius ratio, L/R, the behavior of the shell closely resembles that of a collection of small meridional strips, and the buckling load tends to the classical Euler's formula for a strut ([32, pp. 466]). In this latter case, the effect of boundary conditions becomes crucial. For instance, the critical load for the very short

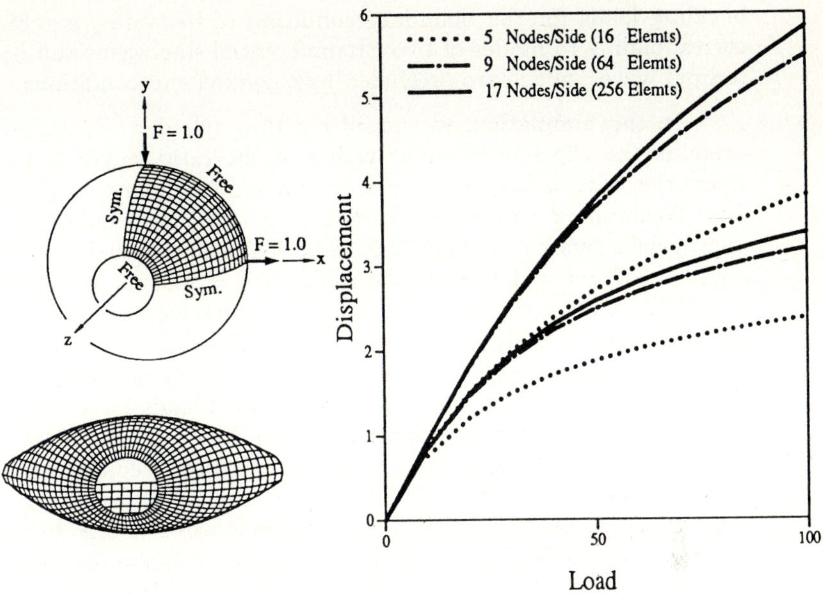

FIGURE 4.4. (a) Initial and deformed mesh configurations for the *large strain* version of the pinched hemisphere problem. The final configuration, involving deformations of $\approx 60\%$ of the radius, is arrived at using *ten* load steps. Deformations are plotted without magnification. (b) Deflection-Load plots for the different mesh configurations used to model the *large deformation* version of the pinched hemisphere problem. In contrast with the linear problem, the displacements under the pinching loads are not equal. The three top curves correspond to the displacements under the vertical loads, while the lower curves represent the displacements under the horizontal loads.

simply-supported cylinder, which is associated with a reticulated mode shape with one half sine wave in the axial direction and several waves in the circumferential direction, will be much smaller than the buckling load corresponding to the clamped case. The buckling mode shapes for the clamped short cylinder case are axisymmetric.

For long cylinders, an approximate solution for the critical load which is independent of length, referred to herein as the classical solution, is given by (see e.g., [32, pp. 465])

$$P_{cl} \approx \frac{Et^2}{R\sqrt{3(1-\nu^2)}} \, . \tag{4.2}$$

This approximation is valid only for long cylinders with fixed end boundary conditions (clamped or simply-supported). The eigen-modes associated with these critical loads can be axisymmetric, reticulated, or "diamond" shaped. Reference [10], however, shows analytically that the buckling loads for the boundary condition of free ends, is $0.38 \times P_{cl}$, corresponding to modes of two circumferential sine-waves and no longitudinal waves, which are precluded by the fixed end conditions.

In this simulation, we consider a thin cylinder ($R/t \approx 400$) and examine the effect of different values of the ratio length L to radius R on the critical load for the different end boundary conditions. A 20×20 mesh is used to model one-eighth of the cylinder, using the appropriate symmetry conditions. The material properties of the shell are $E = 567$, $\nu = 0.3$, the cylinder radius is $R = 100$, and the thickness is $t = 0.247$. The results obtained in the simulation for several values of the ratio L/R are presented in Figure 4.5 normalized relative to the classical value of P_{cl} given by (4.2). From this Figure, it is readily apparent that the present formulation predicts higher critical loads for short clamped cylinders than short simply-supported cylinders and cylinders with free ends. For long cylinders, the present formulation predicts that critical loads for the clamped and simply-supported cases will converge to the classical approximation, and the free end case will converge to the value reported by [10]. Hence, the computed results are in complete agreement with the classical behavior of cylindrical shells, as described above.

4.3.2. *Snap Through of a Hinged Cylindrical Panel.* A shallow shell in the form of a cylindrical panel with small curvature is pinned and subjected to a central load, as shown in Figure 4.6. Increasing values of the applied load lead, eventually, to snap-through of the panel and reversal of its curvature. This problem is simulated using a 4×4 mesh on one quadrant using symmetry conditions for two different values of the thickness: $R/t = 200$, (Figure 4.6(a)) and $R/t = 400$ (Figure 4.6(b)). In the case of $R/t = 200$, the load-deflection path contains two limit points, and displacement control can be used to solve the equations successfully. In the $R/t = 400$ case, however, the path "kicks in" after the first limit-point, displacement-control thus fails and use of arc-length control becomes necessary. The results shown in Figures 4.6(a),(b), obtained with rather large load steps, are in complete agreement with those reported in the literature; see i.e., [11] and [7].

4.3.3. *Post-Buckling of an L-Shaped Plate Strip Under an In-Plane Bending Load.* We consider the stability analysis of an L-shaped flat plate which is clamped on one end and subjected to an in-plane load on the other, as shown in Figure 4.7. This problem has been investigated by [4], and [28,29] to assess the performance of nonlinear rod elements. The material properties for this problem are: Young's modulus $E = 71240$, Poisson's ratio $\nu = 0.31$, the edge length is $l = 240$, the edge width is $w = 30$, and the thickness is $t = 0.6$. A 68 element mesh is used to evaluate the post-buckling response of the square bracket frame. The bifurcation point is arrived at by employing an extended system of

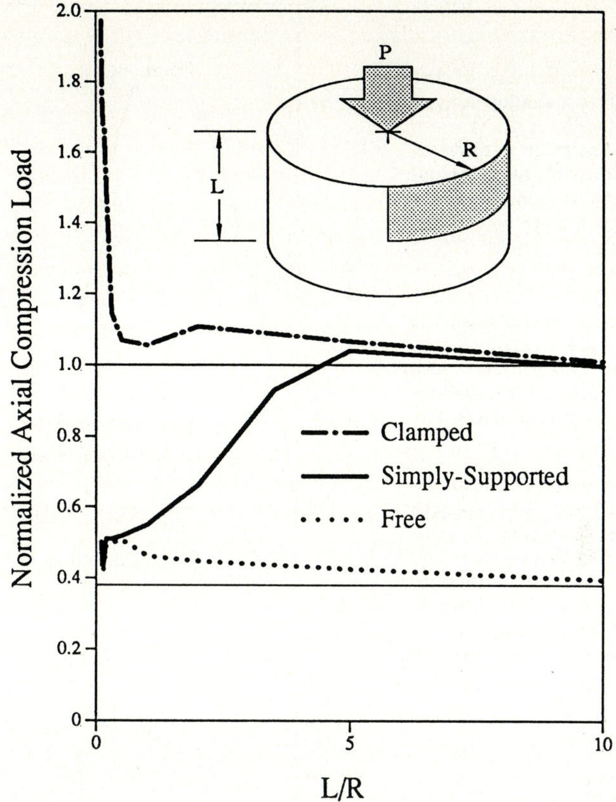

FIGURE 4.5. Effect of length-to-radius ratio L/R on the critical buckling load of an axially compressed cylinder under different boundary conditions. Results are obtained using linearized buckling analysis and normalized with respect to the classical approximation (4.2). A 20×20 mesh is employed to discretize the shaded area using the appropriate symmetry boundary conditions.

equations, as described in [33], combined with a mode-switching technique at the bifurcation point, as in [34]. Thus, the stability analysis is performed with no added initial imperfection to the structure. The post-buckling response is shown in Figure 4.7 where a plot of the load versus the deflection in the out-of-plane direction is given. We observe that this load-defection curve remains flat within a wide range of out-of-plane deflections, up to a value of $\delta z \approx 50$ where a pronounced stiffening of the curve is observed. These results are in good agreement with the general response pattern reported [27], and [4].

FIGURE 4.6.

(a) Problem definition of the snap-through of a shallow hinged cylindrical panel. A 4 × 4 mesh is used to discretize a quarter-section of the problem, employing symmetry boundary conditions. Two cases are considered by varying the thickness: $R/t = 200$ and $R/t = 400$.

(b) Load-deflection path for the $R/t = 200$ case. Displacement control using two different step levels are employed to demonstrate the robustness of the formulation.

(c) Load-deflection path for the $R/t = 400$ case. Arc-Length control was necessary in this case, due to the limit points and "kick-ins" present in the solution path. Again two different arc-length levels are employed to demonstrate the robustness of the formulation.

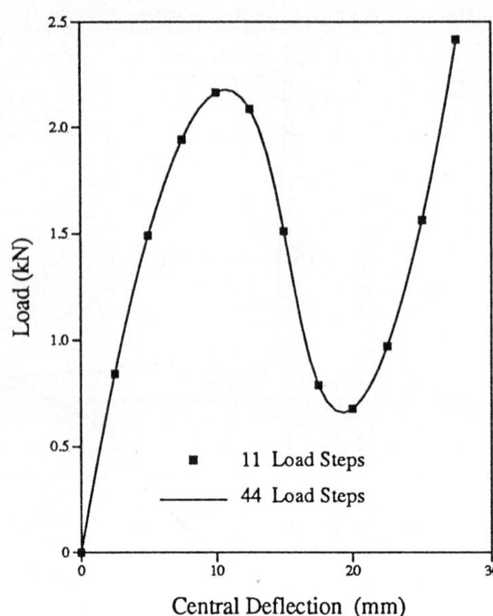

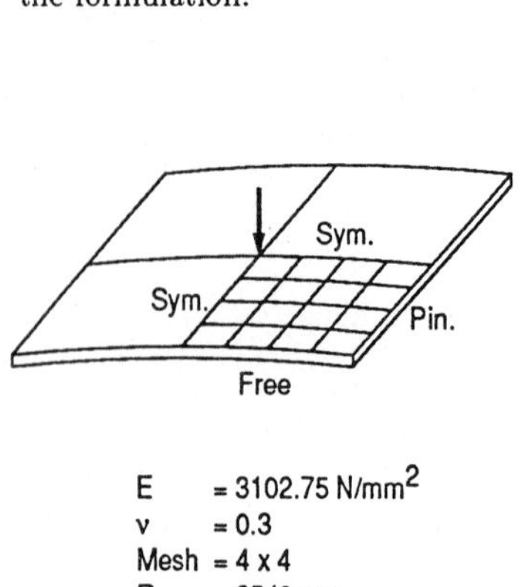

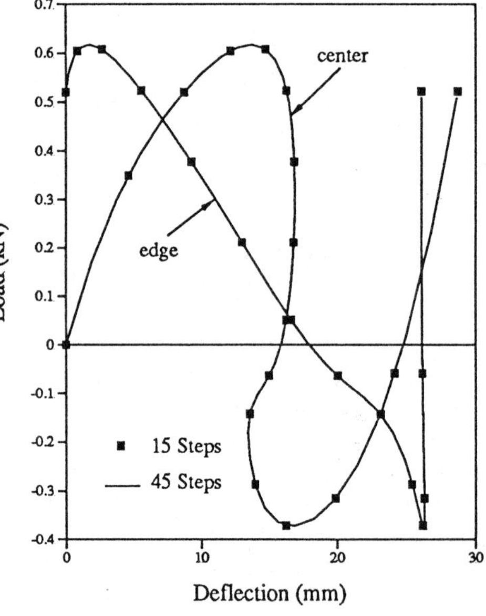

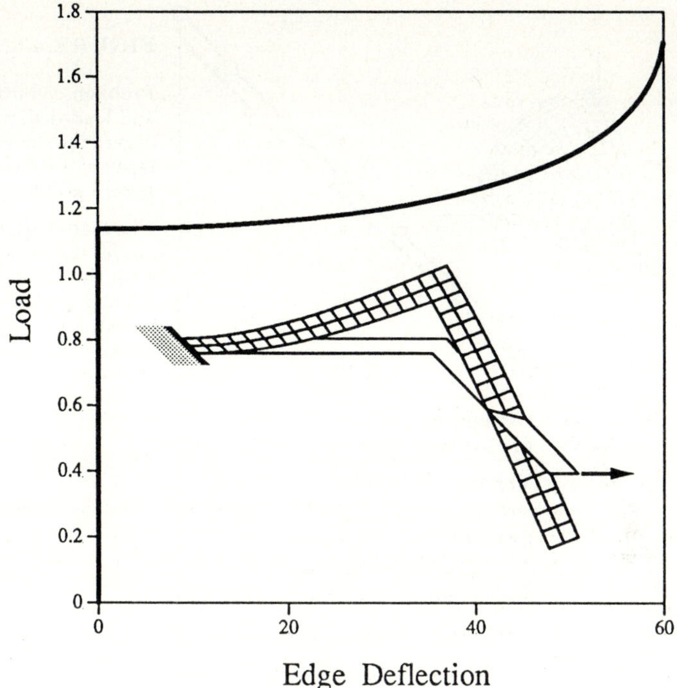

Edge Deflection

FIGURE 4.7. Load-deflection curve in the post-buckling regime for the clamped square bracket using a 68 element mesh. The applied load is plotted against the out-of-plane deflection.

4.4. Thickness Change

By careful examination of the weak form of the momentum balance equations, [26] shows that loads applied to the surface of a shell structure affect not only the displacement of the mid-surface, but also the deformation of the director. The following example (the collapse of a clamped hemispherical shell of a rubber material) demonstrates the *localized effects* of thickness change and the importance of the correct *surface loading*. We denote by *correct* loading that loading which takes into account the effects of surface loading and by *incorrect* loading that loading which is simply applied to the mid-surface.

4.4.1. *Collapse of a Rubber Sphere.*   A hemisphere is clamped along the circumference and subjected to a point load at the pole as shown in Figure 4.8. Load control was used to drive the top of the

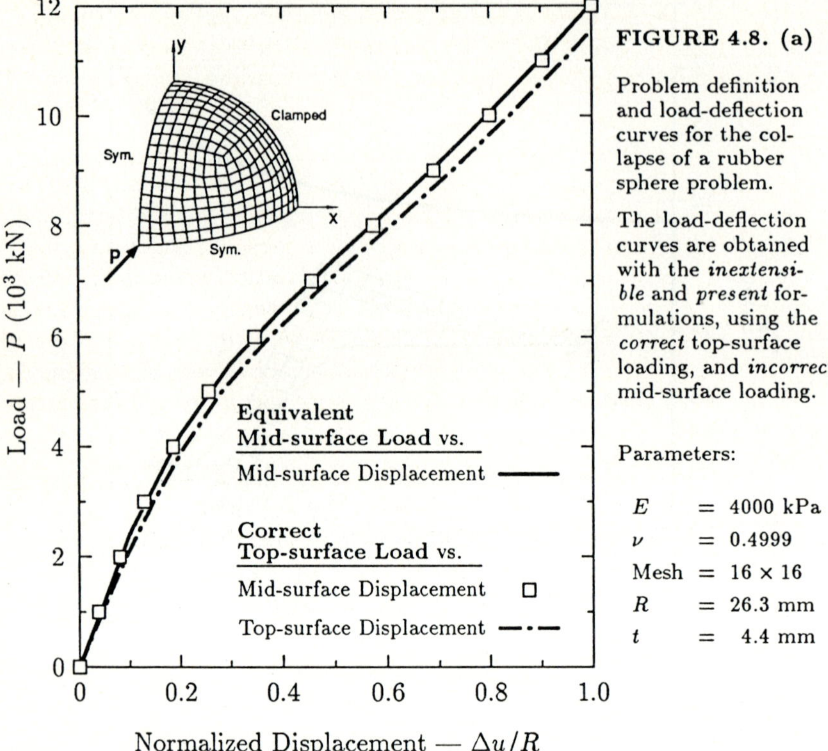

Load — P (10^3 kN)

Equivalent
Mid-surface Load vs.
Mid-surface Displacement

Correct
Top-surface Load vs.
Mid-surface Displacement □
Top-surface Displacement — · —

Normalized Displacement — $\Delta u/R$

FIGURE 4.8. (a)

Problem definition and load-deflection curves for the collapse of a rubber sphere problem.

The load-deflection curves are obtained with the *inextensible* and *present* formulations, using the *correct* top-surface loading, and *incorrect* mid-surface loading.

Parameters:

E	=	4000 kPa
ν	=	0.4999
Mesh	=	16 × 16
R	=	26.3 mm
t	=	4.4 mm

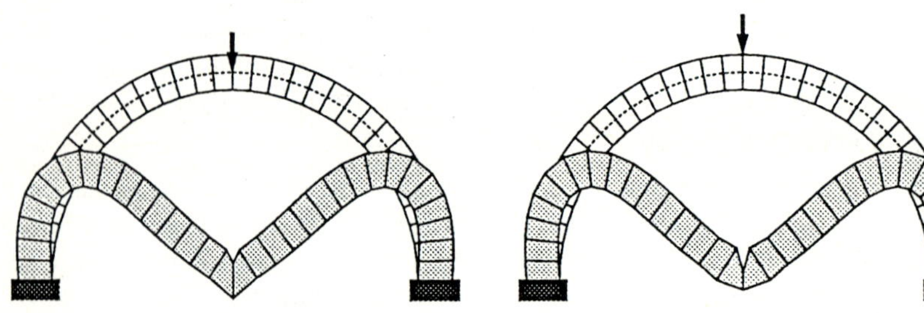

FIGURE 4.8. (b)

Sectional view of the deformed configuration subject to the *incorrect* mid-surface loading. Virtually no thickness change is observed, and the bottom surface (spuriously) exhibits a "kink" under the point load.

FIGURE 4.8. (c)

Sectional view of the deformed configuration subject to the *correct* top-surface loading. Approximately 52% thickness change is observed, and the bottom surface no longer exhibits the "kink" under the point load.

hemisphere the distance of the radius to complete collapse in 10 steps. The formulation presented in the previous sections is used to analyze this problem using the *incorrect* mid-surface loading, and the *correct* top-surface loading (which results in loading of the thickness stretch variable).

The load-deflection plots of the inextensible formulation of [24] and the results of the present formulation are presented in Figure 4.8 (a). The loads resulting from both loadings of the present formulation seem to coincide with those of the inextensible formulation when plotted with reference to the *mid-surface* displacement. However, when these loads are plotted against the *top-surface* displacement (i.e., the true point of application of the load), a significant change is observed for the *correct* loading. This further indicates the degree of localization of thickness change effects.

The final deformed configuration using the *incorrect* loading is shown in Figure 4.8 (b), whereas Figure 4.8 (c) depicts the deformed configuration when the *correct* loading of the present formulation is employed. The former case produces almost no thickness change, whereas the latter yields the anticipated change in thickness localized to the point of application of the load.

ACKNOWLEDGEMENTS

We are indebted to P. Wriggers for his generous involvement in the implementation of the continuation methods employed in this paper. We wish to thank professors S.S. Antman, T.J.R. Hughes, J. Marsden, C.R. Steele and R.L. Taylor for many helpful discussions. Support was provided by N.S.F. grant 2-DJA-491 with Stanford University. This support is gratefully acknowledged.

REFERENCES

1. Ahmad, S., B.M. Irons, and O.C. Zienkiewicz, [1970], "Analysis of thick and thin shell structures by curved finite elements," *Int. J. Num. Meth. Engng.*, **2**, pp. 419-451.

2. Antman, S.S., [1976a], "Ordinary differential equations of nonlinear elasticity I: Foundations of the theories of non-linearly elastic rods and shell," *Arch. Rat. Mech. Anal.* **61**, Vol.4, p.307-351.

3. Antman, S.S., [1976b], "Ordinary differential equations of nonlinear elasticity II: Existency and regularity theory for conservative boundary value problems," *Arch. Rat. Mech. Anal.* **61**, Vol.2, p.353-393.

4. Argyris, J.H., H. Balmer, J.St. Doltsinis, P.C. Dunne, M. Haase, M. Hasse, M. Kleiber, G.A. Malejannakis, J.P. Mlejenek, M. Muller, and D.W. Scharpf, [1979], "Finite Element Method — The Natural Approach", *Comput. Meths. Appl. Mech. Engng.*, **17/18**, 1 106.

5.   Bathe, K.J., and E.N. Dvorkin, [1984], "A continuum mechanics based four-node shell element for general non-linear analysis," *Int. J. Computer-Aided Engng. Software*, Vol. 1.

6.   Belytschko, T., H. Stolarski, W.K. Liu, N. Carpenter, & J. S-J. Ong, [1985], "Stress Projection for Membrane and Shear Locking in Shell Finite elements," *Comp. Meth. Appl. Mech. Engng.*, Vol 51, pp 221-258.

7.   Bergan, P.G., Horrigmoe, G., Krakeland, B., and Soreide, T.H., [1978], "Solution techniques for nonlinear Finite Element problems," *Int. J. Num. Meth. Engng.*, **12**, pp.1677-1696.

8.   Bushnell, D., [1985], *Computerized Buckling Analysis of Shells*, Mechanics of Elastic Stability, Vol. **9**, Martinus Nijoff Publishers, Boston.

9.   Ericksen, J.L., and Truesdell, C., [1958], "Exact theory of stress and strain in rods and shells," *Arch. Rat. Mech. Anal.*, Vol. 1, No. 4, pp. 295-323.

10.   Hoff, N.J., and T.C. Soong [1965], "Buckling of Circular Cylindrical Shells in Axial Compression," *Int. Journ. Mech. Sci.*, **7**, pp. 489-520.

11.   Horrigmoe, G., [1977], "Finite Element Instability Analysis of Free-Form Shells," Report 77-2, *Division of Structural Mechanics, Norwegian Institute of Technology, University of Trondheim, Norway*.

12.   Hughes, T.J.R., and W.K. Liu, [1981a], "Nonlinear finite element analysis of shells: Part I-Three-dimensional shells," *Comp. Meth. Appl. Mech. Engng.*, **26**, 331-362.

13.   Hughes, T.J.R., and W.K. Liu, [1981b], "Nonlinear finite element analysis of shells: Part II -Two-dimensional shells," *Comp. Meth. Appl. Mech. Engng.*, **27**, 167-182.

14.   Keller, H.B., [1977], "Numerical Solution of Bifurcation and Nonlinear eigenvalue problems," *Applications of Bifurcation Theory*, P. Rabinowitz, ed., Academic Press, New York, pp. 359-384.

15.   Naghdi, P.M., [1972], "The theory of shells," in *Handbuch der Physik*, Vol Via/2, Mechanics of Solids II, C. Truesdell Ed., Springer-Verlag, Berlin.

16.   Parks, K.C., and G.M. Stanley, [1986], "A curved C^0 shell element based on assumed natural-coordinate strains," *J. Appl. Mech.*, Vol. 53, No. 2, pp. 278-290.

17.   Reissner, E., [1964], "On the Form of Variationally Derived Shell Equations," *J. Appl. Mech.*, Vol. 31, pp. 233-238.

18.   Reissner, E., [1974], "Linear and Nonlinear Theories of Shells, in *Sechler Anniversary Volume*, pp.29-44, Prentice Hall, New York.

19.   Rheinboldt, W.C., [1974], "Methods for Solving Systems of Nonlinear Equations," *CBMS Regional Conference Series in Applied Mathematics*, **14**, Society for Industrial and Applied Mathematics, Philadelphia.

20.   Rheinboldt, W.C., [1986], *Numerical Analysis of Parametrized Nonlinear Equations*, Wiley Interscience, NewYork.

21.   Schweizerhof, K.H., and P. Wriggers, [1986], "Consistent Linearization of Path Following Methods in Nonlinear FE Analysis," *Comp. Meth. Appl. Mech. Engng.*, **59**, 261-279

22.   Simo, J.C. and D.D. Fox, [1989], "On a Stress Resultant Geometrically Exact Shell Model. Part I: Formulation and Optimal Parametrization," *Comp. Meth. Appl. Mech. Engng,* **72**, 267-304.

23.   Simo, J.C., D.D. Fox and M.S. Rifai, [1989], "On a Stress Resultant Geometrically Exact Shell Model. Part II: The Linear Theory; Computational Aspects," *Comp. Meth. Appl. Mech. Engng,* to appear.

24.   Simo, J.C., D.D. Fox and M.S. Rifai, [1989], "On a Stress Resultant Geometrically Exact Shell Model. Part III: Computational Aspects of the Nonlinear Theory," *Comp. Meth. Appl. Mech. Engng,* to appear.

25.   Simo, J.C. and J.G. Kennedy, [1989], "On a Stress Resultant Geometrically Exact Shell Model. Part IV: Nonlinear Plasticity. Formulation and Integration Algorithms," *Comp. Meth. Appl. Mech. Engng,* to appear.

26.   Simo, J.C., M.S. Rifai and D.D. Fox, [1989], "On a Stress Resultant Geometrically Exact Shell Model. Part V: Variable Thickness Shells with Through-the-Thickness Stretching," *Comp. Meth. Appl. Mech. Engng,* to appear.

27.   Simo, J.C. and L.V. Quoc, [1986], "A 3-Dimensional Finite Strain Rod Model. Part II: Geometric and Computational Aspects," *Comp. Meth. Appl. Mech. Engng.*, **58**, 79-116.

28.   Simo, J.C., and L. Vu-Quoc, [1987a], "A beam model including shear and torsional warping distorsions based on an exact geometric description of nonlinear deformations," *Int. J. Solids Structures,* Submitted for publication.

29.   Simo, J.C. and L. Vu-Quoc, [1987b], "On the dynamics in Space of rods undergoing large motions -A geometrically exact approach," *Comp. Meth. Appl. Mech. Engng.,* To appear.

30.   Simo, J.C., P. Wriggers, K.H. Schweizerhoff and R.L. Taylor, [1986], "Postbuckling Analysis Involving Inelasticity and Unilateral Constraints," *Int. J. Num. Meth. Engng,* **23**, 779-800.

31.   Taylor, R.L., [1987], "Finite Element Analysis of Linear Shell Problems," *Proceedings of the Mathematics of Finite Elements and Applications,* (MAFELAP 1987), S.R. Whitheman Editor.

32.   Timoshenko S.P., and J.M. Gere, [1961], *Theory of Elastic Stability* , Mc-Graw Hill, New York.

33.   Wriggers, P., and J.C. Simo, [1989]. "A General Purpose Algorithm for Extended Systems in Continuation Methods," *Preprint.*

34.   Wriggers, P., P. Wagner, and C. Miehe. [1988], "A Quadratic Convergent Procedure for the Calculation of Stability Points in Finite Element Analysis," *Comp. Meth. Appl. Mech. Engng,* to appear.

Elastic-Plastic Analysis of Thin Shells and Folded Plate Structures with Finite Rotations

E. Stein, F. Gruttmann and K.H. Lambertz

1. Concepts for describing finite rotations of thin shells

During the last years numerous nonlinear shell theories using different levels of approximations have been developed. These theories are able to model small, moderate, large or finite rotations. Going from moderate to finite rotations a significant difficulty of nonlinear shell theories is associated with the incorporation of rotations into the general shell equations since finite rotations are not commutative and thus do not transform like vectors. Different approaches to finite rotations have been discussed in the literature, see e.g. Reissner [18], Wempner [29], Simmonds and Danielson [22], Libai and Simmonds [10], Atluri [1], Badur and Pietraskiewicz [3] , Makowski and Stumpf [11] and Simo and Fox [23].

However there exists a gap between modern nonlinear shell theories and the computational approach using e.g. numerical methods like finite elements. In the latter nonlinear shell models are essentially based on the so called degenerated approach, see e.g. Ramm [16], Park and Stanley [15]. This method uses the three dimensional theory and introduces the shell assumptions on the finite element level. Another procedure for the developement of nonlinear shell models is based on shell theories using the Kirchhoff–Love hypothesis, see e.g. Harte [8], Nolte [14]. However, this assumption leads in the finite rotation case to numerous nonlinear terms and thus cannot be implemented in an efficient way for the finite rotation case. Recke and Wunderlich [17] used a mixed finite element model and the so called total Lagrangian description for the computation of large rotations of shells. However in their approach the magnitude of rotations is restricted within a load increment. Simo and Fox [23] derive a nonlinear shell theory where they assume inextensibility of the director field. A singularity–free parametrization and update procedures of the rotation fields are developed.

In this paper we like to present a finite element model for shells subjected to finite rotations and deflections. As a basis a nonlinear shear elastic shell theory is used which leads to relatively simple closed form expressions for the strain measures which are computed from the weak form of equilibrium equations for arbitrary shell geometries.

The derivation of a nonlinear shell theory from the three–dimensional continuum theory is based on the introduction of a kinematical assumption. To develop a so-called shear elastic theory we introduce a director which is not normal to the deformed middle surface of the shell. Then by using an orthornormal coordinate system, attached to the moving middle surface, we are able to express derivatives of the rotation tensor by axial vectors. The Green–Lagrangian strain tensor which has been frequently used to derive shell theories yields very complicated expressions for the rotations. Thus it is not very useful for computational models. In this paper the shell strains are computed by a polar decomposition of the material deformation gradient. The principle of virtual work is written in such a way that Cristoffel symbols do not appear explicitly in the formulation. This is very important for the numerical implementation. The work conjugate stresses appear to be the stress resultants of the Biot stress tensor. This tensor is invariant against rigid body motions and thus can be used to formulate constitutive laws.

The finite element model allows the discretisation of arbitrary smooth shell geometries which are obtained by an isoparametric mapping. Thus a broad class of shell problems can be investigated like e.g. cylindrical, spherical, or hypar shells. Since shear strains are considered the shape functions for displacements and rotations only must fulfil C^0-continuity. The independent rotational degrees of freedom are described by different representations.

Associated with shear–elastic shell formulations based on displacement methods as well as with the degenerated concept are locking phenomena. Since our main goal is the formulation of shell elements undergoing finite rotations we use standard formulations to overcome locking. Methods which have been implemented are the reduced integration technique,see e.g. Malkus and Hughes [12], and the so called hour–glass stabilization, see e.g. Belytschko et. al. [4]. Better finite elements based on mixed methods should be used.

The resulting nonlinear equations are then solved by a Newton procedure which is coupled — depending on the problem at hand — with an arc–length scheme, see e.g. Schweizerhof, Wriggers [21].

The advantages of the developed finite element can be observed in non–linear applications with large deflections and rotations. Finite rotations may be calculated using only one load increment if elastic material response is assumed. Thus, limitations in the load step size are only due to the global convergence properties of the solution algorithm or associated with stability points.

The examples which have been investigated show good agreement with solutions given in the literature which often have been obtained with considerably more load increments and thus with more computational effort.

2. Kinematics in shellspace and of the middle surface

In this section the kinematics of thin shells are described. We assume a smooth, continuous, and differentiable middle surface Ω. The boundary of the shell will be denoted by C. The shell thickness h is small compared to the smallest curvature radius R.

We define the orthornormal basis t_i (i=1,2,3) with the associated coordinates s_i in the initial configuration. The director t_3 is normal to the undeformed shell middle surface thus the vectors t_1 and t_2 are tangent vectors. Further, the orthornormal basis a_i is introduced in the current configuration. Since transverse shear strains are allowed the vector a_3 is no longer normal to the deformed shell middle surface. Since t_i and a_i are orthonormal vectors we can give the following transformation

$$t_i(s_\alpha) = R_0(s_\alpha)e_i \quad , \quad a_i(s_\alpha) = R(s_\alpha)t_i(s_\alpha) \quad \alpha = 1,2 \quad , \qquad (2.1)$$

where e_i denotes the fixed cartesian basis. In chapter 4 the orthogonal tensors R_0 and R are described in matrix notation with different rotational degrees of freedom. To derive the material deformation gradient it is necessary to form the derivatives of the basis a_i

$$\frac{\partial a_i}{\partial s_\alpha} = a_{i,\alpha} = \Omega_\alpha a_i = \omega_\alpha \times a_i. \qquad (2.2)$$

with respect to the surface coordinates s_α. Therefore we introduce a skew-symmetric tensor Ω_α which is defined by

$$\Omega_\alpha = \frac{\partial(RR_0)}{\partial s_\alpha}(RR_0)^T \quad , \qquad (2.3)$$

and the associated axial vector $\boldsymbol{\omega}_\alpha$. In a similar way the time derivative and the variation of the base vector $\mathbf{a}_i$

$$\frac{\partial \mathbf{a}_i}{\partial t} = \dot{\mathbf{a}}_i = \mathbf{W}\mathbf{a}_i = \mathbf{w} \times \mathbf{a}_i \quad ,$$
$$\delta \mathbf{a}_i = \delta \mathbf{W}\mathbf{a}_i = \delta \mathbf{w} \times \mathbf{a}_i$$

(2.4)

with the skew–symmetric tensors $\mathbf{W} = \dot{\mathbf{R}}\mathbf{R}^T$, $\delta \mathbf{W} = \delta \mathbf{R}\mathbf{R}^T$ and the associated axial vector $\mathbf{w}$ and $\delta \mathbf{w}$ are derived.

Using a Reissner-Mindlin kinematic the position vector $\mathbf{x}$ of a point in the shell space in the current configuration is given by

$$\mathbf{x} = \mathbf{x}_0 + \zeta \mathbf{a}_3 \qquad -\frac{h}{2} \le \zeta \le +\frac{h}{2}.$$

(2.5)

with the classical assumption of inextensibility of the director vektor $\mathbf{a}_3$. Here,

$$\mathbf{x}_0 = \mathbf{X}_0 + \mathbf{u} = x_i\,\mathbf{e}_i = (X_i + u_i)\,\mathbf{e}_i$$

(2.6)

is the position vector of a point on the shell middle surface and $\zeta = s_3$ is the coordinate in the thickness direction of the shell. The assumption of inextensibility is valid for thin shells with $h/R \ll 1$ and is also justified by the small strain assumption. To describe the deformation of the shell it is necessary to derive the material deformation gradient $\mathbf{F}$ which is defined by $d\mathbf{x} = \mathbf{F}\,d\mathbf{X}$. Making use of the axial vectors $\mathbf{F}$ can be computed with regard to the shell kinematics as

$$\mathbf{F} = (\mathbf{x}_{0,1} + \boldsymbol{\omega}_1 \times \zeta \mathbf{a}_3) \otimes \mathbf{e}_1 + (\mathbf{x}_{0,2} + \boldsymbol{\omega}_2 \times \zeta \mathbf{a}_3) \otimes \mathbf{e}_2 + \mathbf{a}_3 \otimes \mathbf{e}_3 .$$

(2.7)

With (2.4) the variation of the deformation gradient leads to

$$\delta \mathbf{F} = (\delta \mathbf{x}_{0,1} + \delta \boldsymbol{\omega}_1 \times \zeta \mathbf{a}_3 + \boldsymbol{\omega}_1 \times \zeta(\delta \mathbf{w} \times \mathbf{a}_3)) \otimes \mathbf{e}_1$$
$$+ (\delta \mathbf{x}_{0,2} + \delta \boldsymbol{\omega}_2 \times \zeta \mathbf{a}_3 + \boldsymbol{\omega}_2 \times \zeta(\delta \mathbf{w} \times \mathbf{a}_3)) \otimes \mathbf{e}_2$$
$$+ (\delta \mathbf{w} \times \mathbf{a}_3) \otimes \mathbf{e}_3 .$$

(2.8)

This expression is used in the next section for the formulation of the internal virtual work. Using (2.7) the Green-Lagrangian strain tensor $\mathbf{E} = \frac{1}{2}(\mathbf{F}^T\mathbf{F} - \mathbf{1})$ could be obtained in a material description. However, this leads to very complicated terms and thus in this paper a different approach is followed. The shell strains are derived by a polar decomposition of the material deformation gradient.

3. Principle of virtual work

3.1 Dynamic field equations and boundary conditions

In this section we define stress resultants and stress couple resultants from the three-dimensional theory and develop the dynamic field equations equations and the boundary conditions of the shell middle surface. The stress resultants $\mathbf{n}_\alpha$ and stress couple resultants $\mathbf{m}_\alpha$

$$\mathbf{n}_\alpha := N_{\alpha i}\mathbf{a}_i = \int_{-\frac{h}{2}}^{\frac{h}{2}} \mathbf{p}_\alpha\, d\zeta \quad , \quad \mathbf{m}_\alpha := \bar{M}_{oi}\mathbf{a}_i = \int_{-\frac{h}{2}}^{\frac{h}{2}} (\mathbf{x} - \mathbf{x}_0) \times \mathbf{p}_\alpha\, d\zeta \tag{3.1}$$

are computed by integration of the stress vector $\mathbf{p}_\alpha$ with respect to the coordinate ζ. Here $\mathbf{p}_\alpha$ is defined by $\mathbf{P}(s_1, s_2, \zeta) = \mathbf{p}_i(s_1, s_2, \zeta) \otimes \mathbf{e}_i$, where $\mathbf{P}$ denotes the non-symmetric first Piola-Kirchhoff stress tensor. Following classical shell theories we assume $\bar{M}_{\alpha 3} = \mathbf{m}_\alpha \cdot \mathbf{a}_3 = 0$. The back transformation of $\mathbf{n}_\alpha$ and $\mathbf{m}_\alpha$ with the rotation tensors $\mathbf{R}$ and $\mathbf{R}_0$ yields

$$\mathbf{N}_\alpha := N_{\alpha i}\mathbf{e}_i = \mathbf{R}_0^T\mathbf{R}^T\mathbf{n}_\alpha \quad , \quad \mathbf{M}_\alpha := \bar{M}_{\alpha i}\mathbf{e}_i = \mathbf{R}_0^T\mathbf{R}^T\mathbf{m}_\alpha \quad . \tag{3.2}$$

Since $\mathbf{R}$ is obtained by a polar decomposition of the material deformation gradient, $\mathbf{N}_\alpha$ and $\mathbf{M}_\alpha$ are integrals of the non-symmetric Biot stress tensor defined by $\mathbf{R}^T\mathbf{P}$.

Starting with the extended principle of virtual work for a three-dimensional body B

$$\int_{(B)} (\mathbf{P} \cdot \delta\mathbf{F} - \rho_0\,\mathbf{b} \cdot \delta\mathbf{u})\, dV - \int_{(\partial B_\sigma)} \bar{\mathbf{t}}_0 \cdot \delta\mathbf{u}\, dA = \int_{(B)} \rho_0\ddot{\mathbf{x}} \cdot \delta\mathbf{u}\, dA \tag{3.3}$$

with the applied external loads $\bar{\mathbf{t}}_0$ and $\rho_0\mathbf{b}$ where ρ_0 is the density of the body in the initial configuration. Recalling the stress resultants and the stress couple resultants defined in (3.1) and (3.2) and the virtual deformation gradient (2.8), the principle of virtual work can be transformed into a two-dimensional form, see Gruttmann [6]. After some manipulations the internal virtual work is presented as

$$\delta\mathcal{W} = \int_{(B)} \mathbf{P} \cdot \delta\mathbf{F}\, d\zeta d\Omega$$

$$= \sum_{\alpha=1}^{2} \int_{(\Omega)} [\mathbf{n}_\alpha \cdot (\delta\mathbf{x}_{0,\alpha} - \delta\mathbf{w} \times \mathbf{x}_{0,\alpha}) + \mathbf{m}_\alpha \cdot \delta\mathbf{w}_{,\alpha}]\, d\Omega \tag{3.4}$$

$$= \sum_{\alpha=1}^{2} \int_{(\Omega)} [\mathbf{N}_\alpha \cdot \delta(\mathbf{R}_0^T\mathbf{R}^T\mathbf{x}_{0,\alpha}) + \mathbf{M}_\alpha \cdot \delta(\mathbf{R}_0^T\mathbf{R}^T\boldsymbol{\omega}_\alpha)]\, d\Omega \,.$$

The virtual work of the external loads is given by

$$\delta \mathcal{A} = \int\limits_{(\Omega)} (\bar{\mathbf{q}} \cdot \delta\mathbf{x}_0 + \bar{\mathbf{m}} \cdot \delta\mathbf{w}) \, d\Omega \tag{3.5}$$

with the applied surface tractions $\bar{\mathbf{q}}$ and $\bar{\mathbf{m}}$. The partial integration of (3.3) employing (3.4) and (3.5) yields

$$\sum_{\alpha=1}^{2} \{ -\int\limits_{(\Omega)} [(\mathbf{n}_{\alpha,\alpha} + \bar{\mathbf{q}} - \bar{\rho}\ddot{\mathbf{x}}_0) \cdot \delta\mathbf{x}_0] \, d\Omega$$

$$-\int\limits_{(\Omega)} [(\mathbf{m}_{\alpha,\alpha} + \mathbf{x}_{0,\alpha} \times \mathbf{n}_\alpha + \bar{\mathbf{m}} - I_\rho \dot{\mathbf{w}}) \cdot \delta\mathbf{w}] \, d\Omega \tag{3.6}$$

$$+\int\limits_{(C)} (\mathbf{n}_\alpha \cdot \delta\mathbf{x}_0 + \mathbf{m}_\alpha \cdot \delta\mathbf{w}) \, dC \} = 0.$$

Here we set $\bar{\rho} = \int\limits_{-\frac{h}{2}}^{\frac{h}{2}} \rho_0 \, d\zeta$ and the surface inertia is defined by $I_\rho = \int\limits_{-\frac{h}{2}}^{\frac{h}{2}} \rho_0 \zeta^2 \, d\zeta$. With the fundamental lemma of variational calculus we get the dynamic field equations of the shell

$$\left. \begin{aligned} \mathbf{n}_{1,1} + \mathbf{n}_{2,2} + \bar{\mathbf{q}} &= \bar{\rho}\ddot{\mathbf{x}}_0 \\ \mathbf{m}_{1,1} + \mathbf{m}_{2,2} + \mathbf{x}_{0,1} \times \mathbf{n}_1 + \mathbf{x}_{0,2} \times \mathbf{n}_2 + \bar{\mathbf{m}} &= I_\rho \dot{\mathbf{w}} \end{aligned} \right\} \text{in } \Omega \tag{3.7}$$

and the boundary conditions

$$\mathbf{n}_\alpha = \mathbf{0} \quad , \quad \mathbf{m}_\alpha = \mathbf{0} \qquad \text{on } C. \tag{3.8}$$

These equations are well known in the literature, see e.g. Reissner [19], Libai and Simmonds [10], Badur and Pietraszkiewics [3], Simo and Fox [23]. Reissner obtained these equations directly from the equilibrium of the stress resultants at an infinitesimal shell element whereas Badur and Pietraskiewicz arrive at equations (3.7) and (3.8) by using a three–dimensional Cosserat theory.

With the definitions

$$\begin{aligned} a_{\alpha i} &= (\mathbf{R}_0^T \mathbf{R}^T \mathbf{x}_{0,\alpha}) \cdot \mathbf{e}_i = \mathbf{x}_{0,\alpha} \cdot \mathbf{a}_i \\ \bar{b}_{\alpha i} &= (\mathbf{R}_0^T \mathbf{R}^T \boldsymbol{\omega}_\alpha) \cdot \mathbf{e}_i = \boldsymbol{\omega}_\alpha \cdot \mathbf{a}_i \end{aligned} \tag{3.9}$$

and the assumption $\bar{\mathbf{m}} \cdot \mathbf{a}_3 = 0$ we obtain the equilibrium of stress couple resultants with respect to the basis vector $\mathbf{a}_3$ in the static case

$$\sum_{\alpha=1}^{2}(\mathbf{m}_{\alpha,\alpha} + \mathbf{x}_{0,\alpha} \times \mathbf{n}_\alpha + \bar{\mathbf{m}}) \cdot \mathbf{a}_3 = e_{\gamma\beta}(N_{\alpha\beta}a_{\alpha\gamma} + \bar{M}_{\alpha\beta}\bar{b}_{\alpha\gamma}) = 0 \quad , \tag{3.10}$$

where $e_{\gamma\beta}$ is the surface permutation tensor.

3.2 Derivation of the strain measures

The polar decomposition of the membrane strains $a_{\alpha\beta}$ yields

$$\begin{aligned}
a_{\alpha\beta} &= R_{\alpha\gamma}\varepsilon_{\gamma\beta} & \varepsilon_{\gamma\beta} &= \varepsilon_{\beta\gamma}, \\
\bar{b}_{\alpha\beta} &= R_{\alpha\gamma}\kappa_{\gamma\beta} & \kappa_{\gamma\beta} &\neq \kappa_{\beta\gamma}.
\end{aligned} \tag{3.11}$$

The components $R_{\alpha\beta}$ may be written in a rotation matrix

$$(R_{\alpha\beta}) = \begin{pmatrix} cos\gamma & sin\gamma \\ -sin\gamma & cos\gamma \end{pmatrix} \quad , \quad tan\,\gamma = \frac{a_{12} - a_{21}}{a_{11} + a_{22}} \tag{3.12}$$

where $tan\,\gamma$ follows with the symmetry of the membrane strains ($\varepsilon_{12} = \varepsilon_{21}$). For ease of representation we will change the notation of the stress couples. By the cross product in (3.1) the indices of the components of $\bar{M}_{\alpha\beta}$ are defined. The first index denotes the normal of the cross section and the second index the direction of the stress couple. However, it is useful to specify the stress couples in the same way as the associated stresses. This transformation is a simple permutation which can be written by $M_{\alpha\beta} = \epsilon_{\gamma\beta}\bar{M}_{\alpha\gamma}$. In this case the work conjugate strains are $b_{\alpha\beta} = -\mathbf{a}_{3,\alpha} \cdot \mathbf{a}_\beta$. Using the polar decomposition (3.11) we are now able to compute the shell strains

$$\begin{aligned}
\tilde{\varepsilon}_{\alpha\beta} &= R_{\gamma\alpha}a_{\gamma\beta} - \delta_{\alpha\beta} = \tilde{\varepsilon}_{\beta\alpha} \quad , \\
\varepsilon_{\alpha3} &= a_{\alpha3} \quad , \\
\tilde{\kappa}_{\alpha\beta} &= R_{\gamma\alpha}b_{\gamma\beta} - \tilde{\kappa}^0_{\alpha\beta} \neq \tilde{\kappa}_{\beta\alpha} \quad ,
\end{aligned} \tag{3.13}$$

where $\tilde{\kappa}^0_{\alpha\beta}$ are the curvatures of the initial configuration. These strains are invariant against rigid body motions.

We are now able to reformulate the internal virtual work (3.4) as

$$\delta \mathcal{W} = \int\limits_{(\Omega)} (N_{\alpha\beta}\delta a_{\alpha\beta} + N_{\alpha 3}\delta a_{\alpha 3} + \bar{M}_{\alpha\beta}\delta \bar{b}_{\alpha\beta})d\Omega$$

$$= \int\limits_{(\Omega)} (\tilde{N}_{\beta\gamma}\delta\tilde{\varepsilon}_{\gamma\beta} + N_{\alpha 3}\delta\varepsilon_{\alpha 3} + \tilde{M}_{\beta\gamma}\delta\tilde{\kappa}_{\gamma\beta})d\Omega \qquad (3.14)$$

$$+ \int\limits_{(\Omega)} [\delta\gamma\, \epsilon_{\gamma\beta}(N_{\alpha\beta}a_{\alpha\gamma} + \bar{M}_{\alpha\beta}\bar{b}_{\alpha\gamma})]d\Omega$$

with stress resultants $\tilde{N}_{\beta\gamma} = N_{\alpha\beta}R_{\alpha\gamma}$ and $\tilde{M}_{\beta\gamma} = M_{\alpha\beta}R_{\alpha\gamma}$. The third integral in (3.14) vanishes since (3.10) holds.

Remarks:

(i) Following classical shell theories the constitutive law in chapter 4 is formulated with the symmetric part of the curvature strains $\tilde{\kappa}_{\alpha\beta}$, see e.g. [3],[23].

(ii) The strain measures for two-dimensional straight beams subjected to finite rotations can be derived as a special case from the shell strains (3.13). This leads to stretch, shear and curvature measures which conform to those reported in Reissner [19].

(iii) The classical linear Reissner–Mindlin theory for plates and shells may be derived from this non linear theory by applying standard consistent linearization procedures which are described in e.g. Marsden and Hughes [13], or Hughes and Pister [9].

3.3 Computation of the orthogonal matrices

For the calculation of the strain components we need explicit expressions of the orthogonal matrices and their associated axial vectors. Since our formulation in the preceeding sections only involves the rotation tensors and axial vectors the representation of the rotations does not affect the general equations. Thus within this framework we can incorporate different representations of finite rotations. Another approach may be found in e.g. Simo [23], who used a singularity-free parametrization with two degrees of freedom to describe the rotations and update procedures for the rotation field.

Now the rotation tensor $\mathbf{R}_0$ given by (2.1) will be derived. Note, that the base vectors

$$\mathbf{g}_1 = \frac{\partial \mathbf{X}_0}{\partial \xi} \qquad , \qquad \mathbf{g}_2 = \frac{\partial \mathbf{X}_0}{\partial \eta} \tag{3.15}$$

are obtained by partial differentiation of the position vector of the undeformed shell middle surface $\mathbf{X}_0(\xi,\eta)$ with respect to the convective coordinates ξ und η. The base vectors $\mathbf{g}_\alpha$ then define a tangential surface to the undeformed shell middle surface. Both vectors $\mathbf{g}_1$ and $\mathbf{g}_2$ are neither unit vectors nor perpendicular. However, we are able to compute an orthorgonal basis $\mathbf{t}_i$ (i=1,2,3) as follows

$$\mathbf{t}_3 = \frac{\mathbf{g}_1 \times \mathbf{g}_2}{\|\mathbf{g}_1 \times \mathbf{g}_2\|} \quad , \quad \mathbf{t}_1 = \frac{\mathbf{g}_1}{\|\mathbf{g}_1\|} \quad , \quad \mathbf{t}_2 = \mathbf{t}_3 \times \mathbf{t}_1 \quad . \tag{3.16}$$

The matrix $\mathbf{R}_0$ contains the vectors $\mathbf{t}_i$ (i=1,2,3) columnwise and thus is determined uniquely by $\mathbf{t}_i$.

Next we discuss the evaluation of the rotation tensor $\mathbf{R}$. With two angles β_1 and β_2 $\mathbf{R}$ is given by a matrix multiplication of orthogonal matrices

$$\mathbf{R}(\beta_1,\beta_2) = \mathbf{R}_2(\beta_2)\mathbf{R}_1(\beta_1) \quad ,$$

$$\mathbf{R}_1(\beta_1) = \begin{pmatrix} 1 & 0 & 0 \\ 0 & cos\beta_1 & -sin\beta_1 \\ 0 & sin\beta_1 & cos\beta_1 \end{pmatrix} \quad , \quad \mathbf{R}_2(\beta_2) = \begin{pmatrix} cos\beta_2 & 0 & -sin\beta_2 \\ 0 & 1 & 0 \\ sin\beta_2 & 0 & cos\beta_2 \end{pmatrix} \quad . \tag{3.17}$$

With this formulation spatial rotations are restricted to 90^0. The use of (3.17) in (2.2) and (2.4) leads to explicit expressions of the associated axial vectors, see [6].

Another representation with quaternions is written in the paper of Stein, Lambertz and Plank [26]. To get a finite element formulation of folded plate structures, the exponential map $\mathbf{R} = e^\Theta$, is introduced , with Θ as a skew-symmetric matrix describing an infinitesimal rotation. The components of the axial vector ϑ asociated with Θ are used as degrees of freedom in the finite element formulation. Using the four quaternions (also called Euler-parameters) the exponential map is given by $q_0 = cos\frac{1}{2}\|\vartheta\|$ and $\mathbf{q} = \frac{\vartheta}{\|\vartheta\|}sin\frac{1}{2}\|\vartheta\|$ and $\mathbf{R}$ can be written as

$$\mathbf{R} = 2 \begin{pmatrix} q_0^2 + q_1^2 - \frac{1}{2} & q_1q_2 - q_3q_0 & q_1q_3 + q_2q_0 \\ q_2q_1 + q_3q_0 & q_0^2 + q_2^2 - \frac{1}{2} & q_2q_3 - q_1q_0 \\ q_3q_1 - q_2q_0 & q_3q_2 + q_1q_0 & q_0^2 + q_3^2 - \frac{1}{2} \end{pmatrix} \quad . \tag{3.18}$$

Using the matrices $\mathbf{R}_0$ and $\mathbf{R}$ we are now able to compute the shell strains explicitly.

4. Constitutive equations for small strains and numerical treatment

4.1 Layer model

Besides the kinematical relations and the equilibrium equations we have to formulate a constitutive law to determine the deformations of the shell. In a pure mechanical theory we neglect thermal influences on the deformation process, and thus we are lead to an isothermal constitutive law. The numerical treatment of folded plate structures and shells uses the integrals of the stresses. If we assume an inelastic material response a numerical integration of the stresses with respect to the coordinate ζ is necessary. The most useful approach is the so-called layer model, see Figure 1.

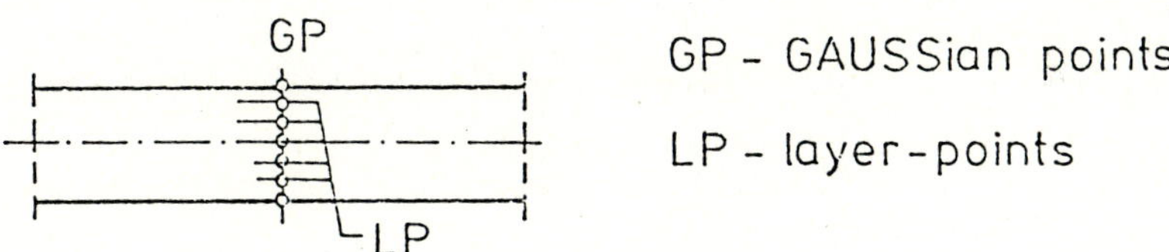

Fig. 1. Layer model

The strains of a layer point i with the coordinate ζ^i are given by

$$\varepsilon^i_{\alpha\beta} = \tilde{\varepsilon}_{\alpha\beta} + \zeta^i \tilde{\kappa}_{\alpha\beta}. \tag{4.1}$$

Now we can define the stress vector $\boldsymbol{\sigma} = \{\sigma_{11}, \sigma_{22}, \sigma_{12}, \sigma_{13}, \sigma_{23}\}^i$ and the strain vektor $\boldsymbol{\varepsilon} = \{\varepsilon_{11}, \varepsilon_{22}, \varepsilon_{12}, \varepsilon_{13}, \varepsilon_{23}\}^i$ of the layer point i. Within the small strain assumption the strain rates can be decomposed in an elastic and an inelastic part

$$\dot{\boldsymbol{\varepsilon}} = \dot{\boldsymbol{\varepsilon}}^{el} + \dot{\boldsymbol{\varepsilon}}^{pl}, \tag{4.2}$$

where the index i of the layer point is dropped for simplicity.

4.2 Elastic-plastic v.Mises type deformation

The elastic part of the material law may be written

$$\dot{\boldsymbol{\varepsilon}}^{el} = \mathbf{C}\dot{\boldsymbol{\sigma}} \tag{4.3}$$

with the elasticity matrix $\mathbf{C}$.

The classical v.Mises yield criterion with nonlinear isotropic and kinematic hardening is formulated

$$F(\sigma, \alpha.e_v) = \frac{3}{2}(\sigma - \alpha)^T \mathbf{D}(\sigma - \alpha) - \kappa^2(e_v) \leq 0. \qquad (4.4)$$

Here, α is the backstress tensor and $\mathbf{D}$ is defined by

$$\sigma^D = \mathbf{D}\sigma \quad , \qquad (4.5)$$

where σ^D denotes the deviator of σ. The yield stress κ is a function of the equivalent plastic strain e_v. The time derivative of ϵ_v is given by

$$\dot{e}_v = \sqrt{\frac{2}{3}\dot{\varepsilon}^{pl} \cdot \dot{\varepsilon}^{pl}}. \qquad (4.6)$$

The rate of the back stresses follows by Pragers rule

$$\dot{\alpha}^D = \frac{2}{3}H'_\alpha \dot{\varepsilon}^{pl} \qquad (4.7)$$

with the derivative of the kinematic part of hardening H'_α with respect to e_v. Finally the associated flow rule describes the plastic strain rates

$$\dot{\varepsilon}^{pl} = \dot{\lambda}\sigma^D \qquad (4.8)$$

where $\dot{\lambda}$ is the the plastic multiplier.

4.3 Projection method and linearization of the stresses

The calculation of the inelastic stresses is achieved by a projection method, which can be seen as an implicit integration of the plastic strain rates (4.8). Therefore the so-called trial stresses

$$\sigma_{tr} = \sigma_n + \mathbf{C}\Delta\varepsilon \qquad (4.9)$$

are introduced. In an incremental process denotes σ_n the stress vector of the last increment. The projection of the trial stresses σ_{tr} onto the yield surface is described by the saddle point problem

$$L(\sigma, \lambda) = I(\sigma, \lambda) + \lambda F(\sigma) \longrightarrow \text{stat} \quad ,$$
$$I(\sigma) = (\sigma_{tr} - \sigma)^T \mathbf{C}^{-1}(\sigma_{tr} - \sigma) \qquad (4.10)$$

with the Lagrange parameter λ and the yield condition $F(\boldsymbol{\sigma})$ (see e.g. [25]). The solution of (4.10) yields the stresses

$$\boldsymbol{\sigma}(\mathbf{E}, \lambda) = \boldsymbol{\alpha}_n + \bar{\mathbf{C}}(\lambda)\mathbf{E} \quad,$$

$$\bar{\mathbf{C}}(\lambda) = (\mathbf{C}^{-1} + \frac{3\lambda}{2(1 + \lambda H'_\alpha)}\mathbf{D})^{-1} \quad, \tag{4.11}$$

$$\mathbf{E} = \mathbf{C}^{-1}\boldsymbol{\sigma}_n + \Delta\boldsymbol{\varepsilon} \quad.$$

as a function of λ. The explicit presentation of λ is not possible, it follows from the iterative solution of the yield function (4.4) with the stresses (4.11). The linearization of the stresses with respect to the strains i.e. the so-called consistent tangent matrix yields, see [5],[25],

$$\mathbf{C}_T = \frac{\partial\boldsymbol{\sigma}}{\partial\boldsymbol{\varepsilon}} = \bar{\mathbf{C}} - \frac{\bar{\mathbf{C}}(\boldsymbol{\sigma}^D - \boldsymbol{\alpha}^D)(\boldsymbol{\sigma}^D - \boldsymbol{\alpha}^D)^T\bar{\mathbf{C}}^T}{\frac{4}{9}\kappa^2 y' + (\boldsymbol{\sigma}^D - \boldsymbol{\alpha}^D)^T\bar{\mathbf{C}}(\boldsymbol{\sigma}^D - \boldsymbol{\alpha}^D)} \quad,$$

$$y' = \frac{(H'_\alpha + \kappa' + \lambda H''_\alpha\kappa)(1 + \lambda H'_\alpha)}{1 - \lambda\kappa' - \lambda^2 H''_\alpha\kappa} \quad. \tag{4.12}$$

This linearization process is necessary to get a quadratic rate of convergence by solving the nonlinear finite element equations with Newtons method.

The numerical integration (e.g. Simpson integration) of the projected stresses (4.11) yields

$$\tilde{N}_{\alpha\beta} = h\sum_{i=1}^{n} w^i\sigma_{\alpha\beta},$$

$$\tilde{M}_{\alpha\beta} = h\sum_{i=1}^{n} w^i\zeta^i\sigma_{\alpha\beta}, \tag{4.13}$$

$$N_{\alpha 3} = h\sigma_{\alpha 3}.$$

Here w^i is the weighting factor and n is the total number of layer points. In a similar way the tangential stiffness matrix is integrated (see [26]).

5. Finite element discretization

In this section the finite element formulation is developed for the shell theory presented above. The displacement and rotation fields are approximated using the isoparametric approach. Within this method the same shape functions are specified for the interpolation of geometry and displacement fields. Thus, by employing the isoparametric finite element approximation we are able to describe arbitrary shell geometries. The shape functions have to be constructed in such a form that they fulfil the boundary and transition conditions according to the variational formulation used. This is an essential requirement for the convergence of the method. Since shear deformations are considered in the theory presented we only need C^0-continuity for the displacement and rotations at the element boundaries. The approximation of geometry, displacements and rotations are given by

$$X_i = \sum_{k=1}^{n} N_k(\xi, \eta) X_{ik} \qquad i = 1, 2, 3 \quad ,$$

$$u_i = \sum_{k=1}^{n} N_k(\xi, \eta) v_{ik} \quad , \tag{5.1}$$

$$\vartheta_i = \sum_{k=1}^{n} N_k(\xi, \eta) \vartheta_{ik} \quad .$$

Explicit formulaes for the shape functions $N_k(\xi, \eta)$ may be found in e.g. Zienkiewicz [30] for four–or nine–node elements. The vector of the nodal degrees of freedom $\mathbf{v}_k$ is represented by

$$\mathbf{v}_k = \{v_{1k}, v_{2k}, v_{3k}, \vartheta_{1k}, \vartheta_{2k}, \vartheta_{3k}\}. \tag{5.2}$$

If (3.17) is used, only two rotational degrees of freedom are necessary. Now the finite element formulation for the static equilibrium of the shell can be derived in matrix formulation. The approximations (5.1) are inserted into the principle of virtual work (3.3) and together with (3.4) and (3.5) one gets

$$\mathbf{G}^h(\mathbf{v}) = \bigcup_{i=1}^{n_e} (\int_{\Omega_e} \mathbf{B}^T(\mathbf{v}) \boldsymbol{\sigma}^h \, d\Omega - \bar{\mathbf{P}}). \tag{5.3}$$

$\boldsymbol{\sigma}^h$ are the approximated stress resultants and stress couple resultants of the shell which are given in terms of kinematical quantities through the constitutive law (see chapter 4), and $\bar{\mathbf{P}}$ denotes the vector of the applied external loads. Ω_e is the element area and the operator $\cup$ denotes the standard assembly process of the single element matrices into a global structure. The matrix $\mathbf{B}$ which contains the finite element approximations follows with (3.4). For explicit presentations of $\mathbf{B}$ see [6], [7] and [26]. Equation (5.3) contains transcendental functions and thus is highly nonlinear. This is different from a shell theory with moderate rotations which leads to a cubically nonlinear system of equations for the displacements. Since an isoparametric representation of the shell geometry is adopted, no transformations of tangent matrices and residuals from local to global coordinates are necessary.

The solution of the nonlinear equations (5.3) is obtained by Newton's method. For this purpose we have to construct the tangent stiffness of $\mathbf{G}^h$ by computing the directional derivative of (5.3). This leads to the following incremental solution scheme

$$D\,\mathbf{G}^h(\mathbf{v}_i)\,\Delta\mathbf{v}_{i+1} = -\mathbf{G}^h(\mathbf{v}_i)$$
$$\mathbf{v}_{k+1} = \mathbf{v}_i + \Delta\mathbf{v}_{i+1}. \tag{5.4}$$

The tangential stiffness matrix is given by

$$D\,\mathbf{G}_e^h(\mathbf{v}_e) = \bigcup_{i=1}^{n_e}(\int_{(\Omega_e)} (\mathbf{B}^T\mathbf{D}\mathbf{B} + \mathbf{K}_g)\,d\Omega) \tag{5.5}$$

where $\mathbf{D}$ follows by numerical integration of $\mathbf{C}_T$ (4.12) with respect to the coordinate ζ (see e.g.[26]). The integral of $\mathbf{K}_g$ is called the initial stiffness matrix, which is obtained by taking the derivative of $\mathbf{B}$. The element tangent matrices $D\,\mathbf{G}_e^h(\mathbf{v})$ and the residual vectors $\mathbf{G}_e^h(\mathbf{v})$ are computed using Gauss quadrature. To avoid locking effects a selective reduced integration is applied, see e.g. Malkus and Hughes [12]. However, this procedure does not necessarily lead to tangent matrices with correct rank which may produce singular tangent matrices depending on the boundary conditions. By adding of stabilization matrices the correct rank of the matrices can be regained. For a discussion of these procedures, see e.g. Belytschko et.al. [4] or Stein, Wagner, Wriggers [28]. This method has been used in the presented shell element in cases where the reduced integration caused singular tangent stiffnesses.

6. Computer implementation

The implementations have been done in the program systems FEAP (Finite Element Analysis Program) and in INA-SP (INelastic Analysis of Shells and Plates).

6.1 FEAP

FEAP has been developed at the University of California (Department of Civil Engineering) by R.L. Taylor. A description of the program may be found in Zienkiewicz [30]. The computer system has been developed on virtual memory computers operating in a UNIX environment. FEAP consists of several general modules:

1. Problem control

2. Problem definition and mesh input

3. Element library

4. Problem solution

5. Graphics output

The problem solution module is centered arround a unique macro programming language concept in which the solution algorithm is written by the user. Accordingly, with this unique capability, each user may construct a solution strategy which meets his specific need. There are sufficient macro instructions included in the system for many applications in structural mechanics.

6.2 INA-SP

This program was developed at the Institut für Baumechanik und Numerische Mechanik of the University of Hannover during a research project in the DFG-research group on "Nonlinear computatations in structural engineering" (organized in the DFG-Schwerpunkt "Nichtlineare Berechnungen im Konstruktiven Ingenieurbau"). The system uses a lot of subroutines from the DFGBIB (see e.g.[27]), a public-domain software pool including finite element programs written by the members of the above mentioned research group. INA-SP depends on a mesh independent geometrical description and includes modules for adaptive mesh refinement with a-priori and a-posteriori criteria, respectively. Several shell and facet elements are implemented, and material nonlinear behaviour is treated by different layer models. Graphic output, based on GKS, completes the program.

7. Examples

7.1 Geometrical nonlinear examples

In this section we compare the developed finite element formulation with known solutions in the literature. Due to the ability of the shell element to model finite rotations the solution range of some examples could be extended.

7.1.1 Cylindrical shell without end diaphragms

The cylindrical shell shown in Figure 2 is loaded by point forces at two opposite sides of the cylinder. Since the shell is very thin, $R/h = 52.69$, large parts of the shell undergo rigid body motions. Due to the boundary conditions, hour–glass modes appear caused by the reduced integration. They are stabilized as described in chapter 5. Thus this example provides a severe test for the finite element formulation.

One quarter of the shell has been discretized by 32 four node quadrilaterals. The nonlinear behaviour has been studied by Harte [8] who employed different shell theories to calculate moderate deflections of the shell up to 17 times of the shell thickness. The load deflection curve calculated with the shell element proposed in this paper is in very good agreement with the results reported by Harte.

However, we are able to compute finite deflections and rotations of this shell up to the entire stretching of the cylinder into a plane plate. For the limiting case the maximum deflection can be computed as $u_3 = (\frac{\pi}{2} - 1)R$. This solution is approached for a load going to infinity, see load deflection curve in Figure 3. A deformed configuration of the shell associated with this case is depicted in Figure 4.

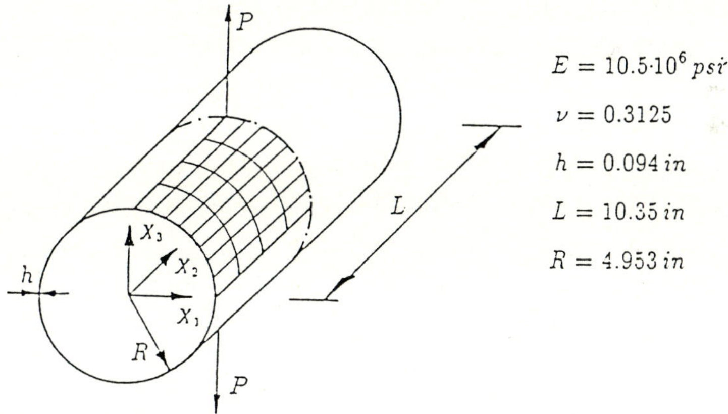

Fig. 2. Cylindrical shell without end diaphragms

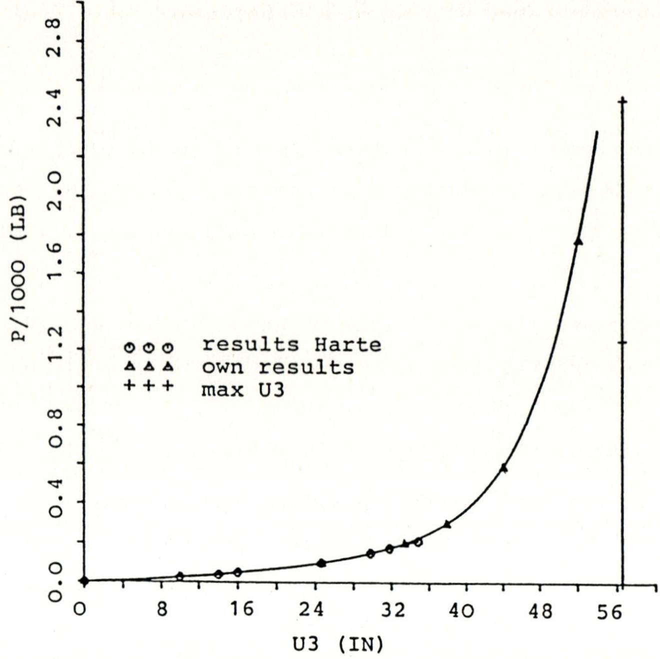

Fig. 3. Load deflection curve of a cylindrical shell without end diaphragms

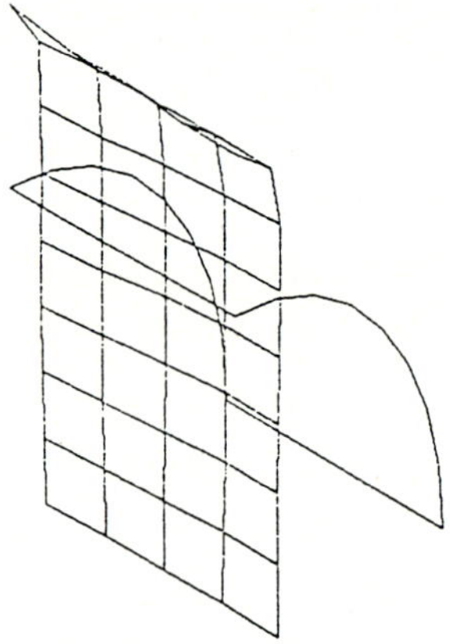

Fig. 4. Deformed finite element mesh of a cylindrical shell

7.1.2 Clamped polygonal beam (Right angle frame)

This example has been analyzed with three-dimensional nonlinear beam elements by Argyris [2] and Simo and Vu-Quoc [24]. With a width to height ratio of 1/50 the beam is almost a thin plate. Thus it should be discretized by shell elements. A relatively coarse mesh with only 44 elements is employed to model this structure. Geometry and data can be found in Figure 5. The structure is loaded in-plane by a point load at its end. After passing a stability point the structure displays an out-of-plane deflection which can be initialized by applying a perturbation load.

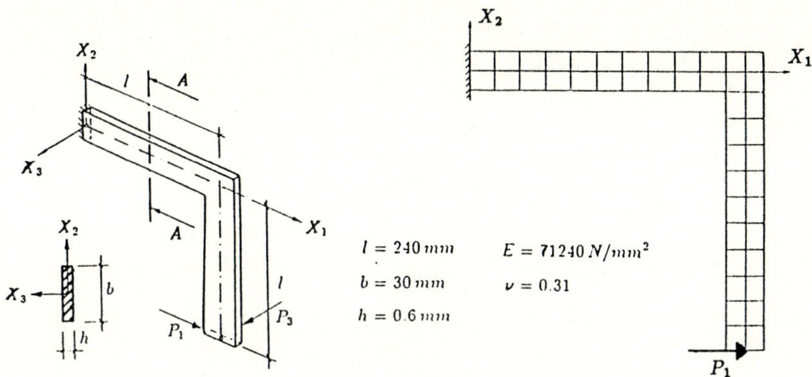

Fig. 5. Clamped frame under point load

In [2] the stability loads are computed for discretizations with beam as well as with plate elements. The plate discretization leads to a critical load of $P_k = 1.1453\,N$ which is 5.3 % higher then the beam solution $P_k = 1.088\,N$. Applying a perturbation load of $P_3 = 0.001\,P_1$ the stability problem is reduced to a pure stress problem and can therefore be solved using Newton's method together with an arc-length scheme.

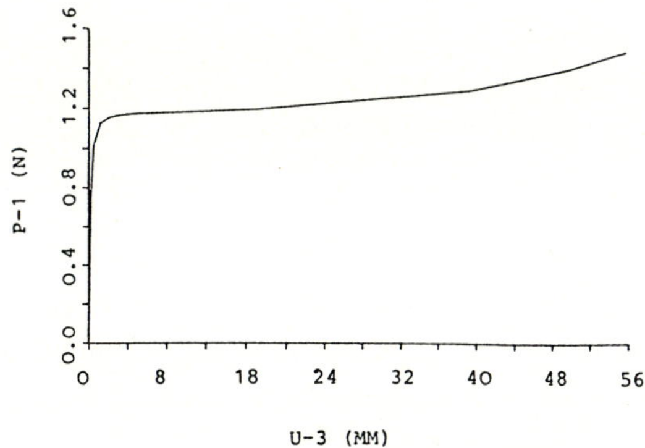

Fig. 6. Load deflection curve of the frame

Figure 6 shows the load deflection curve of the displacement u_3 with respect to the load P_1. A solution using nonlinear plate or shell elements is not available in the literature. The step size of the load increments is dictated in this example by the arc–length method. Deformed configurations are depicted in Figure 7. These plots show that the structure undergoes large rotations and deflections in the secondary equilibrium path after passing the stability point.

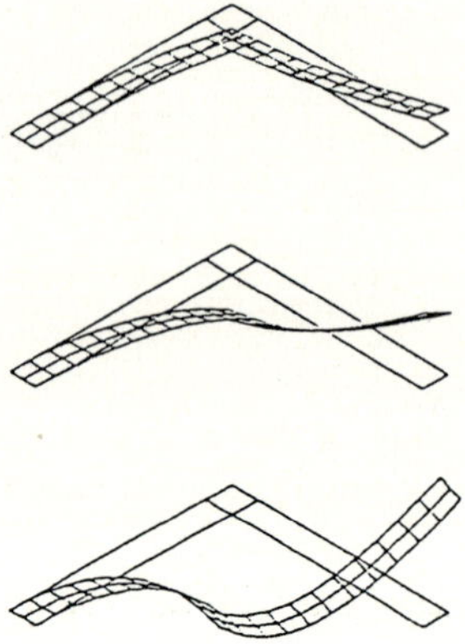

Fig. 7. Deformed meshes of the frame

7.2 Geometrical and material nonlinear deformation process

For the ultimate load analysis of thin walled structures, geometrical as well as material nonlinear behaviour must be regarded. In the following examples the numerical results of the nonlinear formulation and developed algorithmus are first compared with experiments and then the application to a structure of practical interest is shown.

7.2.1 Plate girder subjected to patch loading

For comparison with experimental results and the confirmation of the theoretical model, a series of ultimate load tests with plate girders at the Technical University of Braunschweig were very suitable [20]. The experiments were performed using plate girders with very slender webs subjected to symmetric patch loading. The test stand arrangement is shown in Figure 8 , where the analyzed part of the girder is marked by a FE-mesh. The experimental ultimate load was 248 kN. Numerical studies were performed with up to 350 elements and gave a lower bound of 244 kN and an upper bound of 249 kN (see [25]). For comparison a purely elastic calculation resulted in a critical load of 975 kN. In Figure 9 the lateral web displacement in the middle of the girder is shown for the test girder with $t_w = 4.14mm$ and a second one with $t_w = 15mm$. From Figure 10 it is obviously the development of two plastic hinges in the web that causes the failure of the girder (In the upper flange there are no inelastic deformations). The failure of the girder with $t_w = 15mm$ is due to three hinge mechanism in the upper flange, see the yield zone in Figure 11.

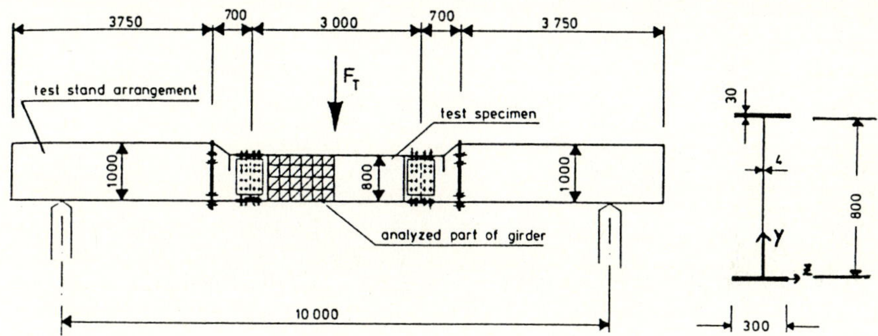

Fig. 8. Plate girder subjected to patch loading

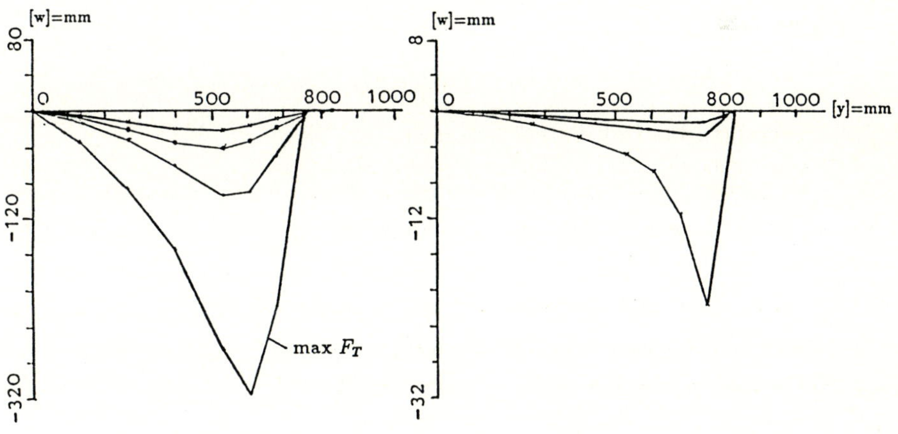

$t_w = 4.14mm$

$t_w = 15mm$

Fig. 9. Lateral displacements of the web

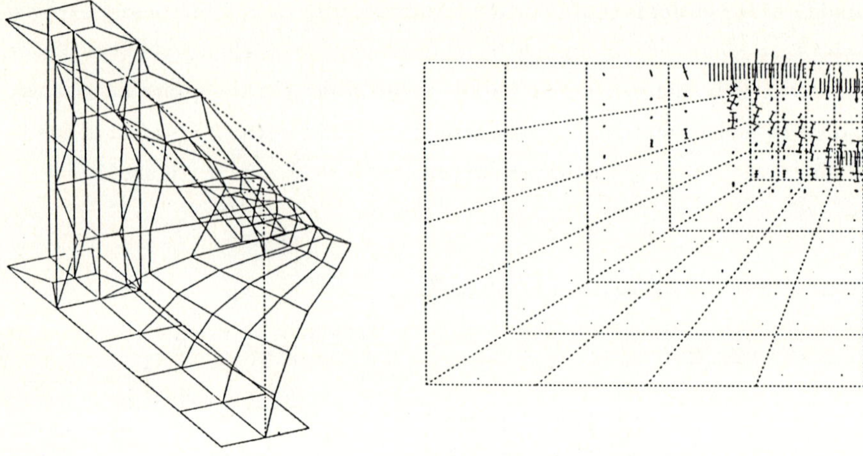

Fig. 10. a) Deformed structure   b) Plastified part of the web

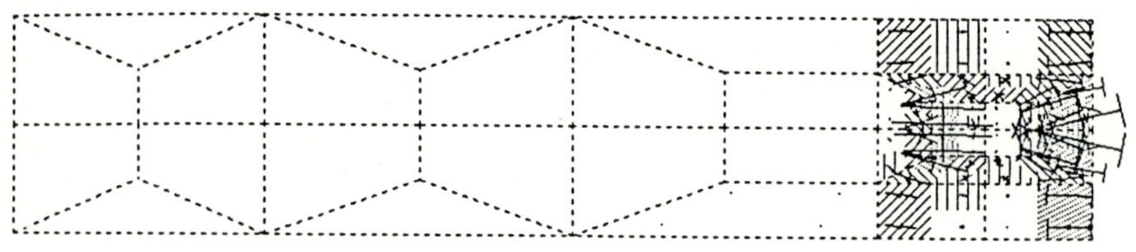

Fig. 11. Three hinge mechanism in the upper flange

7.2.2 Ultimate load of a steel frame

In Figure 12 the dimensions of a steel frame and the construction of the corner are shown. The frame is subjected to a uniform loading on its main girder. Figure 13 shows the displacements of different points normal to the plane of the frame with increasing loads. The ultimate load concerning plastified haunch edges, see Figure 14 is obtained at a load level of 2.4 times the load for linear elastic structural analysis.

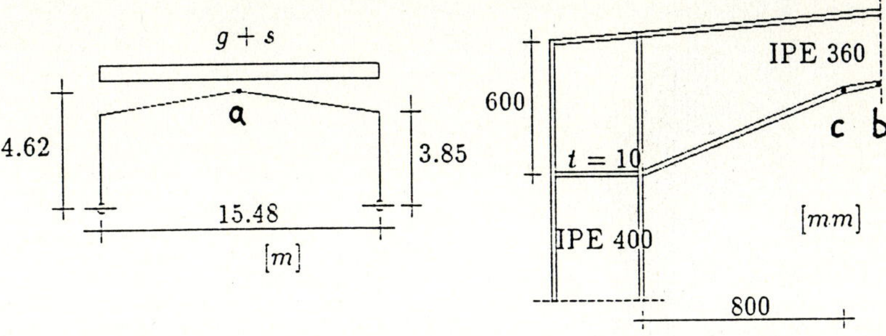

Fig. 12. Steel frame subjected to a uniform loading

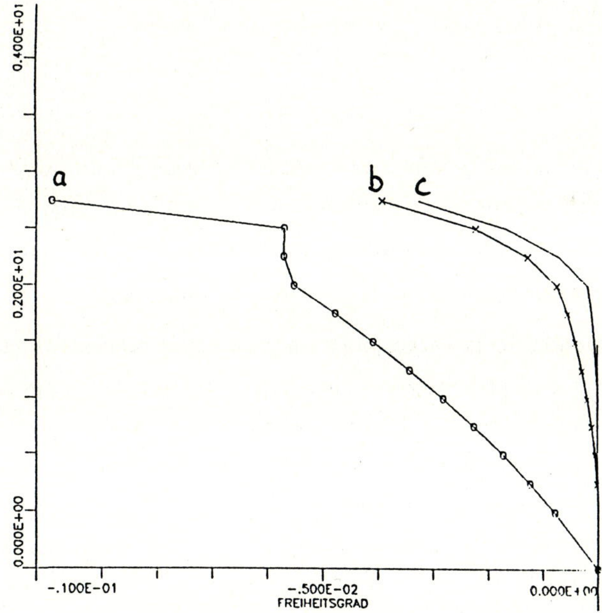

Fig. 13. Load deflection curve of the steel frame

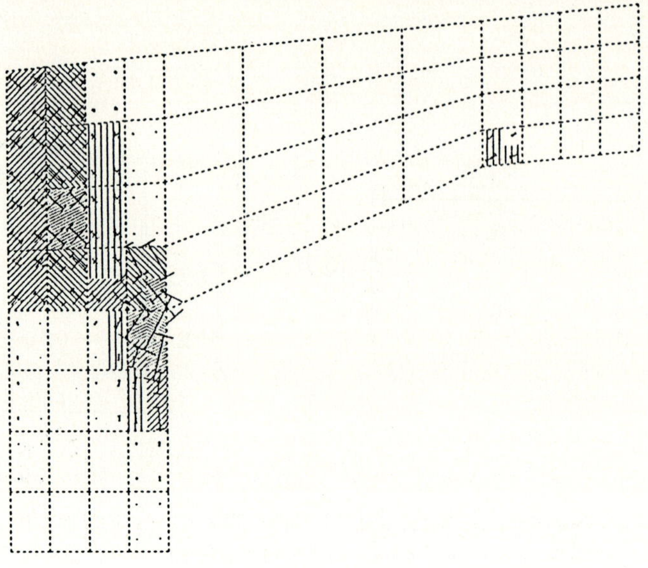

Fig. 14. Plastified zones of the steel frame

8. Conclusion

In this paper a bending theory for thin elastic shells undergoing finite rotations has been developed. For this purpose the principle of virtual work is formulated in terms of the Biot stress resultants. The shell strains are derived by a polar decomposition of the material deformation gradient.

The finite element formulation is based on the isoparametric concept which allows the discretization of arbitrary shell geometries. Since finite rotations are modeled correctly by the developed finite element formulation load increments are only limited by the numerical solution algorithms for the case that elastic material response is assumed. Thus the numerical process for the computation of problems with large or finite rotations is very effective. Furthermore, the consistent tangential stiffness matrix for geometrical nonlinear and elastic-plastic deformations with hardening is derived analytically so that quadratic convergence of the Newton-Raphson iteration process is provided. In this paper we assume small elastic and plastic strains. A couple of instructive examples with finite rotations and plastified zones of thin walled structures are given.

References

1 Atluri, S.N.: Alternative stress and conjugate strain measures, and mixed variational formulations involving rigid rotations, for computational analysis of finitely deformed solids, with application to plates and shells; Computers & Structures 18 (1984), 93-116

2 Argyris, J.H.; Balmer, H.; Doltsinis, J.St.; Dunne, P.C.; Haase, M.; Kleiber, M., G. A.; Mlejnek, H.-P.; Müller, M.; Scharpf, D. W. : Finite element method - the natural approach; Comput. Meth. Appl. Mech. Engng. 17/18 (1979), 1-106

3 Badur, J.; Pietraszkiewicz, W. : On geometrically non-linear theory of elastic shells derived from pseudo-cosserat continuum with constrained microrotations; Lecture notes in engineering : Finite rotations in structural mechanics (Proceedings of the Euromech Colloquium 197, Jabłonna, Poland, 1985); editor : W. Pietraszkiewicz; Springer-Verlag 1986

4 Belytschko, T.; Bachrach, W.E. : Efficient implementation of quadrilaterals with high-coarse mesh accuracy; Comput. Meths. Appl. Mech. Engng. (54), 279-301

5 Gruttmann, F.; Stein, E. : Tangentiale Steifigkeitsmatrizen bei Anwendung von Projektionsverfahren in der Elastoplastizitätstheorie; Ingenieur-Archiv 58 (1988), 15-24

6 Gruttmann, F.: Theorie und Numerik schubelastischer Schalen mit endlichen Drehungen unter Verwendung der Biot-Spannungen; Forschungs- und Seminarberichte aus dem Bereich der Mechanik der Universität Hannover, Bericht Nr. F88/1, Hannover 1988.

7 Gruttmann, F. ; Stein, E.; Wriggers, P. : Theory and numerics of thin elastic shells with finite rotations; Ingenieur-Archiv 59 (1989), 54-67

8 Harte, R. : Doppelt gekrümmte finite Dreieckelemente für die lineare und geometrisch nichtlineare Berechnung allgemeiner Flächentragwerke; Mitteilung Nr. 82-10, Nov. 1982; Institut für Konstruktiven Ingenieurbau, Ruhr-Universität Bochum

9 Hughes, T.J.R.; Pister, K.S.: Consistent linearization in mechanics of solids and structures; Comp. & Struct., 8 (1978), 391-397

10 Libai, A.; Simmonds, J.G.: Nonlinear elastic shell theory; Advances in Applied Mechanics, 23 (1983), Academic Press, Inc.

11 Makowski,J.; Stumpf, H.: Finite strains and rotations in shells, Lecture notes in engineering : Finite rotations in structural mechanics (Proceedings of the Euromech Colloquium 197, Jabłonna, Poland, 1985); editor : W. Pietraszkiewicz; Springer-Verlag 1986

12 Malkus, D.S.; Hughes, T.R.J.: Mixed finite element methods - reduced and selective integration techniques : a unification of concepts; Comp. Meth. Appl. Mech. Engng., 15 (1978), 63-81

13 Marsden, J.E.; Hughes, T.J.R.: Mathematical foundations of elasticity; Prentice-Hall, Englewood Cliffs, 1983

14 Nolte, L.P.: Beitrag zur Herleitung und vergleichende Untersuchung geometrisch nichtlinearer Schalentheorien unter Berücksichtigung großer Rotationen, Mitt. Inst. f. Mechanik, No. 39, Ruhr-Universität Bochum (1983).

15 Park, K.C.; Stanley, G.M.: A curved C^0 shell element based on assumed natural-coordinate strains, Trans. ASME, 53 (1986), 278-290

16 Ramm, E. : Geometrisch nichtlineare Elastostatik und finite Elemente; Bericht Nr. 76-2, Institut für Baustatik der Universität Stuttgart, 1976

17 Recke, L.; Wunderlich, W.: Rotations as primary unknowns in the nonlinear theory of shells and corresponding finite element models, Lecture notes in engineering : Finite rotations in structural mechanics (Proceedings of the Euromech Colloquium 197, Jabłonna, Poland, 1985); editor : W. Pietraszkiewicz; Springer-Verlag 1986

18 Reissner, E.: A note on two-dimensional, finite-deformation theories of shells, Int. J. Non lin. Mech. 17 (1982), 217-221

19 Reissner, E.: On one-dimensional finite strain beam theory, the plane problem, Journal Appl. Math. Phys., 23, (1972), 795-804

20 Scheer, J.: Dehnungsmessungen an Versuchsträgern zum Vergleich für theoretische nichtlineare Berechnungen; Technical report 6092, Institut für Stahlbau, TU Braunschweig, 1983

21 Schweizerhof, K.; Wriggers, P.: Consistent linearizations for path following methods in nonlinear FE analysis, Comp. Meth. Appl. Mech. Engng. 59 (1986), 261-279

22 Simmonds, J.G.; Danielson, D.A.: Non-linear shell theory with finite rotations and stress-function vectors; J. Appl. Mech., Trans. ASME, Serie E (39), 1972, 1085-1090

23 Simo, J.C.; Fox, D.D.: On a stress resultant geometrical exact shell model. Part I: Formulation and optimal parametrization; Comp. Meth. Appl. Mech. Engng. 72 (1989), 267-304

24 Simo, J.C.; Vu-Quoc, L.: Three-dimensional finite-strain rod model;   Part II: Computational aspects; Computer Methods in Applied Mechanics and Engineering 58 (1985), p. 79-116, North-Holland

25 Stein, E.; Lambertz, K.H.; Plank, L.: Ultimate load analysis of thin walled steel structures with elasto-plastic properties using FEM – Theoretical, algorithmic and numerical investigations. Lecture notes in engineering : Finite rotations in structural mechanics (Proceedings of the Euromech Colloquium 197, Jabłonna, Poland, 1985); editor : W. Pietraszkiewicz; Springer-Verlag 1986

26 Stein, E.; Lambertz, K.H.; Plank, L.: Verfahren zur Traglastberechnung dünnwandiger Strukturen mit elastoplastischen Deformationen; in: Nichtlineare Berechnungen im konstruktiven Ingenieurbau; editor: E. Stein; Springer-Verlag 1989

27 Stein, E.; Lambertz, K.H.; Plank, L.; Wunderlich W.; Cramer H.; Redanz W.: Die DFGBIB des Schwerpunktes "Nichtlineare Berechnungen im KIB"– eine Sammlung kompatibler Programmbausteine für die Forschung; in: Nichtlineare Berechnungen im konstruktiven Ingenieurbau; editor: E. Stein; Springer-Verlag 1989

28 Stein, E.; Wagner, W.; Wriggers, P.: Concepts of modeling and discretization of elastic shells for nonlinear finite element analysis ,

in : Proceedings of MAFELAP 1987 Conference, editor: J.R. Whiteman, Academic Press, London (1988), 205-232

29 Wempner,G.: Finite elements ,finite rotations and small strains; Int. J. Solids and Structures 5 (1969), 117-153

30 Zienkiewicz, O.C.: The finite element method, Mc Graw Hill, 1977.

On the Optention of the Tangent Matrix for Geometrically Nonlinear Analysis Using Continuum Based Beam/Shell Finite Elements

E. OÑATE[*], E. DVORKIN[**], M.E. CANGA[**] and J. OLIVER[*]

* E.T.S. Ingenieros de Caminos, Canales y Puertos.
Universidad Politécnica de Catalunya, Barcelona, Spain.
** Facultad de Ingeniería.
Universidad de Buenos Aires, Buenos Aires, Argentina.

Summary

The incremental finite element equations for geometrically non linear problems are obtained via the full incremental form of the principle of virtual displacements using a Generalized Lagrangian approach. This leads to the expression of the tangent matrix in a straight forward manner and an example of application for 2D elasticity is presented. For large displacements/large rotations beam/shell problems the incremental equations are derived using a quadratic approximation for the increment of the reference vectors in terms of the nodal rotation increments. It is shown how this approach leads to a complete tangent matrix which in the examples analyzed seems to be competitive with respect to the simplified form obtained by linearizing the changes in the reference vectors.

INTRODUCTION

The incremental equations for solving geometrically non linear structural problems can be derived in a variety of ways. The most popular alternative is probably to directly linearize the non linear virtual work equations. This leads to the obtention of the classic "tangent" stiffness matrix traditionally used for solving these kind of problems. This procedure can be also interpreted in several more "rigorous" or "intuitive" equivalent ways like: second variation, total differential, linear incrementation, Taylor expansion limited to linear terms, direct consequence of Newton–Raphson method for solving non linear equations, etc. This approach has been used by an extensive number of authors, both in the context of the Total Lagrangian (TL) and Updated Lagrangian (UL) descriptions. For references see, for example, the reference list of chapter 19 of [1] and reference [2].

A second approach is to derive the incremental equations by substracting the virtual work equations written for two equilibrium configurations and then linearizing the result. This method was originally suggested by Yaghmai [2] and then followed by others [3]–[7]. It was shown by Frey [8] that the final linearized equations in both approaches are the same if the displacement field is linearly interpolated in the displacement unknowns in a standard finite element form.

However, very little work seems to have been done in exploiting the finite element formulation based in the "full" incremental form obtained by writting the virtual work equations for an incremental deformed configuration. This approach is developed in this work using a Generalized Lagragian description. It will be shown that retaining all the non linear terms leads to a "secant" expression from which the standard tangent matrix can be directly obtained. An example of the simplicity of this derivation for 3D elasticity problems is presented.

Next we will show that the obtention of a complete tangent matrix for continuum based beam/shell finite element analysis demands a quadratic approximation for the changes in the reference vectors in terms of nodal rotations increments. The efficiency of the complete tangent matrix is checked out with some examples of application.

Basic Concepts

We will consider an elastic body with initial volume $^{\circ}V$ at a known deformed configuration corresponding to time t. A *Generalized Lagrangian* description will be used in which strains and stresses are refered to an intermediate reference configuration ^{r}V (Fig. 1) [2]. Thus, the Green–Lagrange strain tensor at time $t + \Delta t$ refered to the configuration ^{r}V can be written as

$$
{}^{t+\Delta t}_{r}\varepsilon_{ij} = \frac{1}{2}({}^{t+\Delta t}_{r}u_{i,j} + {}^{t+\Delta t}_{r}u_{j,i} + {}^{t+\Delta t}_{r}u_{k,i}\,{}^{t+\Delta t}_{r}u_{k,j}) \tag{1}
$$

with

$$
{}^{t+\Delta t}_{r}u_{i,j} = \frac{\partial\,{}^{t+\Delta t}u_i}{\partial\,{}^{r}x_j} \quad ; \quad {}^{t+\Delta t}u_i = {}^{t}u_i + \Delta u_i \quad , \quad i,j = 1,2,3 \tag{2}
$$

where $^{r}x_i$ are the cartesian coordinates of point i in the reference configuration ^{r}V and $^{t}u_i$ and Δu_i are the cartesian components of the displacement and incremental displacement vectors, respectively (see Fig. 1).

The strain increments are obtained as

$$
\Delta\varepsilon_{ij} = {}^{t+\Delta t}_{r}\varepsilon_{ij} - {}^{t}_{r}\varepsilon_{ij} = {}_{r}e_{ij} + {}_{r}\eta_{ij} \tag{3}
$$

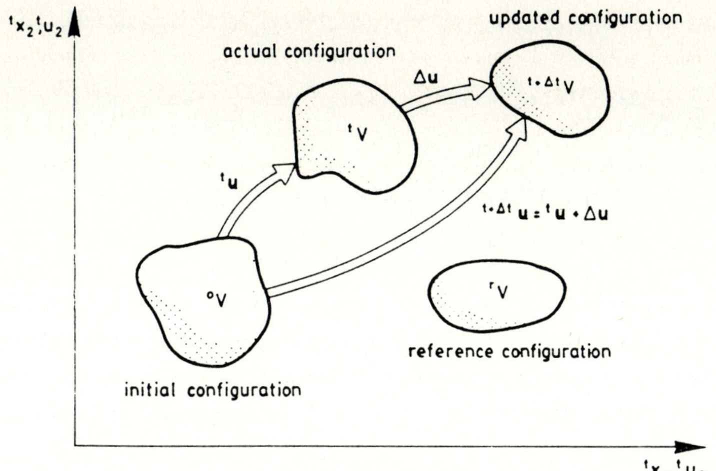

Figure 1   Initial, actual, updated and reference configurations.

where $_r e_{ij}$ and $_r \eta_{ij}$ are first and second order strain increments given by

$$_r e_{ij} = \frac{1}{2}(_r u_{i,j} + _r u_{j,i} + \underline{^t_r u_{k,i} \, _r u_{k,j} + _r u_{k,i} \, ^t_r u_{k,j}} \atop = 0 \quad for \quad ^r V = \, ^t V \tag{4a}$$

$$_r \eta_{ij} = \frac{1}{2} \, _r u_{k,i} \, _r u_{k,j} \tag{4b}$$

with

$$_r U_{i,j} = \frac{\partial \Delta U_i}{\partial ^r x_j} \tag{5}$$

Note that for an updated Langragian description ($^r V = \, ^t V$) the underlined term in eq.(4a) are zero.

The principle of virtual displacements at time $t + \Delta t$ can be written as [2]

$$\int_{^r V} \delta ^{t+\Delta t}_r \varepsilon^T \, ^{t+\Delta t}_r \sigma \, ^r dV = \int_{^r V} \delta ^{t+\Delta t} \mathbf{u}^T \, ^{t+\Delta t} \mathbf{b} \, ^r dV \tag{6}$$

where

$$\delta_r^{t+\Delta t}\boldsymbol{\epsilon} = [\delta_r^{t+\Delta t}\epsilon_{11}, \delta_r^{t+\Delta t}\epsilon_{22}, \delta_r^{t+\Delta t}\epsilon_{33}, 2\delta_r^{t+\Delta t}\epsilon_{12}, 2\delta_r^{t+\Delta t}\epsilon_{13}, 2\delta_r^{t+\Delta t}\epsilon_{23}]^T$$

$$_r^{t+\Delta t}\boldsymbol{\sigma} = [_r^{t+\Delta t}\sigma_{11}, _r^{t+\Delta t}\sigma_{22}, _r^{t+\Delta t}\sigma_{33}, _r^{t+\Delta t}\sigma_{12}, _r^{t+\Delta t}\sigma_{13}, _r^{t+\Delta t}\sigma_{23}]^T$$

$$(7)$$

$$\delta^{t+\Delta t}\mathbf{u} = [\delta^{t+\Delta t}u_1, \delta^{t+\Delta t}u_2, \delta^{t+\Delta t}u_3,]^T$$

$$\mathbf{b} = [b_1, b_2, b_3]^T$$

In eq.(6) $_r^{t+\Delta t}\boldsymbol{\sigma}$ is the second Piola-Kirchhoff stress vector at time $t + \Delta t$ refered to the configuration r. For simplicity, only body forces $^{t+\Delta t}\mathbf{b}$ are considered in eq.(6). Further, since the displacements $^t u_i$ are assumed to be known (hence $\delta^t u_i = 0$)

$$\delta_r^{t+\Delta t}\boldsymbol{\epsilon} = \delta_r\mathbf{e} + \delta_r\boldsymbol{\eta} \tag{8a}$$

$$\delta^{t+\Delta t}\mathbf{u} = \delta(\Delta\mathbf{u}) \tag{8b}$$

On the other hand, the stresses are updated in an incremental form as

$$_r^{t+\Delta t}\boldsymbol{\sigma} = {}_r^t\boldsymbol{\sigma} + \Delta\boldsymbol{\sigma} \tag{9}$$

Equation (5) can thus be rewritten using eqs. (6)–(9) as

$$\int_{rV} [(\delta_r\mathbf{e}^T + \delta_r\boldsymbol{\eta}^T)_r^t\boldsymbol{\sigma} + (\delta_r\mathbf{e}^T + \delta_r\boldsymbol{\eta}^T)\Delta\boldsymbol{\sigma}] \, {}^r dV = \int_{rV} \delta(\Delta\mathbf{u})^{Tt+\Delta t}\mathbf{b} \, {}^r dV \tag{10}$$

For simplicity we will consider a linear hiperelastic constitutive equation defined in the material frame as

$$\Delta\boldsymbol{\sigma} = {}_r^t\mathbf{D}\Delta\boldsymbol{\epsilon} = {}_r^t\mathbf{D}(_r\mathbf{e} + {}_r\boldsymbol{\eta}) \tag{11}$$

where $_r^t\mathbf{D}$ is the constitutive matrix at time t referred to configuration ^rV. [2].

Substituting (11) in (10) yields

$$\int_{rV} [\delta_r\mathbf{e}^T {}_r^t\mathbf{D}_r\mathbf{e} + \delta_r\mathbf{e}^T {}_r^t\boldsymbol{\sigma} + \delta_r\boldsymbol{\eta}^T {}_r^t\boldsymbol{\sigma} + (\delta_r\mathbf{e}^T {}_r^t\mathbf{D}_r\boldsymbol{\eta} + \delta_r\boldsymbol{\eta}^T {}_r^t\mathbf{D}_r\mathbf{e}) +$$

$$+ \delta_r\boldsymbol{\eta}^T {}_r^t\mathbf{D}_r\boldsymbol{\eta}] \, {}^r dV = \int_{rV} \delta(\Delta\mathbf{u})^T {}^{t+\Delta t}\mathbf{b}^r dV \tag{12}$$

The underlined terms in (12) represent second and third order terms in Δu_i. These terms are neglected for the obtention of the tangent matrix. However, their inclusion leads to secant equilibrium expressions from which different solution algorithms can be derived [9], [10].

Example. GNL finite element formulation for 2D elasticity

As an example we will consider the case of a 2D elastic solid discretized in isoparametric solid finite elements with shape functions $N^k(\xi, \eta, \zeta)$[1]. The displacement field can be interpolated in an standard manner as [1]

$$^t\mathbf{u} = \mathbf{N}^t\mathbf{a} \quad , \quad \Delta\mathbf{u} = \mathbf{N}\Delta\mathbf{a} \tag{13}$$

where
$$\mathbf{N} = [\mathbf{N}^1, \mathbf{N}^2, \cdots, \mathbf{N}^n] \quad , \quad \mathbf{N}^k = N^k\mathbf{I}_2$$
$$^t\mathbf{a}^k = [^tu_1^k, {}^tu_2{}^k]^T \quad , \quad \Delta\mathbf{a}^k = [\Delta u_1^k, \Delta u_2^k]^T \tag{14}$$

are the shape function matrix, and displacement vectors of node k and $\mathbf{I}_2$ is the 2×2 unit matrix.

After adequate substitution of (13) in (4), (5) it can be obtained for each element

$$_r\mathbf{e} = {}_r^t\mathbf{B}_L\Delta\mathbf{a}$$
$$_r\boldsymbol{\eta} = \frac{1}{2}\,{}_r^t\mathbf{B}_1(\Delta\mathbf{a})\Delta\mathbf{a} \tag{15}$$

where ${}_r^t\mathbf{B}_L$ and ${}_r^t\mathbf{B}_1$ are first and second order strain matrices, respectively. The form of these matrices is given in Fig. 2.

From (15) we can obtain

$$\delta_r\mathbf{e} = {}_r^t\mathbf{B}_L\delta(\Delta\mathbf{a})$$
$$\delta_r\boldsymbol{\eta} = {}_r^t\mathbf{B}_1(\Delta\mathbf{a})\delta\Delta\mathbf{a} \tag{16}$$
$$\delta_r\boldsymbol{\eta}^T\,{}_r^t\boldsymbol{\sigma} = \delta(\Delta\mathbf{a})^T\,{}_r^t\mathbf{B}_{NL}^T\,{}_r^t\mathbf{S}\,{}_r^t\mathbf{B}_{NL}\Delta\mathbf{a}$$

The expressions of $\mathbf{B}_{NL}$ and ${}_r^t\mathbf{S}$ are shown in Fig. 3.

Substituting eqs. (16) in (12) and retaining linear terms only the standard tangent expression is obtained after simple mathematics as

$$_r^t\mathbf{K}_T\Delta\mathbf{a} = -{}_r^t\boldsymbol{\Psi} \tag{17}$$

Eq.(17) is the basis for obtaining the numerical solution in a standard incremental form. In (17)

$$\,_r^t\mathbf{B}_L = [\,_r^t\mathbf{B}_L^1, \,_r^t\mathbf{B}_L^2, \cdots, \,_r^t\mathbf{B}_L^n] \quad ;$$

$$\,_r^t\mathbf{B}_1 = [\,_r^t\mathbf{B}_1^1, \,_r^t\mathbf{B}_1^2, \cdots, \,_r^t\mathbf{B}_1^n] \quad ;$$

$$\,_r^t\mathbf{B}_L^k = \,_r^t\mathbf{B}_{L_0}[\mathbf{I}_2 + \,_r^t\mathbf{L}^T]$$

$$\,_r^t\mathbf{B}_1^k = \,_r^t\mathbf{B}_{L_0}\,_r^t\hat{\mathbf{L}}^T$$

$$\,_r^t\mathbf{B}_{L_0}^k = \begin{bmatrix} \,_r N_{,1}^k & 0 \\ 0 & \,_r N_{,2}^k \\ \,_r N_{,2}^k & \,_r N_{,1}^k \end{bmatrix} \quad ; \quad \,_r^t\mathbf{L} = \begin{bmatrix} \,_r^t l_{11} & \,_r^t l_{12} \\ \,_r^t l_{21} & \,_r^t l_{22} \end{bmatrix}$$

$\,_r^t\hat{\mathbf{L}}$ like $\,_r^t\mathbf{L}$ with $\,_r^t\hat{l}_{ij}$ instead of $\,_r^t l_{ij}$

$$\,_r N_{,i}^k = \frac{\partial N^k}{\partial \,^r x_i} \quad , \quad \,_r^t l_{ij} = \frac{\partial \,^t u_i}{\partial \,^r x_j} \quad , \quad \,_r^t \hat{l}_{ij} = \frac{\partial (\Delta u_i)}{\partial \,^r x_j}$$

Figure 2   Matrices $\,_r^t\mathbf{B}_L$ and $\,_r^t\mathbf{B}_1$ for 2D elasticity problems.

$$\,_r^t\mathbf{B}_{NL} = \begin{bmatrix} \,_r^t\bar{\mathbf{B}}_{NL} & \bar{\mathbf{0}} \\ \bar{\mathbf{0}} & \,_r^t\bar{\mathbf{B}}_{NL} \end{bmatrix} \quad ; \quad \,_r^t\bar{\mathbf{B}}_{NL} = \begin{bmatrix} \,_r N_{,1}^1 & 0 & \,_r N_{,1}^2 & 0 \cdots & \,_r N_{,1}^n \\ \,_r N_{,2}^1 & 0 & \,_r N_{,2}^2 & 0 \cdots & \,_r N_{,2}^n \end{bmatrix}$$

$$\,_r^t\mathbf{S} = \begin{bmatrix} \,_r^t\hat{\mathbf{S}} & 0 \\ 0 & \,_r^t\hat{\mathbf{S}} \end{bmatrix} \quad ; \quad \,_r^t\hat{\mathbf{S}} = \begin{bmatrix} \,_r^t\sigma_{11} & \,_r^t\sigma_{12} \\ \,_r^t\sigma_{21} & \,_r^t\sigma_{22} \end{bmatrix}$$

$$\mathbf{0} = \begin{bmatrix} 0 & 0 \\ 0 & 0 \end{bmatrix} \quad ; \quad \bar{\mathbf{0}} = \begin{Bmatrix} 0 \\ 0 \end{Bmatrix}$$

Figure 3   Matrices $\,_r^t\mathbf{B}_{NL}$ and $\,_r^t\mathbf{S}$ for 2D elasticity problems.

$$\,^t_r\boldsymbol{\Psi} \;=\; \int_{^rV} \,^t_r\mathbf{B}^T_L \,^t_r\boldsymbol{\sigma}\,^rdV \;-\; \int_{^rV^e} \mathbf{N}^{T\,t+\Delta t}\mathbf{b}\,^rdV \tag{18}$$

is the residual force vector and $\,^t_r\mathbf{K}_T$ is the tangent matrix given by

$$\,^t_r\mathbf{K}_T \;=\; \int_{^rV} \,^t_r\mathbf{B}^T_L \,^t_r\mathbf{D} \,^t_r\mathbf{B}_L \,^rdV \;+$$

$$+\; \int_{^rV} \,^t_r\mathbf{B}^T_{NL} \,^t_r\mathbf{S} \,^t_r\mathbf{B}_{NL} \,^rdV \;=\; \,^t_r\mathbf{K}_L \;+\; \,^t_r\mathbf{K}_\sigma \tag{19}$$

where $\,^t_r\mathbf{K}_L$ and $\,^t_r\mathbf{K}_\sigma$ are the well known symmetric initial displacement and initial stress matrices [1], [2]. Again note that the expression of these matrices for Total and Update formulations are directly obtained simply by making $^rV = \,^0V$ and $^rV = \,^tV$. Note that in the latter case $\,^t_r\mathbf{B}_L = \,^t_r\mathbf{B}_{L_0}$ and $\,^t_r\mathbf{K}_L$ coincides with the standard stiffness matrix for infinitesimal theory [1], [2].

The full non linear form of eq.(12) is the basis for obtaining the *secant matrix*. The expression of this matrix can be written in a variety of ways and for most cases is non–symmetric [9]. A symmetric form of the secant matrix has been recently obtained by Oñate [10].

Formulation for continuum based beam and shell finite elements

We will consider jointly the analysis of 3–D beams of rectangular cross section and shells (Fig. 4) using degenerated solid finite elements [1] [2]. The vector of incremental displacements for both type of problems can be written as

Beams:
$$\Delta\mathbf{u} \;=\; \sum_{k=1}^{n} \mathbf{N}^k\left(\Delta\mathbf{u}^k + \frac{\eta}{2}a^k\Delta\mathbf{v}^k_s + \frac{\tau}{2}b^k\Delta\mathbf{v}^k_t\right) \tag{20}$$

Shells:
$$\Delta\mathbf{u} \;=\; \sum_{k=1}^{n} \mathbf{N}^k\left(\Delta\mathbf{u}^k + \frac{\tau}{2}b^k\Delta\mathbf{v}^k_t\right) \tag{21}$$

In eqs. (20) and (21) $\mathbf{N}^k$ is the shape function matrix of node k, (a^k, b^k) and (b^k) are the (constant) dimensions of the beam cross section and the shell thickness of node k, respectively and $\Delta\mathbf{v}^k_a(a = s, t)$ are the incremental changes of the nodal reference vectors (see Fig. 4). Since the rotations are assumed to be large, vectors are defined by non–conmutative orthogonal transformations and therefore

$$\Delta\mathbf{v}^k_a = \,^{t+\Delta t}\mathbf{v}^k_a - \,^t\mathbf{v}^k_a = [^{\Delta t}\mathbf{R}^k - \mathbf{I}]^t\mathbf{v}^k_a \quad , \quad a = s, t \tag{22}$$

90

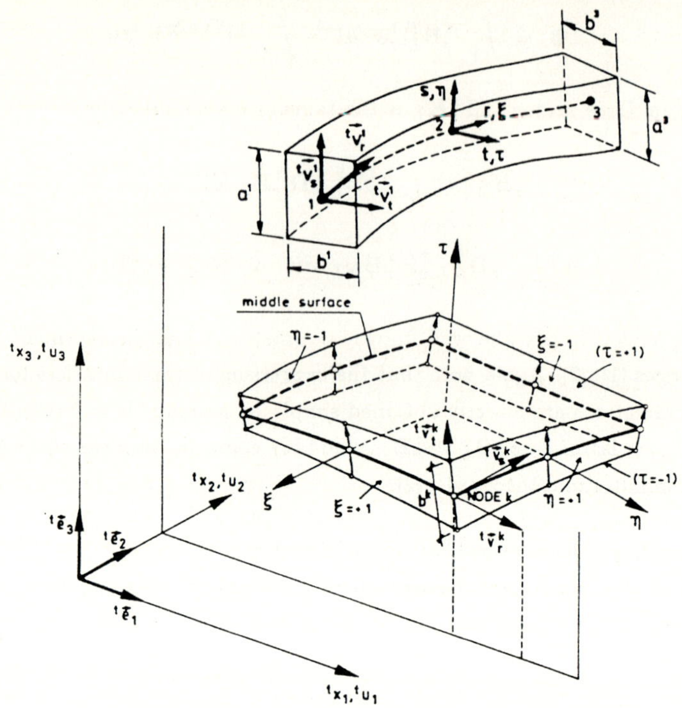

Figure 4 Geometrical configurations of 3D continuum based beam/shell finite elements.

In (22) $^{\Delta t}\mathbf{R}^k$ is the incremental rotation matrix of node k given in exact form by [11]

$$^{\Delta t}\mathbf{R}^k = e^{\Delta \mathbf{S}^k} = \mathbf{I}_3 + \Delta\mathbf{S}^k + \frac{1}{2!}(\Delta\mathbf{S}^k)^2 + \frac{1}{3!}(\Delta\mathbf{S}^k)^3 + \cdots \tag{23}$$

where

$$\Delta\mathbf{S}^k = \begin{bmatrix} 0 & -\Delta\theta_z^k & \Delta\theta_y^k \\ \Delta\theta_z^k & 0 & -\Delta\theta_x^k \\ -\Delta\theta_y^k & \Delta\theta_x^k & 0 \end{bmatrix} \tag{24}$$

is the standard skew symmetric matrix with axial vector $\Delta\boldsymbol{\theta}^k = [\Delta\theta_x^k, \Delta\theta_y^k, \Delta\theta_z^k]$, $\Delta\theta_i^k$ being the so called pseudo incremental nodal rotations about the ith global axe. Closed form expressions for the rotation matrix $^{\Delta t}\mathbf{R}^k$ can be ound in [11].

Note that for simplicity we have defined here the pseudo nodal rotations $\Delta\theta_i^k$ about *global* axes for both the beam and shell cases. However, as it is well known for smooth shells a definition of $\Delta\theta_i^k$ about local axes is more adequate [2].

The infinitesimal rotation case is simply obtained by neglegecting non linear terms in (13). Thus

$$^{\Delta t}\mathbf{R}^k = \mathbf{I}_3 + \Delta \mathbf{S}^k \qquad (25)$$

and consequently in this case

$$\Delta \mathbf{v}_a^k = \Delta \mathbf{S}^k \, {}^t\mathbf{v}_a^k \qquad\qquad a = s,t \qquad (26)$$

i.e. if the rotations are infinitesimal vectors are transformed via skew–symmetric transformations [11].

Eq.(22) is not practical for deriving the incremental equilibrium equations. An alternative frequently used is to use the linearized form (26). This, however leads to incomplete forms of the tangent matrix, as it will be shown next.

Let us choose a second order approximation or $\Delta \mathbf{v}_a^k$ as

$$\Delta \mathbf{v}_a^k = [\mathbf{I} + \frac{1}{2}\Delta \mathbf{S}^k]\Delta \mathbf{S}^k \, {}^t\mathbf{v}_a^k =$$

$$= [\mathbf{I} + \frac{1}{2}\Delta \mathbf{S}^k][{}^t\mathbf{V}_a^k]^T \Delta \boldsymbol{\theta}^k \qquad\qquad a = s,t \qquad (27)$$

where ${}^t\mathbf{V}_a^k$ is the skew symmetric matrix of node k with axial vector ${}^t\mathbf{v}_a^k$.

Substituting (27) in (20) yields

$$\Delta \mathbf{u} = \sum_{k=1}^{n}[\mathbf{N}_1^k + \mathbf{N}_2^k(\Delta\theta_i^k)]\Delta \mathbf{a}^k \qquad (28)$$

with

$$\mathbf{N}_1^k = N^k[\mathbf{I}_3, \frac{\eta}{2}a^k \, {}^t\mathbf{V}_s^k + \frac{\tau}{2}b^k \, {}^t\mathbf{V}_t^k]$$

$$\mathbf{N}_2^k(\Delta\theta_i^k) = \frac{1}{2}N^k\Delta \mathbf{S}^k[0, \frac{\eta}{2}a^k \, {}^t\mathbf{V}_s^k + \frac{\tau}{2}b^k \, {}^t\mathbf{V}_t^k]$$

$$\Delta \mathbf{a}^k = \left\{ \begin{array}{c} \Delta \mathbf{u}^k \\ \Delta \boldsymbol{\theta}^k \end{array} \right\} \qquad (29)$$

Note that the underlined terms in (29) are zero for the case of shells.
From (28) we can obtain the first and second order strain increments as

92

$$_r e = [^t_r\mathbf{B}_L + {}^t_r\mathbf{B}_L^*(\Delta\boldsymbol{\theta})]\Delta a$$

$$_r\eta = \frac{1}{2}[^t_r\mathbf{B}_1(\Delta a) + {}^t_r\mathbf{B}_1^*(\Delta\boldsymbol{\theta}^2) + {}^t_r\mathbf{B}_1^{**}(\Delta\boldsymbol{\theta}^3)]\Delta a \tag{30}$$

and

$$\delta_r e = [^t_r\mathbf{B}_L + {}^t_r\hat{\mathbf{B}}_L(\Delta\boldsymbol{\theta})]\delta(\Delta a) \tag{31a}$$

$$\delta_r\eta = {}^t_r\mathbf{B}_1(\Delta a)\delta(\Delta a) + \text{H.O.T.} \tag{31b}$$

The higher order terms in (31b) can be "a priori" neglected since they lead to second and third order terms in (12) which need not be considered in the final linearized form of the equilibrium equations.

Substituting the terms shown in eqs.(31) in (12) *and retaining the linear terms only* we obtain the following equilibrium expression

$$\int_{rV}({}^t_r\mathbf{B}_L^{T} {}^t_r\mathbf{D} \, {}^t_r\mathbf{B}_L\Delta a + {}^t_r\mathbf{B}_1 \, {}^t_r\boldsymbol{\sigma} + \underline{{}^t_r\hat{\mathbf{B}}_L^{T} {}^t_r\boldsymbol{\sigma}}) \, {}^r dV = -{}^t_r\boldsymbol{\Psi} \tag{32}$$

where ${}^t_r\boldsymbol{\Psi}$ is the residual force vector of (18). It can be shown that [10], [12]

$$^t_r\mathbf{B}_1^{T} {}^t_r\boldsymbol{\sigma} = {}^t_r\mathbf{K}_\sigma\Delta a$$

$$^t_r\hat{\mathbf{B}}_1 \, {}^t_r\boldsymbol{\sigma} = {}^t_r\hat{\mathbf{K}}_\sigma\Delta a \tag{33}$$

where ${}^t_r\mathbf{K}_\sigma$ and ${}^t_r\hat{\mathbf{K}}_\sigma$ are both symmetric initial stress matrices. Eq.(33) can thus be finally written as

$$[^t_r\mathbf{K}_L + {}^t_r\mathbf{K}_\sigma + \underline{{}^t_r\hat{\mathbf{K}}_\sigma}]\Delta a \equiv$$

$$\equiv {}^t_r\mathbf{K}_T\Delta a = -{}^t_r\boldsymbol{\Psi} \tag{34}$$

where

$$\mathbf{K}_T = {}^t_r\mathbf{K}_L + {}^t_r\mathbf{K}_\sigma + \underline{{}^t_r\hat{\mathbf{K}}_\sigma} \tag{35}$$

is the tangent matrix. Note that the underlined terms in eqs. (32), (34) and (35) represent the difference with the formulation obtained using the linear approximation for Δv_a^k of eq.(26). Matrix ${}^t_r\hat{\mathbf{K}}_\sigma$ is therefore a direct consequence of the quadratic approximation of eq.(27). It is worth also noting that the use of a higher approximation for Δv_a^k does not contribute new linear terms in eq.(32). Therefore, we can conclude that the expression of ${}^t_r\mathbf{K}_T$ of eq.(35) represents the complete tangent matrix. The detailed expression of ${}^t_r\mathbf{K}_T$ can be found in [10], [12].

Examples

The advantages of using the full tangent matrix of eq. (35) is checked out with some examples of application for the solution of simple 2D beam problems. The two node linear beam element with one point reduced integration is used in all the examples [1], [2]. In each example we show the number of iterations needed to attain convergence using the tangent matrix of eq. (35) and that obtained with a linear approximation for $^t\mathbf{v}_a^k$ (i.e. non including matrix $^t_r\hat{\mathbf{K}}_\sigma$). This later approach will be termed here "incomplete tangent approach".

From the numbers presented in Figs. 5–7 it is clear that the complete tangent matrix is computationaly advantageous and it should be recommended for practical use.

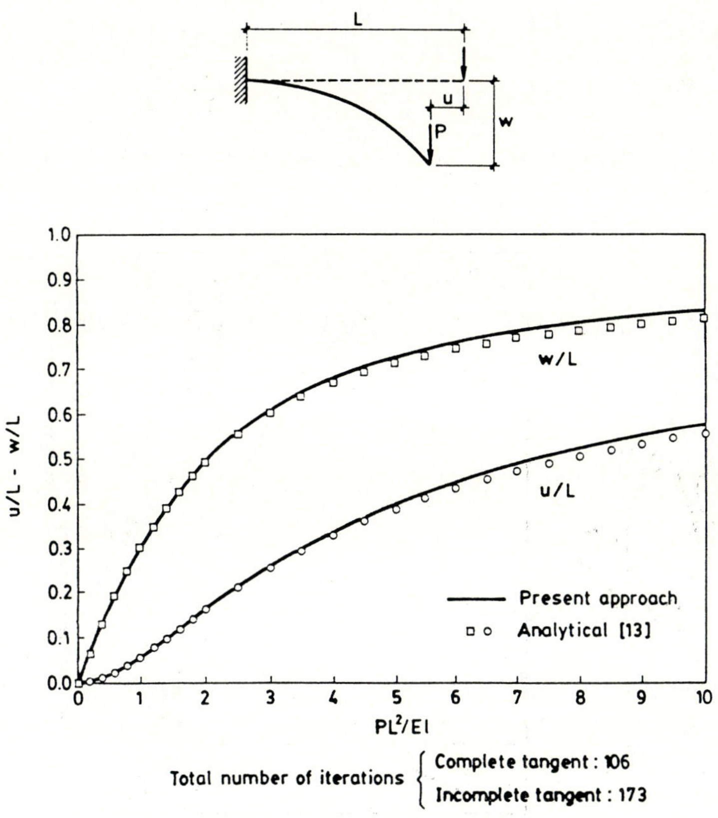

Figure 5 Clamped beam under point load. 10 linear beam elements used. For material parameters see [13].

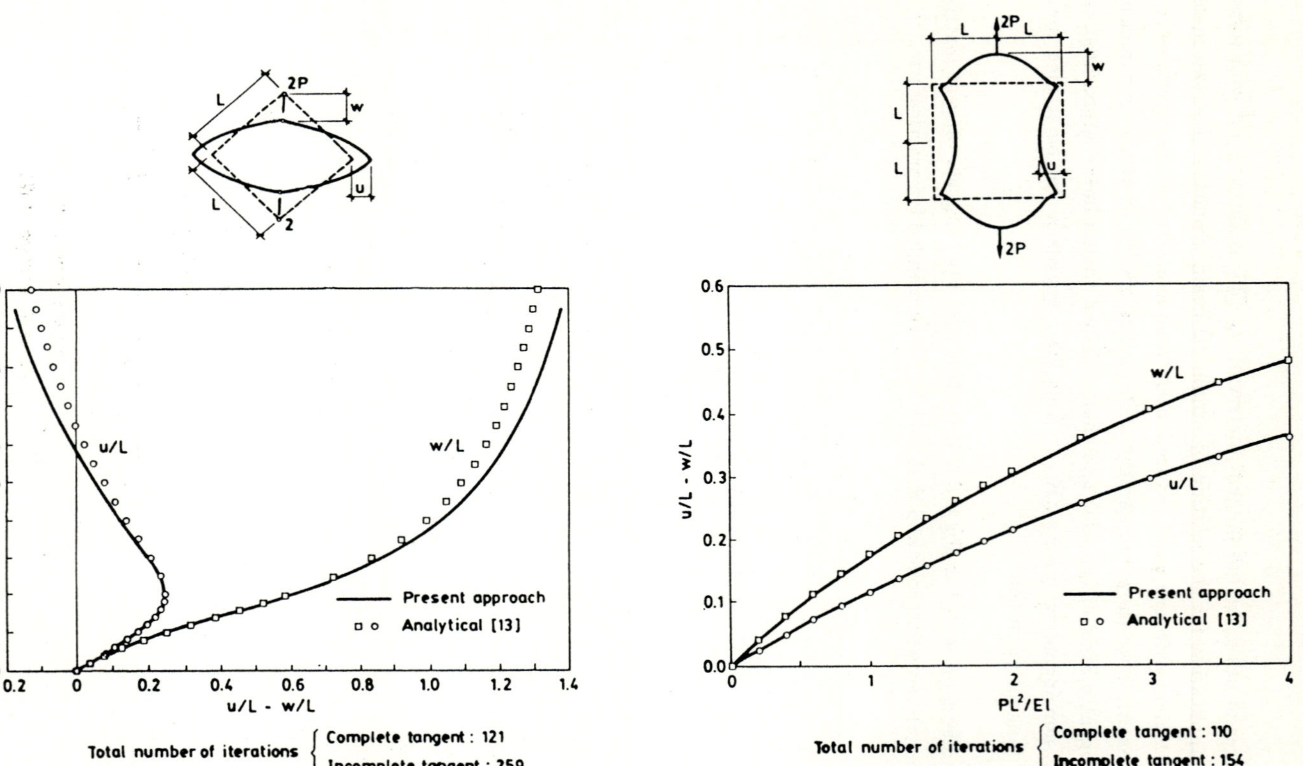

Figure 6   Square frame under symmetric loading. 10 linear beam elements used in 1/4 frame for symmetry. For material parameters see [13].

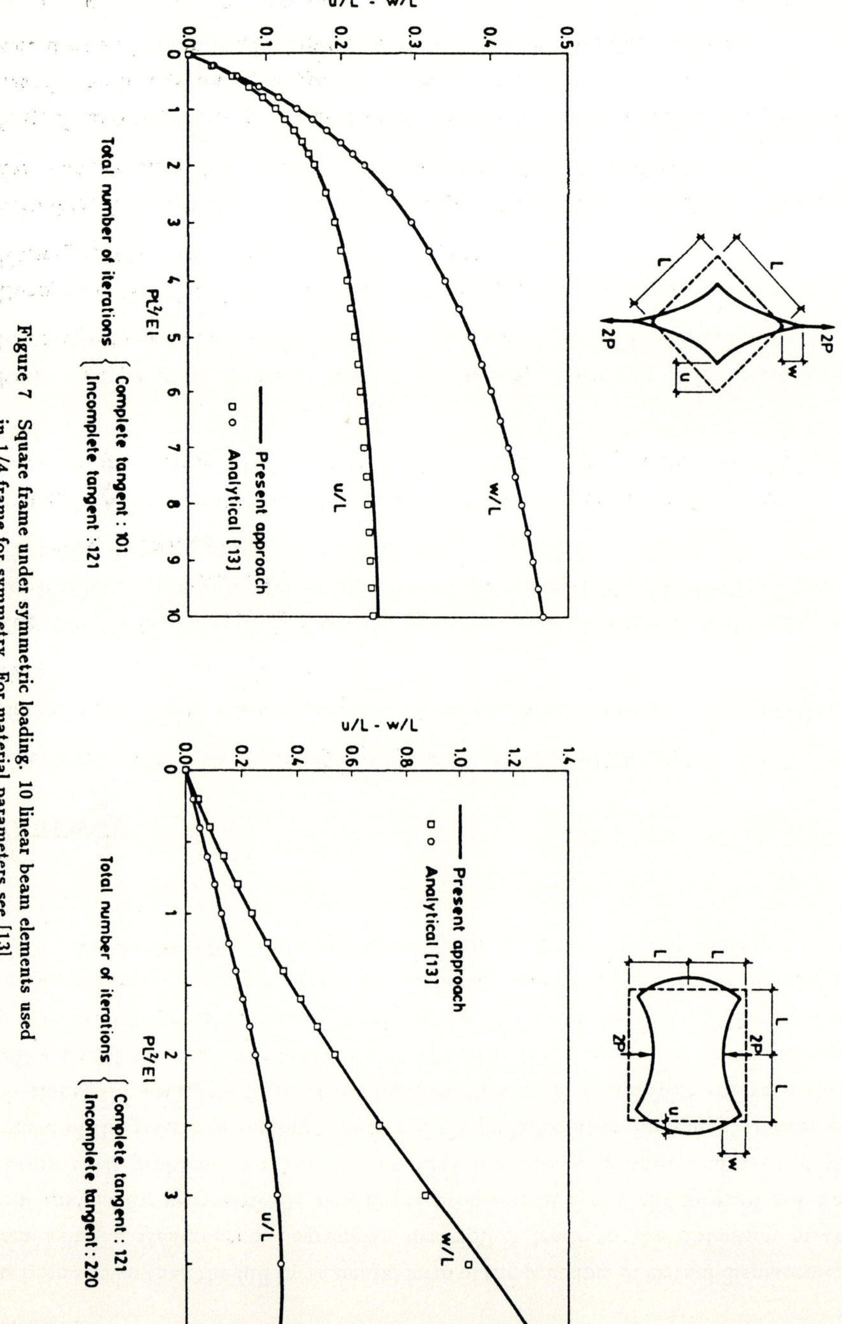

Figure 7   Square frame under symmetric loading. 10 linear beam elements used in 1/4 frame for symmetry. For material parameters see [13].

Conclusions

It has been shown that the full incremental form of the principle of virtual displacement written using a Generalized Lagrangian description leads to the obtention of the tangent matrix for geometrically non linear problems in a straight forward manner. For beam/shell problems we have shown that the cuadratic approximation of the increment of the reference vectors in terms of the nodal rotation increment is essential for obtaining the complete form of the tangent matrix. The examples analyzed show the efficiency of the complete tangent matrix versus the "incomplete" form obtained using a linearized approximation of the changes in the reference vectors. The full form of the incremental equilibrium equations can be also exploited for deriving alternative "secant" discretized equilibrium expressions which could be useful for some particular cases.

REFERENCES

1. Zienkiewicz, O.C.: *The finite element method.* McGraw-Hill, 1975.

2. Bathe, K.J.: *Finite Element Procedures in Non Linear Analysis.* Prentice Hall, 1982.

3. Yaghmai, S.: Incremental analysis of large deformations in mechanics of solids with applications to axisymmetric shells of revolution. *Report No. SESM 68-17*, Univ. California, Berkeley, 1968.

4. Larsen, P.K.: Large displacement analysis of shells of revolution including creep, plasticity and viscoelasticity. *Report SESM 71-22*, Univ. California, Berkeley, 1971.

5. Bathe, K.J.; Ramm, E. and Wilson, E.L.: Instability analysis of free form shells by finite elements. *Int. J. Num. Meth. Engng.*, **9**, 2, pp. 353–386, 1975.

6. Horrigmoe, G.: Non linear finite element models in solid mechanics. *Report 76-2*, Norwegian Inst. Tech., Univ. Trondheim, 1970.

7. Mondkar, D.P. and Powell, G.H.: Finite element analysis of non linear static and dynamic response. *Int. J. Num. Meth. Engng.*, **11**, 3, pp. 499–520, 1977.

8. Frey, F. and Cescotto, S.: Some new aspects of the incremental total lagrangian description in non linear analysis in *"Finite Element in Non Linear Mechanics"*, edited by P. Bergan et al. Tapir Publishers, Univ. of Trondheim, 1978.

9. Oñate, E.; Oliver, J.; Miquel-Canet, J. and Suárez, B.: A finite element formulation for geometrically non linear problems using a secant matrix:

Applications to 3-D Trusses in *Computational Mechanics.*, edited by G.Yagawa y S.N. Atluri. Springer-Verlag, Tokyo, 1986.

10. Oñate, E.: On the obtention of secant and tangent matrices for the analysis of geometrically non linear problems using finite elements. *Internal Report, E.T.S. Ingenieros de Caminos*, Technical University of Catalonia, Spain, 1988.

11. Argyris, J.: An excursion into large rotations. *Comp. Meth. Appl. Mech. Engng.*, **32**, pp. 85–155, 1982.

12. Dvorkin, E.; Oñate, E. and Oliver, J.: On a non linear formulation for curved Timoshenko beam elements considering large displacements/rotation increments. *Int, J. Num. Meth. Engng.*, 1988.

13. Mattiason, K.: Numerical results from large deflection beam and frame problems analyzed by means of elliptic integrals. *Int. J. Num. Meth. Engng.*, **17**, 145–153, (1981).

Part 2
Instability and Nonlinear Responses of Shells

Fundamentals of Numerical Algorithms for Static and Dynamic Instability Phenomena of Thin Shells

W.B. KRÄTZIG

Institute for Statics and Dynamics
Faculty of Civil Engineering
Ruhr-University, Bochum, Germany

Summary

The paper designs a general stability theory for discretized
shell structures based on step-wise numerical solution techniques. To this purpose the nonlinear principle of virtual work
is transformed into its incremental subprinciple and finally
discretized. The resulting equation for Kelvin-Voigt-material,
usually denoted as tangential equation of motion, turns out to
be a sufficient and suitable basis for the numerical evaluation
of arbitrary nonlinear responses including their instability
phenomena.

1. Nonlinear Equation of Motion and 1st Variation

Let us consider the response of an arbitrary discontinuum S
mapped into a 2n-dimensional normed vector space R^{2n} with n-
dimensional subspaces R_V^n, R_P^n of time-dependent nodal displacements V and loads P :

$$V = \{V_1\ V_2\ \ldots\ V_i\ \ldots\ V_n\}\ ,$$
$$P = \{P_1\ P_2\ \ldots\ P_i\ \ldots\ P_n\}\ . \tag{1.1}$$

R shall be structured by the inner product

$$(P,V)\ =\ P^T \cdot V\ =\ P_1 V_1 + P_2 V_2 + \ldots P_i V_i + \ldots P_n V_n\ . \tag{1.2}$$

Any load-deflection path $V(P,t)$ of S in R^{2n} shall be described by the **nonlinear equation of motion**

$$M \cdot \ddot{V} + G(\dot{V},V,t) = P(t)\ ,\quad (\dot{\ldots}) = d\ldots/dt \tag{1.3}$$

with the global mass matrix M , nodal velocities $\dot{V}$ and accelerations $\ddot{V}$, the matrix function G of internal nodal
forces due to viscous damping as well as elasto-plastic re-

storing mechanisms, and with $\mathbf{P}$ as load vector. $\mathbf{G}$ may be subject to arbitrary constraints, e.g. yield conditions; it shall be assumed as continuous, Gateaux-differentiable and including its derivatives bounded in R^{2n} . Under similar restrictions for $\mathbf{P}$ any solution $\mathbf{V}(P,t)$ of (1.3) satisfies a Lipschitz condition and thus depends uniquely on the initial conditions $\mathbf{V}_o, \dot{\mathbf{V}}_o$ at time t_o .

Nonlinear responses $\mathbf{V}(P,t)$ of (1.3) may exhibit various instability phenomena, e.g. bifurcating, diverging or snapping. In order to construct incremental-iterative procedures for the detection of those responses including their instabilities we decompose (1.1)

$$\mathbf{V} = \overset{\circ}{\mathbf{V}} + \bar{\mathbf{V}} + \overset{+}{\mathbf{V}} , \qquad \mathbf{P} = \bar{\mathbf{P}} + \overset{+}{\mathbf{P}} \tag{1.4}$$

into initial imperfections $\overset{\circ}{\mathbf{V}}$, the fundamental state $\bar{\mathbf{V}}, \bar{\mathbf{P}}$ and infinitesimal perturbations $\overset{+}{\mathbf{V}}, \overset{+}{\mathbf{P}}$ of the adjacent state. Analogous decompositions follow for $\dot{\mathbf{V}}, \ddot{\mathbf{V}}$. Forming the 1. variation of (1.3) with respect to the fundamental state

$$\mathbf{M} \cdot \delta \ddot{\bar{\mathbf{V}}} + \left.\frac{\partial \mathbf{G}}{\partial \dot{\mathbf{V}}}\right|_{\bar{\mathbf{V}}} \cdot \delta \dot{\bar{\mathbf{V}}} + \left.\frac{\partial \mathbf{G}}{\partial \mathbf{V}}\right|_{\bar{\mathbf{V}}} \cdot \delta \bar{\mathbf{V}} = \delta \bar{\mathbf{P}} , \tag{1.5}$$

we derive under observation of $\delta \bar{\mathbf{V}} = \overset{+}{\mathbf{V}}, \ \delta \dot{\bar{\mathbf{V}}} = \overset{+}{\dot{\mathbf{V}}}, \ \delta \ddot{\bar{\mathbf{V}}} = \overset{+}{\ddot{\mathbf{V}}}, \ \delta \bar{\mathbf{P}} = \overset{+}{\mathbf{P}}$ and of mass conservation $(\dot{\mathbf{M}} = 0)$ the following linearized equation of motion for $\overset{+}{\mathbf{V}}(\overset{+}{\mathbf{P}},t)$:

$$\mathbf{M} \cdot \overset{+}{\ddot{\mathbf{V}}} + \left.\frac{\mathbf{G}}{\dot{\mathbf{V}}}\right|_{\bar{\mathbf{V}}} \cdot \overset{+}{\dot{\mathbf{V}}} + \left.\frac{\partial \mathbf{G}}{\partial \mathbf{V}}\right|_{\bar{\mathbf{V}}} \cdot \overset{+}{\mathbf{V}} = \overset{+}{\mathbf{P}} = \mathbf{P} - \bar{\mathbf{P}} , \tag{1.6}$$

Herein the jacobians

$$\left.\frac{\partial \mathbf{G}}{\partial \dot{\mathbf{V}}}\right|_{\bar{\mathbf{V}}} = \left.\frac{\partial \mathbf{G}_i}{\partial \dot{\mathbf{V}}_j}\right|_{\bar{\mathbf{V}}_j} , \qquad \left.\frac{\partial \mathbf{G}}{\partial \mathbf{V}}\right|_{\bar{\mathbf{V}}} = \left.\frac{\partial \mathbf{G}_i}{\partial \mathbf{V}_j}\right|_{\bar{\mathbf{V}}_j} , \qquad i,j = 1 \ldots n \tag{1.7}$$

abbreviate tangential damping and stiffness properties of $\mathbf{S}$ in $\bar{\mathbf{V}}$. Re-substituting of $\bar{\mathbf{P}} = \mathbf{F}_I = \mathbf{M} \cdot \ddot{\bar{\mathbf{V}}} + \mathbf{G}(\bar{\mathbf{V}}, \dot{\bar{\mathbf{V}}}, \bar{\mathbf{V}})$ into (1.6) furnishes the final form of the **tangential equation of motion**

$$\mathbf{M} \cdot \overset{+}{\ddot{\mathbf{V}}} + \mathbf{C}_T \cdot \overset{+}{\dot{\mathbf{V}}} + \mathbf{K}_T \cdot \overset{+}{\mathbf{V}} = \mathbf{P} - \mathbf{F}_I = \mathbf{P} - \mathbf{M} \cdot \ddot{\bar{\mathbf{V}}} - \mathbf{G}(\bar{\mathbf{V}}, \overset{\circ}{\mathbf{V}}, \bar{\mathbf{V}}) \tag{1.8}$$

with the following matrices:

M global mass matrix,

C_T tangential damping matrix,

K_T tangential stiffness matrix,

$F_I = M \cdot \ddot{\bar{V}} + G(\ddot{\bar{V}}, \dot{\bar{V}}, \bar{V})$ vector of internal d'Alembert's, viscous, elasto-plastic nodal forces.

2. Phase Transformation and Stability Theorems

In order to connect the evaluation of stability properties of nonlinear time-dependent processes (1.3) with the theory of 1st order differential equations, we transform (1.8) into the 2n-dimensional phase space $X = \{\dot{V}, V\}$. Assuming regularity for M , det $M \neq 0$, we receive:

$$\frac{d}{dt}\begin{bmatrix} \dot{\bar{V}} \\ \hline \bar{V} \end{bmatrix} = \begin{bmatrix} -M^{-1} \cdot C_T & -M^{-1} \cdot K_T \\ \hline I & 0 \end{bmatrix} \cdot \begin{bmatrix} \dot{\bar{V}} \\ \hline \bar{V} \end{bmatrix} + \begin{bmatrix} M^{-1} \cdot (P - F_I) \\ \hline 0 \end{bmatrix}$$

$$\dot{\bar{X}} = A \cdot \bar{X} + B , \qquad (2.1)$$

with A of order $(2n \times 2n)$, non-hermitean, but with $(n \times n)$ hermitean submatrices. The linear differential equation (2.1) holds in an exact sense only for (infinitesimally small) 1st variations $\delta \bar{X}$. In this case the nonlinearity condition [1]

$$\|B\| = \|M^{-1} \cdot (P - F_I)\| \to 0 \quad \text{as} \quad \delta \bar{X} = \bar{X} \to 0 \qquad (2.2)$$

is satisfied and the remaining homogeneous form

$$\dot{\bar{X}} = A \cdot \bar{X} \quad \text{for} \quad \bar{X} \to 0 \qquad (2.3)$$

of (2.1) qualifies as **Ljapunow's 1st approximation** [2] to describe the instability properties of (1.3)[*].

On the other hand (2.1) also serves as basis of iterative solution processes for (1.3), following the general concept of suc-

---

[*] In [2] one will find only a heuristic justification for the use of the 1st approximation; mathematical proofs, however, seem to exist only for special forms of A .

cessive approximations, where solution points are traced for
$B \to 0$. For the future response computations we thus assume
the availability of suitable stable time integration schemes
in combination with iterative algorithms, such that we are able
to evaluate in a step-wise manner time-dependent response-paths
from (2.1).

To decide on the stability of an arbitrary point of a computed
response $\bar{X}(\bar{X}_o,t)$ we apply a perturbation $\overset{+}{X}$ with initial
conditions:

$$\|\overset{+}{X} (\overset{+}{X}_o,\bar{X}_o,t_o,t_o)\| < \delta(\varepsilon) > 0 . \tag{2.4}$$

Due to **Ljapunow's stability theorems** [2] $\bar{X}$ is identified as

- stable or weakly stable, if for $t>t_o$: $\|\overset{+}{X} (\overset{+}{X}_o,\bar{X}_o,t_o,t)\| \leq \varepsilon$,
- stable in the limit: $\approx \varepsilon$,
- unstable: $> \varepsilon$,
- convergent, if for $t \to \infty$: $\lim \|\overset{+}{X} (\overset{+}{X}_o,\bar{X}_o,t_o,t)\| = 0$.

$$\tag{2.5}$$

A stable and convergent response is called asymptotically
stable.

Geometrically interpreted, one thinks of $\bar{X}$ as a curve in the
2n-dimensional state space surrounded by an initial tube with
radius δ . Due to (2.5) any solution $\overset{+}{X}$ which penetrates this
tube once must thereafter remain in case of stability within a
slightly larger tube with radius ε .

Later on the stability theorems (2.5) will be cast into numerical procedures. Finally we turn to the evaluation of stability properties of nonlinear time-invariant responses. Their
treatment will be based directly on the tangential stiffness
equation

$$K_T \cdot \overset{+}{V} = P - {}_IF = P - G(\overset{\circ}{V},\bar{V}) , \tag{2.6}$$

an algebraic relation, derived by $\overset{\cdot\cdot}{V} = \overset{\cdot}{V} = 0$ from (1.8).

3. Basic Equations of Nonlinear Shell Dynamics

Shell structures are primarily modelled as continua. Let us consider an arbitrary shell, imbedded into the E3, in its initial state with middle surface domain $\overset{\circ}{F}$ and boundary $\overset{\circ}{C}$. General convective curvilinear coordinates $\Theta^\alpha:\Theta^1,\Theta^2$ determine points on F, and the following **fields of variables** are defined on F, respectively along C:

$p=p(\Theta^\alpha,u,t)$ prescribed loads on F, related to $d\overset{\circ}{F}$;

$u=u(\Theta^\alpha,t)$ displacement variables on F;

$f=f(\overset{\circ}{\rho},\ddot{u},t)$ d'Alembert's inertia forces with $\overset{\circ}{\rho}$ the mass density function of $\overset{\circ}{F}$;

$t=t(\Theta^\alpha,u,t)$ boundary force variables, related to $d\overset{\circ}{C}$;

$r=r(\Theta^\alpha,t)$ boundary displacement variables along C;

$\sigma=\sigma(\Theta^\alpha,t)$ Piola-Kirchhoff stress variables of 2nd kind;

$\varepsilon=\varepsilon(\Theta^\alpha,t)$ Green-Lagrange strain variables;

$\dot{\varepsilon}=\dot{\varepsilon}(\Theta^\alpha,t)$ corresponding strain velocities.

These column variables consisting of tensor components permit the formulation of arbitrary nonlinear shell theories [4] with the following

field equations: $\forall \{p, u, f, \sigma, \varepsilon\} \in F$

$$-(p+f) = D_e\cdot\sigma = (D_{eL} + D_{eN}(u))\cdot\sigma, \qquad (3.1)$$

$$\varepsilon = D_k\cdot u = (D_{kL} + \tfrac{1}{2} D_{kN}(u))\cdot u, \qquad (3.2)$$

constitutive relation:

$$\sigma = \overset{\circ}{\sigma} + E(\varepsilon,\dot{\varepsilon},t)\cdot\varepsilon + D(\varepsilon,\dot{\varepsilon},t)\cdot\dot{\varepsilon}, \qquad (3.3)$$

boundary conditions: $\forall \{u,\sigma\} \in F$, $\{r,t\} \in C$

$$t = \overset{\circ}{t} = R_t\cdot\sigma = (R_{tL} + R_{tN}(u))\cdot\sigma \quad \text{along} \quad \overset{\circ}{C}_t, \qquad (3.4)$$

$$r = \overset{\circ}{r} = R_r\cdot u \quad \text{along} \quad \overset{\circ}{C}_r. \qquad (3.5)$$

Herein D_e and D_k represent the dynamic and kinematic field operators. Both consist of linear differential operators D_{eL},

D_{kL} and nonlinear parts $D_{eN}(u)$, $D_{kN}(u)$, which are linear functionals of elements of u or their covariant derivatives. An indispensable property for the future discretization is the symmetry of $D_{kN}(u)$:

$$D_{kN}(u) \cdot v = D_{kN}(v) \cdot u . \tag{3.6}$$

R_t and R_r are recognized as dynamic and kinematic trace operators. The proposed material law (3.3) of a nonlinear Kelvin-Voigt model with residual stresses $\overset{\circ}{\sigma}$, the elasticity tensor E and the viscousity tensor D , seems to be sufficiently general for our aims. Finally it should be stressed that every consistently formulated shell theory can be broken down as presented; for further details the reader is referred to the example of a large rotation theory in [5].

In conclusion the **functional of virtual work**

$$\int\limits_{\overset{\circ}{F}} \rho \; \overset{\cdot\cdot}{u}^{T} \cdot \delta u \; d\overset{\circ}{F} + \int\limits_{\overset{\circ}{F}} \sigma^{T} \cdot \delta\varepsilon \; d\overset{\circ}{F} = \int\limits_{\overset{\circ}{F}} \overset{\circ}{p}^{T} \cdot \delta u \; d\overset{\circ}{F} + \int\limits_{\overset{\circ}{C}_t} \overset{\circ}{t}^{\;T} \cdot \delta r \; d\overset{\circ}{C}_t$$
$$\forall \; \{\overset{\circ}{p},u,\sigma,\varepsilon\} \in F , \quad \{\overset{\circ}{t},r\} \in C_t \tag{3.7}$$

describes - in a weak formulation - dynamic equilibrium (3.1, 3.4) under the constraints (3.2, 3.5)

$$\delta\varepsilon = (D_{kL}+D_{kN}(u)) \cdot \delta u \text{ on } F , \quad \delta r = R_r \cdot \delta u \text{ along } C_r . \tag{3.8}$$

Together with (3.3), (3.7, 3.8) are regarded as functional analogue of (1.3).

4. Incremental Principle of Virtual Work and Tangential Equation of Motion

In order to derive the incremental form of (3.7) we decompose the vector fields of $u,\dot{u},r$

$$u = \overset{\circ}{u}+\bar{u}+\overset{+}{u} = \overset{\circ}{u}+\bar{u}+\delta\bar{u} \rightarrow \delta\bar{u} = \overset{+}{u} ,$$
$$\dot{u} = \dot{\bar{u}}+\dot{\overset{+}{u}} , \tag{4.1}$$
$$r = \overset{\circ}{r}+\bar{r}+\overset{+}{r} = \overset{\circ}{r}+\bar{r}+\delta\bar{r} \rightarrow \delta\bar{r} = \overset{+}{r} ,$$

according to Fig. 1 into imperfections $\overset{\circ}{u},\overset{\circ}{r}$, finite variables $\overline{u},\dot{\overline{u}},\overline{r}$ of the fundamental motion as well as increments $\overset{+}{u}=\delta\overline{u},\dot{\overset{+}{u}},$ $\overset{+}{r}=\delta\overline{r}$ of the adjacent motion. $\overset{+}{u},\dot{\overset{+}{u}},\overset{+}{r}$ are assumed as small and thus again interpreted as 1st variations of the fundamental motion.

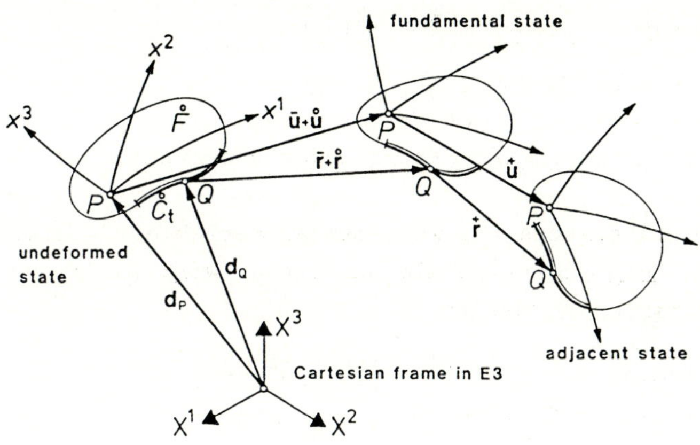

Fig. 1. Different states of shell deformation

Similarly we decompose strains and strain velocities of the fundamental motion

$$\varepsilon = \overline{\varepsilon} + \delta\overline{\varepsilon} + \frac{1}{2!}\delta^2\overline{\varepsilon} = \overline{\varepsilon} + \overset{+}{\varepsilon} + \frac{1}{2}\overset{++}{\varepsilon} \tag{4.2}$$

$$= (D_{kL} + \tfrac{1}{2}D_{kN}(2\overset{\circ}{u}+\overline{u}))\cdot\overline{u} + (D_{kL}+D_{kN}(\overset{\circ}{u}+\overline{u}))\cdot\overset{+}{u} + \tfrac{1}{2}D_{kN}(\overset{+}{u})\cdot\overset{+}{u} ,$$

$$\dot{\varepsilon} = \dot{\overline{\varepsilon}} + \delta\dot{\overline{\varepsilon}} + \frac{1}{2!}\delta^2\dot{\overline{\varepsilon}} = \dot{\overline{\varepsilon}} + \dot{\overset{+}{\varepsilon}} + \frac{1}{2}\dot{\overset{++}{\varepsilon}} \tag{4.3}$$

$$= (D_{kL} + D_{kN}(\overset{\circ}{u}+\overline{u}))\cdot\dot{\overline{u}} + (D_{kL} + D_{kN}(\overset{\circ}{u}+\overline{u}))\cdot\dot{\overset{+}{u}}$$

$$+ D_{kN}(\dot{\overline{u}})\cdot\overset{+}{u} + D_{kN}(\overset{+}{u})\cdot\dot{\overset{+}{u}} ,$$

where multiple crosses signify the polynominal degree of the respective quantity in terms of $\overset{+}{u},\dot{\overset{+}{u}}$. We further find from (4.1,4.2) for the variations δ with respect to $\overset{+}{u}$:

$$\delta u = \delta\overset{+}{u}\,, \quad \delta r = \delta\overset{+}{r}\,,$$

$$\delta\varepsilon = \delta(\bar\delta\bar\varepsilon) + \frac{1}{2!}\delta(\bar\delta^2\bar\varepsilon) = \delta\overset{+}{\varepsilon} + \frac{1}{2}\delta\overset{++}{\varepsilon} \tag{4.4}$$

$$= (D_{kL} + D_{kN}(\mathring{u}+\bar u))\cdot\delta\overset{+}{u} + D_{kN}(\overset{+}{u})\cdot\delta\overset{+}{u}\,.$$

Finally we decompose the force variables

$$p = \bar p + \overset{+}{p}\,, \quad f = \bar f + \overset{+}{f} = -\mathring\rho(\ddot{\bar u}+\ddot{\overset{+}{u}})\,,$$

$$t = \bar t + \overset{+}{t}\,, \quad \sigma = \mathring\sigma + \bar\sigma + \overset{+}{\sigma} + \overset{++}{\sigma}\,, \tag{4.5}$$

in the same manner, the latter under observation of (3.3,4.2).

If we now substitute the incremental variables (4.1,4.2,4.4)
and (4.5) into the principle of virtual work (3.7), we receive
after suitable reordering

$$\int\limits_{\mathring F} \mathring\rho\,\ddot{\overset{+}{u}}{}^{T}\cdot\delta\overset{+}{u}\;d\mathring F + \int\limits_{\mathring F}\Big[(\mathring\sigma+\bar\sigma)^{T}\cdot\tfrac{1}{2}\delta\overset{++}{\varepsilon} + \overset{+}{\sigma}{}^{T}\cdot\delta\overset{+}{\varepsilon}\,\Big]\;d\mathring F$$

$$= \delta\Big[\int\limits_{\mathring F} \bar p^{\circ T}\cdot\delta\bar u\;d\mathring F - \int\limits_{\mathring F} \mathring\rho\,\ddot{\bar u}\cdot\delta\bar u\;d\mathring F + \int\limits_{\mathring C_t} \bar t^{\circ T}\cdot\delta\bar r\;d\mathring C - \int\limits_{\mathring F} (\mathring\sigma+\bar\sigma)^{T}\cdot\delta\bar\varepsilon\;d\mathring F\Big]$$

$$+ \int\limits_{\mathring F} \overset{+}{p}{}^{\circ T}\cdot\delta\overset{+}{u}\;d\mathring F + \int\limits_{\mathring C_t} \overset{+}{t}{}^{\circ T}\cdot\delta\overset{+}{r}\;d\mathring C \tag{4.6}$$

the **incremental principle of virtual work**. Like its origin it
holds independently of the material law applied. Obviously by
comparison with (3.7), the centered variation δ[...] vani-
shes for all fundamental motions $\bar u,\dot{\bar u},\bar r$ in dynamic equilibrium, and the last line can be cancelled for unexcited neighbor-
ing motions $\overset{+}{p} = \overset{+}{t} = 0$.

The complete incremental principle of virtual work (4.6) contains as functional analogue of (1.8) all necessary physical
information for the evaluation of arbitrary neighboring states,
starting from a fundamental state in dynamic equilibrium. In
the course of the discretization of (4.6) the constitutive re-
lation (3.3) and the strain fields $\varepsilon,\dot\varepsilon,\delta\varepsilon$ due to (4.2,4.3,
4.4) are substituted. In the resulting virtual work functional
powers of arbitrary order of $\bar u,\dot{\bar u}$ are retained, higher than

quadratic ones in $\overset{+}{u},\overset{\cdot}{u}$ are cancelled. Further all displacement, velocity and acceleration fields are discretized in an uniform manner

$$u^P = \Omega^P v^P \rightarrow D_{kL} \cdot u^P = D_{kL}^P v^P, \quad D_{kN}(u^P) \cdot u^P = D_{kN}^P(v^P)v^P \quad (4.7)$$

with u^P the displacements, Ω^P the shape functions and v^P the nodal degrees of freedom of the p-th element. This process delivers finally as stationarity condition of (4.6) the **element tangential equation of motion** [6], in which the functionals of Table 1 have been defined.

The assemblage process of all elements then leads to the **global tangential equation of motion** with the same structure as the element one:

$$M \cdot \overset{\cdot\cdot}{\overline{V}} + C_T \cdot \overset{\cdot}{\overline{V}} + K_T \cdot \overset{+}{\overline{V}} = P - F_I : \qquad (4.8)$$

$$M \cdot \overset{\cdot\cdot}{\overline{V}} + (C_V + C_{uL} + C_{uN}) \cdot \overset{\cdot}{\overline{V}} + (K_e + K_\sigma^\circ + K_{\sigma L} + K_{\sigma N} + K_{\sigma L}^\cdot +$$

$$K_{\sigma N}^\cdot + K_{uL} + K_{uN} + K_{uL}^\cdot + K_{uN}^\cdot) \cdot \overset{+}{\overline{V}} = P - F_\sigma^\circ - F_\sigma - F_\sigma^\cdot - F_m .$$

Herein the following abbreviations have been introduced in agreement with Table 1 (functional indices L: linear in $\overline{V}$, N: quadratic in $\overline{V}$ or bilinear in $\overline{V},\overline{V}$):

M mass matrix,
C_v viscous damping matrix,
C_u initial displacement damping matrix,
K_e elastic stiffness matrix,
K_σ° residual stress matrix,
$K_\sigma(K_\sigma^\cdot)$ elastic (viscous) initial stress matrix,
$K_u(K_u^\cdot)$ elastic (viscous) initial displacement matrix,
$F_\sigma^\circ(F_\sigma,F_\sigma^\cdot,F_m)$ residual (elastic, viscous, D'ALEMBERT's) internal nodal forces.

By comparison of (1.8), (4.8) and Table 1 we finally review the transformation process of the basic nonlinear shell equations (3.1 to 3.7) into the tangential equation of motion for a discretized Kelvin-Voigt shell model, the starting point of our intended instability investigations.

110

$$m^p = \int_{\mathring{F}^p} \mathring{\rho}\, \Omega^{pT} \cdot \Omega^p\, d\mathring{F}^p$$

$$c_T^p \begin{cases} c_v^p = \int_{\mathring{F}^p} D_{kl}^{pT} \cdot D \cdot D_{kl}^p\, d\mathring{F}^p \\[2mm] c_{uL}^p = \int_{\mathring{F}^p} [D_{kl}^{pT} \cdot D \cdot D_{kN}^p(\mathring{v}^p + \bar{v}^p) + D_{kN}^{pT}(\mathring{v}^p + \bar{v}^p) \cdot D \cdot D_{kl}^p]\, d\mathring{F}^p \\[2mm] c_{uN}^p = \int_{\mathring{F}^p} D_{kN}^{pT}(\mathring{v}^p + \bar{v}^p) \cdot D \cdot D_{kN}^p(\mathring{v}^p + \bar{v}^p)\, d\mathring{F}^p \end{cases}$$

$$k_T^p \begin{cases} k_e^p = \int_{\mathring{F}^p} D_{kL}^{pT} \cdot E \cdot D_{kL}^p\, d\mathring{F}^p \\[2mm] k_\sigma^p \begin{cases} k_{\mathring{\sigma}}^p = \left(\int_{\mathring{F}^p} D_{kN}^{pT}(\overset{+}{v}{}^p) \cdot \mathring{\sigma}\, d\mathring{F}^p \right)_{,\mathring{v}^p} \\[2mm] k_{\sigma L}^p = \left(\int_{\mathring{F}^p} D_{kN}^{pT}(\overset{+}{v}{}^p) \cdot E \cdot (D_{kL}^p + D_{kN}^p(\mathring{v}^p)) \cdot \bar{v}^p\, d\mathring{F}^p \right)_{,\mathring{v}^p} \\[2mm] k_{\sigma N}^p = \left(\tfrac{1}{2} \int_{\mathring{F}^p} D_{kN}^{pT}(\overset{+}{v}{}^p) \cdot E \cdot D_{kN}^p(\bar{v}^p) \cdot \bar{v}^p\, d\mathring{F}^p \right)_{,\mathring{v}^p} \\[2mm] k_{\mathring{\sigma}L}^p = \left(\int_{\mathring{F}^p} D_{kN}^{pT}(\overset{+}{v}{}^p) \cdot D \cdot D_{kL}^p \cdot \dot{v}^p\, d\mathring{F}^p \right)_{,\mathring{v}^p} \\[2mm] k_{\mathring{\sigma}N}^p = \left(\int_{\mathring{F}^p} D_{kN}^{pT}(\overset{+}{v}{}^p) \cdot D \cdot D_{kN}^p(\mathring{v}^p + \bar{v}^p) \cdot \dot{v}^p\, d\mathring{F}^p \right)_{,\mathring{v}^p} \end{cases} \\[2mm] k_u^p \begin{cases} k_{uL}^p = \int_{\mathring{F}^p} [D_{kL}^{pT} \cdot E \cdot D_{kN}^p(\mathring{v}^p + \bar{v}^p) + D_{kN}^{pT}(\mathring{v}^p + \bar{v}^p) \cdot E \cdot D_{kL}^p]\, d\mathring{F}^p \\[2mm] k_{uN}^p = \int_{\mathring{F}^p} D_{kN}^{pT}(\mathring{v}^p + \bar{v}^p) \cdot E \cdot D_{kN}^p(\mathring{v}^p + \bar{v}^p)\, d\mathring{F}^p \\[2mm] k_{\dot{u}L}^p = \int_{\mathring{F}^p} D_{kL}^{pT} \cdot D \cdot D_{kN}^p(\dot{v}^p)\, d\mathring{F}^p \\[2mm] k_{\dot{u}N}^p = \int_{\mathring{F}^p} D_{kN}^{pT}(\mathring{v}^p + \bar{v}^p) \cdot D \cdot D_{kN}^p(\dot{v}^p)\, d\mathring{F}^p \end{cases} \end{cases}$$

$$\bar{p}^p = \int_{\mathring{F}^p} \Omega^{pT} \cdot \bar{p}\, d\mathring{F}^p + \int_{\mathring{C}_t^p} R_r^{pT} \cdot \bar{t}^\circ\, d\mathring{C}_t^p$$

$$\overset{+}{p}{}^p = \int_{\mathring{F}^p} \Omega^{pT} \cdot \overset{+}{p}\, d\mathring{F}^p + \int_{\mathring{C}_t^p} R_r^{pT} \cdot \overset{+}{t}{}^\circ\, d\mathring{C}_t^p$$

$$f_T^p \begin{cases} f_{\mathring{\sigma}}^p = \int_{\mathring{F}^p} \left(D_{kL}^{pT} + D_{kN}^{pT}(\mathring{v}^p + \bar{v}^p) \right) \cdot \mathring{\sigma}\, d\mathring{F}^p \\[2mm] f_\sigma^p = \int_{\mathring{F}^p} \left(D_{kL}^{pT} + D_{kN}^{pT}(\mathring{v}^p + \bar{v}^p) \right) \cdot E \cdot \left(D_{kL}^p + \tfrac{1}{2} D_{kN}^p(\bar{v}^p) + D_{kN}^p(\mathring{v}^p) \right) \cdot \bar{v}^p\, d\mathring{F}^p \\[2mm] f_{\dot{\sigma}}^p = \int_{\mathring{F}^p} \left(D_{kL}^{pT} + D_{kN}^{pT}(\mathring{v}^p + \bar{v}^p) \right) \cdot D \cdot \left(D_{kL}^p + D_{kN}^p(\mathring{v}^p + \bar{v}^p) \right) \cdot \dot{v}^p\, d\mathring{F}^p \\[2mm] f_m^p = \int_{\mathring{F}^p} \mathring{\rho}\, \Omega^{pT} \cdot \Omega^p \cdot \ddot{v}^p\, d\mathring{F}^p \end{cases}$$

Table 1. Components of element tangential equation of motion

5. Stability Analysis of Time-Invariant Processes

For time-invariant unstable responses we limit our attention to single-parametric load systems

$$P = \lambda P^{\circ} \tag{5.1}$$

and thus detect load paths in the R^{n+1} of elements $\{V_n, \lambda\}$. According to Fig. 2 the following problems have to be solved:

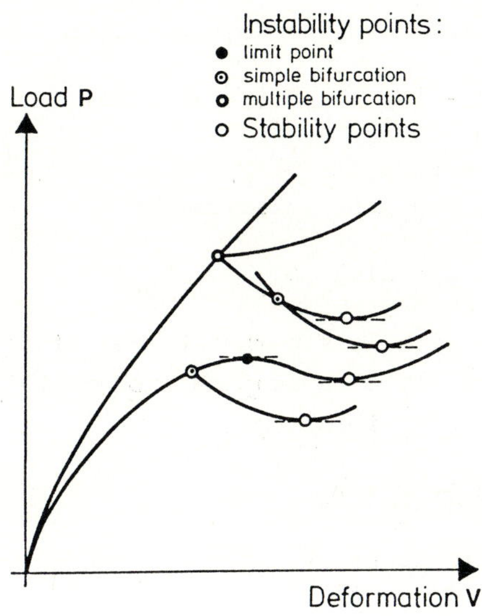

Fig. 2. Instability phenomena of time-invariant responses

- the evaluation of nonlinear primary, secondary and higher-order response-paths;

- the detection of limit points;

- the determination of arbitrary bifurcation points and their multiplicity.

As nodal velocities and accelerations vanish in the static case, (4.8) transforms into the **tangential stiffness relation**

$$\ddot{\vec{V}} = \dot{\vec{V}} = 0 : \quad K_T(\bar{V}) \cdot \vec{V} = P_I - F = \lambda P^{\circ} - F_I \tag{5.2}$$

as basic process description.

Well-established incremental-iterative solution procedures for
(5.2) are available and can be applied for the evaluation of
points of arbitrary nonlinear responses [7,8] including bifurcated branches, if they are endowed with restart capabilities
at each previously computed point.

Instability properties have to be determined as properties of
K_T . Because K_T is of order (nxn) and hermitean for conservative loads, all eigenvalues λ_n (eigenvectors U_n) of K_T
are real (mutually orthogonal), and the decomposition

$$U^T \cdot K_T \cdot U = \Lambda \quad \text{with} \quad U = [U_1 \ldots U_n], \quad \Lambda = \text{diag} \{\lambda_1 \ldots \lambda_n\}$$

$$(5.3)$$

exists. For the same reason the triangularization

$$K_T = L^T \cdot D \cdot L \quad \text{with} \quad D = \text{diag} \{D_{11} \ldots D_{nn}\} \qquad (5.4)$$

can be carried out, usually with less computational effort than
(5.3). For judging the stability of a reached fundamental state
$\{V, \lambda\}$, non-trivial states of equilibrium

$$K_T(\bar{V}) \cdot \overset{+}{V} = \lambda P^\circ - F_I = 0 \rightarrow \overset{+}{V} = \overset{+}{V}_e \qquad (5.5)$$

yield the following classification: $\bar{V}$ is

- stable,   if $\overset{+}{V}$ does not exist, since det $K_T > 0$:
         all λ_i, $D_{ii} > 0$;

- unstable,   if $\overset{+}{V}$ does not exist, < 0:
         at least one λ_i, $D_{ii} < 0$;

- indifferent, if $\overset{+}{V}$ exists, $= 0$:
         at least one λ_i, $D_{ii} = 0$.

The eigenvalues λ_i , occasionally also D_{ii} , are denoted as
stability coefficients: for stable responses all of them are
positive. If one or more are negative, the structure responds
in an unstable manner with respect to the corresponding mode(s)
U_i . The number of negative elements $\{\lambda_i, D_{ii}\}$, usually known as
degree of instability, informs about possible unstable responses
in $\bar{V}$.

Indifferent (neutral) states of equilibrium are response points $\bar{V}$, in which the degree of instability changes. If we denote the computed eigenvector(s) of (5.5) in $\bar{V}$ by $\overset{\leftrightarrow}{V}_e(\bar{V})$, the incremental vector of the original path by $\overset{\leftrightarrow}{V}_c(\bar{V})$ and the tangent vector to the load path by $\lambda,_V(\bar{V})$, we are able to distinguish:

- $\overset{\leftrightarrow}{V}_e = \alpha \overset{\leftrightarrow}{V}_c$ and $\lambda,_v = 0$:   limit point, snap through or snapback;

- one $\overset{\leftrightarrow}{V}_e = \alpha\overset{\leftrightarrow}{V}_c$ and $\lambda_v, \neq 0$, or one $\overset{\leftrightarrow}{V}_e \neq \alpha\overset{\leftrightarrow}{V}_c$:

  simple bifurcation point;

- several $\overset{\leftrightarrow}{V}_e = \alpha\overset{\leftrightarrow}{V}_c$ and $\lambda,_v \neq 0$, several $\overset{\leftrightarrow}{V}_e \neq \alpha\overset{\leftrightarrow}{V}_c$:

  multiple bifurcation point.

In a computed bifurcation point $\bar{V}_c, \lambda_c$ the corresponding buckling mode serves as starting vector for the evaluation of the bifurcated branch:

$$V = \bar{V}_c + \varepsilon \overset{\leftrightarrow}{V}_e , \qquad (5.6)$$

where ε denotes a suitable control parameter: $\varepsilon=0$ follows the primary path, whereas for $\varepsilon=\pm\varepsilon^* \neq 0$ one bifurcated branch is selected. The iteration of (5.2) via F_I leads, if the Riks-Wempner-Wessels scheme or a similar one is applied, after a few corrector-steps to the first point of the bifurcated branch:

$$V = \bar{V}_c + \varepsilon \overset{\leftrightarrow}{V}_e + \sum_j \overset{\leftrightarrow}{V}_j . \qquad (5.7)$$

Fig. 3 demonstrates the unstable responses of 2 degrees of freedom of a flat cylindrical panel, axially loaded and immovably supported along its boundaries in the radial direction. We observe the primary path I with degree of instability 0 and one single point of bifurcation along its course. The secondary path II possesses the degree of instability 1 (0) in its descending (ascending) branch returning to stability in a snapback point. Imperfect paths for imperfections of 5% of the wall thickness complete the response picture.

114

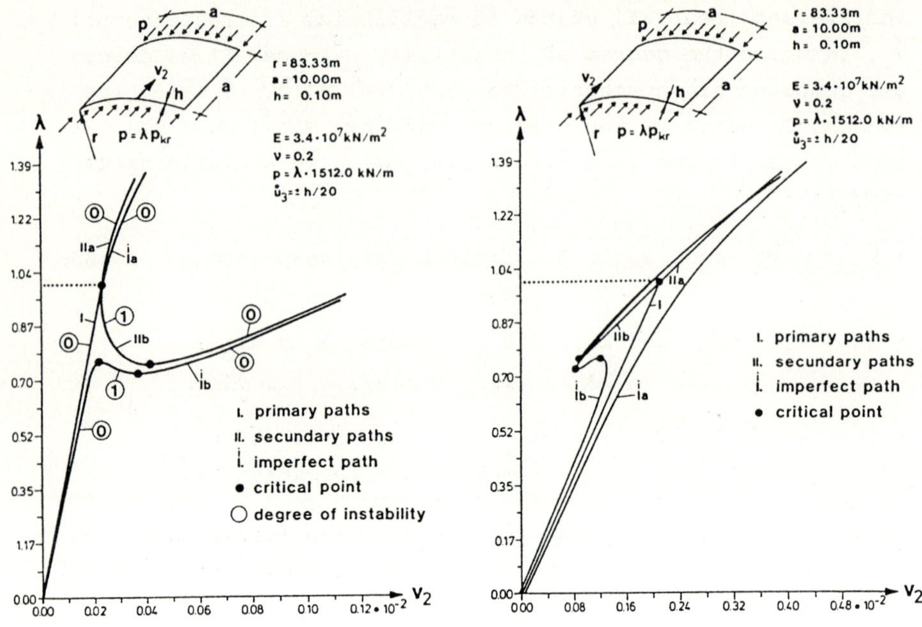

Fig. 3. Unstable responses of a cylindrical panel

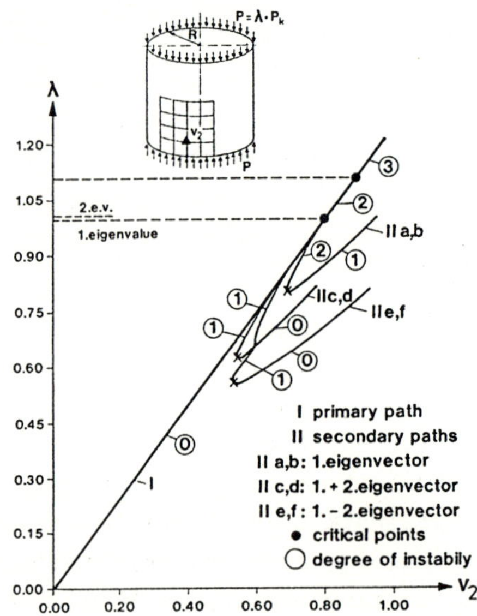

Fig. 4. Unstable response of axially loaded cylindrical shell

The well-known stability problem of the axially loaded circular cylindrical shell is pictured in Fig. 4. 3 bifurcation points are closely neighbored on the primary path I, and all secondary paths descend tangentially along the primary one leading to degrees of instability 2 respectively 3 there. Modes corresponding to the 1st and 2nd bifurcation points are combined in order to compute different postbuckling paths. For further details for both figures see [8].

6. Stability Analysis of Time-Dependent Processes

Time-dependent responses will be treated as time-histories such that we trace load-displacement histories in the R^{2n+2} of elements $\{V_n, \dot{V}_n, \lambda, t\}$. Stability assessments are based on the phase transformation (2.1) of the tangential equation of motion, although (4.8) itself may be also integrated directly.

General stability conditions can be derived from the structure of the matrix $\mathbf{A}$ of (2.1) and the definiteness properties of its submatrices as follows [9]:

$$
\begin{array}{lccc}
 & \mathrm{tr}(\mathbf{M}^{-1} \cdot \mathbf{C}_T) & & \det \mathbf{K}_T \\
\bullet \text{ necessary for stability:} & \geq 0 & \text{and} & > 0 \ , \\
\text{asymptotic stability:} & > 0 & & > 0 \ , \\
\bullet \text{ sufficient for instability:} & < 0 & \text{or} & \leq 0 \ .
\end{array}
$$

$$\tag{6.1}$$

More exact criteria, however, have to be based on Ljapunow's general stability theorems (2.5) applying the following procedure:

- Evaluation of a specific response $\bar{V}(t)$ respectively $\bar{X}(t)$, whose stability is a matter of question, through direct numerical integration of (4.8) or (2.1).

- Performance of systematic small perturbations of $\bar{V}(t), \bar{X}(t)$ at suitable time instants through slightly changed initial conditions at t_o

$$\| \overset{\scriptscriptstyle\pm}{X} (\overset{\scriptscriptstyle\pm}{X}_o, \bar{X}_o, t_o, t_o) \| < \delta \tag{6.2}$$

and computation of the future responses $\overset{\star}{\underline{X}}(\overset{\star}{\underline{X}}_o,\bar{\underline{X}}_o,t_o,t)$.

• Stability assessment of $\|\overset{\star}{\underline{X}}(\overset{\star}{\underline{X}}_o,\bar{\underline{X}}_o,t_o,t)\|$ due to (2.5).

In numerical applications the proposed procedure turns out to be extremely time-consuming. An example which demonstrates this work load is presented in Fig. 5 and 6. These illustrations show a flat cylindrical panel under a constant transverse step

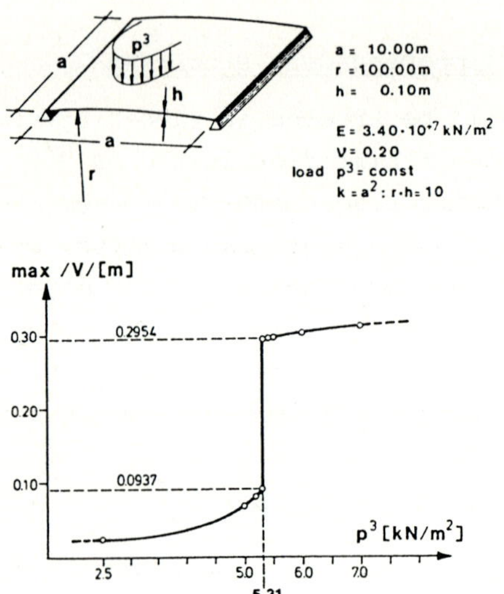

Fig. 5. Cylindrical panel under constant step loading at t_s

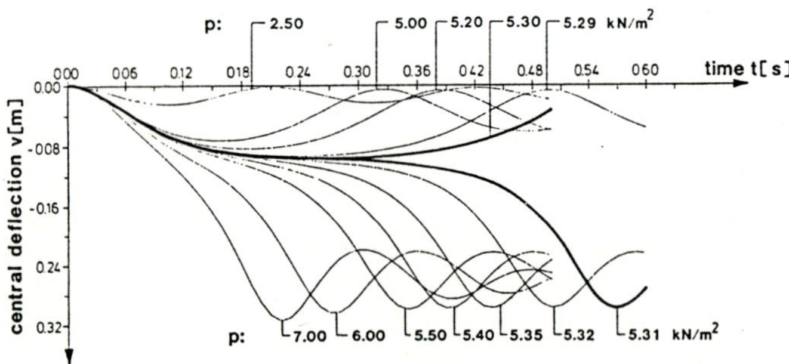

Fig. 6. Central displacement responses of cylindrical panel under constant step loading

loading of increasing intensity p :

$$p^3 = 0 \quad \text{for} \quad t < t_s \, , \qquad p^3 = p \quad \text{for} \quad t \geq t_s \, . \qquad (6.3)$$

All responses for $p \leq 5.30$ kN/m² obviously can be classified as asymptotically stable because they reach, after some transient vibrations, the static deflection $v \approx 5.2$ cm for $t \to 0$. Solutions from $p = 5.31$ kN/m² upwards have to be identified as unstable, if the central deflection v on Fig. 6 exceeds a bound ε , smaller than v_{max} . Because of the dominance of the strongly time-dependent matrices $K_{\sigma L}, K_{\sigma N}$ in K_T the response is neither autonomous nor periodical, as the faster vibrations after the snapping in Fig. 6 demonstrate.

If the response variety of Fig. 6, including its transition to instability, would be unknown, every computed path would have to be subjected to a number of perturbations (6.2) at certain time instants t_o not too far from each other. But finally, only the response for $5.30 < p < 5.31$ would have demonstrated its instability, for instance by a perturbation with small, but final velocity at t_s , as the phase diagram on Fig. 7 shows.

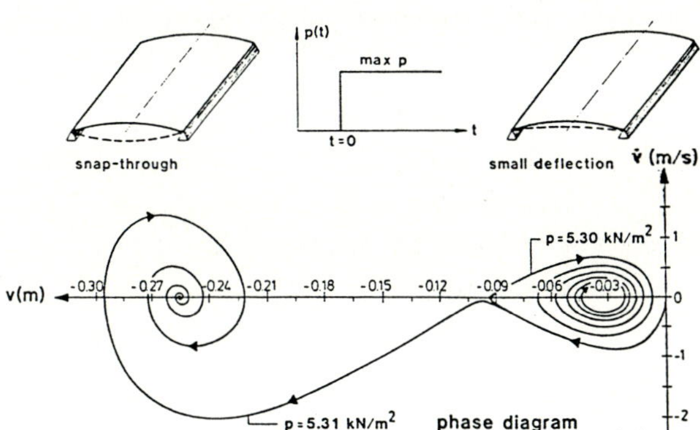

Fig. 7.   Phase diagram of last stable and first unstable response

For limited classes of responses, for which (2.1) forms a regular system of differential equations [1] such as

- autonomous systems: $A = A(t)$,

- non-autonomous, but periodical systems: $A = A(t) = A(t+T)$,

kinetic stability criteria may help to avoid heuristic, time-consuming procedures of the kind mentioned. Instability of a given state of dynamic equilibrium $\bar{V}$ is diagnosed, if for nonlinear, **autonomous systems** non-trivial solutions of (2.1) can be detected, so-called dynamic quasi-bifurcations:

$$P-F_I = 0: \quad B = 0 \quad \rightarrow \quad \dot{\bar{X}} = A \cdot \dot{\bar{X}} \quad \rightarrow \quad \ddot{X}(t) \ . \tag{6.4}$$

Possible solutions of (6.4) can be determined by means of the assumption

$$X = \ddot{X}_O \, e^{\omega t} \quad \text{with} \quad \omega \quad \text{from} \quad \det(A - \omega I) = 0 \ . \tag{6.5}$$

Obviously $\bar{X}(t)$ behaves asymptotically stable (weakly stable), if all characteristic roots ω exhibit negative real parts (are purely imaginary). All other possibilities furnish unstable responses.

The explizit form of the characteristic equation (6.5)

$$\left(\begin{bmatrix} -M^{-1} \cdot C_T & -M^{-1} \cdot K_T \\ \hline I & 0 \end{bmatrix} - \omega \begin{bmatrix} I & 0 \\ \hline 0 & I \end{bmatrix} \right) \cdot \dot{\bar{X}} = \begin{bmatrix} -M^{-1} \cdot C_T - \omega I & -M^{-1} \cdot K_T \\ \hline I & -\omega I \end{bmatrix} \cdot \dot{\bar{X}} = 0 \tag{6.6}$$

can be simplified for vanishing damping $C_T = 0$ to the following **kinetic stability criterion:**

$$\det \begin{bmatrix} -\omega I & -M^{-1} \cdot K_T \\ \hline I & -\omega I \end{bmatrix} = \det \left[\omega^2 I + M^{-1} \cdot K_T \right] = 0 \ , \tag{6.7}$$

$$\det \left[K_T + \omega^2 M \right] = 0 \ .$$

Instability, e.g. the transition from asymptotically stable to weakly stable and finally to unstable responses, is determined by $Re(\omega) = 0 \rightarrow \omega = 0$, when the system loses its vibration capacity. Consequently, as in the case of time-invariant processes, this transition again is characterized by the singularity of the tangent stiffness matrix:

$$\omega = 0 \quad \rightarrow \quad (6.7): \quad \det K_T = 0 \ . \tag{6.8}$$

7. Stability Analysis of Periodical (Rheo-Linear) Systems

As mentioned previously, spezialized stability criteria can
also be developed for **periodical systems**. Assuming again the
forcing process $P(t)$ to remain unchanged during the passage
from the fundamental to the neighboring motion, we have from
(2.1, 4.8):

$$P-F_i = 0 : \qquad M \cdot \ddot{\vec{V}} + C_T \cdot \dot{\vec{V}} + K_T(\bar{V}) \cdot \vec{V} = 0$$

$$\text{or:} \quad \dot{\vec{X}} = A(\bar{X}) \cdot \vec{X} \tag{7.1}$$

with $K_T(\bar{V}) = K_T(t) = K_T(t+T)$, $A(\bar{X}) = A(t) = A(t+T)$ as periodicity condition. If we collect $2n$ fundamental solutions of
(7.1) in $\vec{X}_F$, the monodromy matrix

$$B = \vec{X}_F^{-1}(t) \cdot \vec{X}_F (t+T) \tag{7.2}$$

admits the following **stability criterium:** If the spectral radius κ of B is defined as

$$\kappa(B) = \max_i \{|\lambda_i| : \lambda_i \text{ eigenvalues of } B\} , \tag{7.3}$$

$\vec{V}$ respectively $\vec{X}$ can be classified as

- stable for $\kappa(B) < 1$,
- unstable > 1 ,
- critical $= 1$.

$$\tag{7.4}$$

The limit case $\kappa(B) = 1$ describes the change from stable to
unstable responses for linear fundamental motions. For nonlinear
ones it may, but must not, designate the criterial case of Ljapunow's theory [3].

For linear systems (7.1) simplifies to the **Mathieu matrix differential equation**

$$M \cdot \ddot{V} + C \cdot \dot{V} + (K_e + K_{\sigma S} + K_{\sigma D} \cos \Omega t) \cdot V = 0 , \qquad (7.5)$$

valid for a forcing process

$$P(t) = P_S + P_D \cos \Omega t , \qquad \Omega = 2\pi/T \qquad (7.6)$$

with a static preload P_S and a harmonically pulsating compo-
nent P_D. $K_{\sigma S}$, $K_{\sigma D}$ stand for the initial stress matrices corresponding to P_S, P_D. B is derived from an infinite matrix
series solution, and for the limit $\kappa(B) = 1$, motions with T,
2T can be identified and evaluated from (7.1) by means of finite approximations of infinite eigenvalue problems [10, 12].
For details of the numerical treatment the reader is referred
to [11, 12] leading finally to regions of stable and unstable
responses in a normalized load-frequency space. As an example
Fig. 8 shows the well-known stability map for the central deflection of a cylindrical panel under axially pulsating load
(7.6).

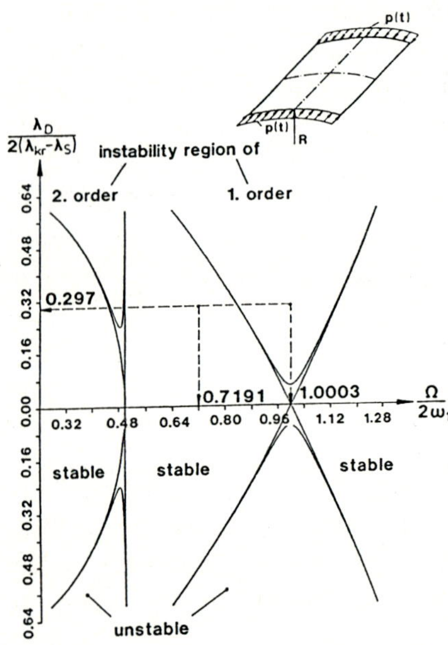

Fig. 8. Stability map for central deflection of cylindrical
         panel

For geometrically nonlinear systems the fundamental matrix

$$\overset{+}{\underset{F}{X}} = \sum_{i=1}^{2n} \overset{+}{\underset{Fi}{X}} = \sum_{i=1}^{2n} \left[\begin{array}{c} \dot{\overset{+}{V}} \\ \hline \overset{+}{V} \end{array} \right]_i \tag{7.7}$$

can theoretically be computed from (7.1), now a **Hill matrix differential equation**, for $2n$ linear independent initial conditions. But even the evaluation of a few solutions for a typical discretized shell structure may overcharge mainframe computers, such that - at least for weakly nonlinear systems - the reduction of $\overset{+}{\underset{F}{X}}$ to a few "eigenmodes" is recommended:

$$\overset{+}{\underset{F}{X}}(t) \approx \sum_{i=1}^{2k} \left[\begin{array}{c} \dot{\psi}_i(t) \cdot \varphi_i \\ \hline \psi_i(t) \cdot \varphi_i \end{array} \right] , \quad k \ll n . \tag{7.8}$$

Herein $\psi(t)$, $\dot{\psi}(t)$ represent a periodical scalar function of time t and its time-derivative, φ_i the "eigenvectors" at time t_o, t_o+T, t_o+2T, ... Using this reduction technique identical stability assessments as in the linear case (7.4) can be applied.

The result of such a 6-mode reduction is demonstrated in the last Figs. for an edge loaded conical shell with $r_o = 1000$ mm, $r_u = 3000$ mm, 5 mm in thickness and 2000 mm in height. The shell is clamped on the lower edge, and for the harmonically pulsating load on the upper rim

$$p(t) = 25.54 + 17.03 \cos 2\pi t/0.019 \text{ kN/m} \tag{7.9}$$

instability is predicted. The fundamental matrix due to (7.8) was reduced by the first 6 modes on Fig. 9, leading to eigenvalues of B :

$$|\lambda_1| = 1.505, \quad |\lambda_2| = 0.834, \quad |\lambda_3| = 0.790, \quad \ldots |\lambda_{12}| = 0.378 .$$

Obviously because of $\kappa = |\lambda_1| = 1.505 > 1$, instability of the response is confirmed. Finally, this unstable response is also computed through step-wise integration of (7.1) and perturbed at $t_p = 0.128$ s by

122

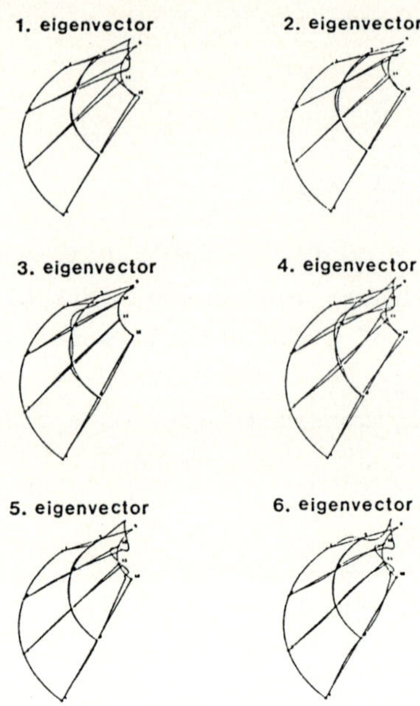

Fig. 9.   Eigenmodes of vibration of a conical shell

$$\psi_i(t_p) \;=\; \dot{\psi}_i(t_p) \;=\; 2.0 \cdot 10^{-5} \;,$$

equally for all modes. Fig. 10 demonstrates the evaluated time-
history of a nodal displacement, Fig. 11 the corresponding
phase diagram [11].

Conclusions

Modern incremental-iterative algorithms enable the numerical
evaluation of arbitrary nonlinear shell responses. Time-invariant as well as time-dependent problems can be treated on the
basis of discretized incremental energy functionals. They contain simultaneously, as demonstrated, all necessary informations
concerning possible instabilities of the mentioned responses.

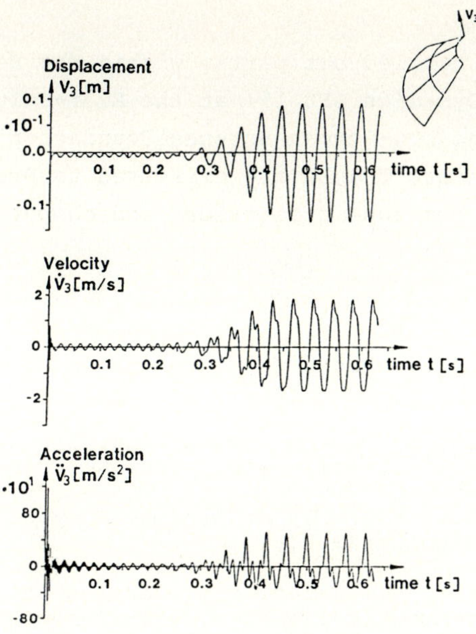

Fig. 10. Time-history of unstable response of nodal deflection

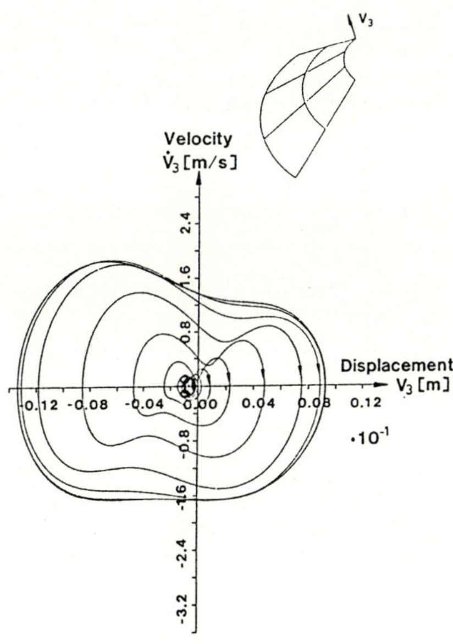

Fig. 11. Phase diagram of unstable response of nodal deflection

Acknowledgement

This research has been carried out partly within the Research Center for Structural Dynamics SFB 151 at the Ruhr-University. Financial support of the DFG-German Science Foundation is gratefully acknowledged. Sincere thanks are expressed to Drs. U. Eckstein and C. Eller for numerous discussions and computations of unstable responses.

References

1. Bellmann, R.: Stability theory of differential equations. Dover Publications, Inc., New York 1969.

2. Ljapunow, A.M.: Problème général de la stabilité du mouvement. Transl.thesis 1893, Ann. math. studies, Princeton 1949.

3. Willems, J.L.: Stability Theory of Dynamical Systems. Th. Nelson & Sons Ltd, London 1970.

4. Krätzig, W.B.; Wittek, U.; Basar, Y.: Buckling of general shells - Theory and numerical analysis. In: Thompson, J.M.T., Hunt, G.W. (ed.): Collapse, 377-394. Cambridge University Press, Cambridge 1983.

5. Basar, Y.; Krätzig, W.B.: Introduction to Finite-Rotation Shell Theories and their Operator Formulation. Contribution in this volume.

6. Krätzig, W.B.: Eine einheitliche statische und dynamische Stabilitätstheorie für Pfadverfolgungsalgorithmen in der numerischen Festkörpermechanik. ZAMM 69 (1989), 203.

7. Ramm, E.: Strategies for Tracing the Nonlinear Response near Limit Points. Wunderlich, W., et al. (ed.): Nonlinear Elastic Analysis in Structural Mechanics, 63-89. Springer-Verlag, Berlin 1980.

8. Eckstein, U.: Nichtlineare Stabilitätsberechnung elastischer Schalentragwerke. Technical Report No. 83-3, Ruhr-University, Institute for Struct. Eng., Bochum 1983.

9. Müller, P.C.: Stabilität und Matrizen. Springer-Verlag, Berlin 1977.

10. Bolotin, V.V.: The Dynamic Stability of Elastic Systems. Russian original 1956, Holden-Day, San Francisco 1964.

11. Eller, C.: Lineare und nichtlineare Stabilitätsanalyse periodisch erregter, visko-elastischer Strukturen. Techn. Report No. 88-2, Ruhr-University, Institute for Struct. Eng., 1988.

12. Basar, Y.; Eller, C.; Krätzig, W.B.: Finite Element Procedures for Parametric Resonance Phenomena of Arbitrary Elastic Shell Structures. Comput. Mechanics 2 (1987), 87-98.

Numerical Aspects of Shell Stability Analysis

E.Riks[1], F.A. Brogan [2] and C.C. Rankin [2]

[1] National Aerospace Laboratory, Anthony Fokkerweg, Amsterdam
[2] Applied Mechanics Lab., Lockheed Missiles & Space Co.Inc.
3251 Hanover Street, Palo Alto, California 94304

Summary

Many shell stability problems can be analyzed following the quasi static approach, i.e. with methods that solve the static equations of equilibrium. This paper focuses on some particular points of this approach. There are cases, however, mode jumping for example, where the methods of statics do not longer suffice and where it becomes necessary to combine the methods of statics with procedures for the integration of the equations of motion. The latter idea is illustrated with an example that was analyzed with the shell finite element code STAGS.

1. Introduction

The notion of stability should be placed in the context of dynamics, although on the level of applications it is seldom treated as such. It has always been convenient to analyze the load carrying capacity of shell structures with static methods, i.e. with methods that determine and analyze the static equilibrium states of a structure without resorting to the equations of motion. This is also the approach that will be discussed here, although we will return to the theme of dynamical response in the closing stages of this overview.

Incremental iterative methods or continuation methods for static analysis are now firmly established in the finite element community [1]. Indeed they play an important role in the nonlinear finite element codes of the present day and the range of applications is growing as is illustrated for instance by the (geometrical) nonlinear crack analysis reported in [2,3]. We feel therefore, that it is not necessary to describe these methods here extensively as this has already been done elsewhere, see for instance [4,5,6].

Instead, we will focus on some particular details of the techniques which are especially useful for the computation and evaluation of the (critical) equilibrium states in the sense of the stability theory. As already mentioned, the paper is closed with a discussion about the need to combine the methods of statics with the methods of dynamics.

2. BY PRODUCTS OF THE CORRECTOR EQUATIONS

2.1 General remarks

The continuation methods considered here are predictor-corrector procedures. The prediction phase is used to advance the solution along a predetermined loading direction. Prediction is followed by corrector operations to determine the solution accurately. In the analysis of stability problems the corrector equations usually belong to the class of iteration methods that are called 'Quasi Newton' or 'Newton like' methods. Characteristic of these methods is that the systems matrix is - to some approximation -, equal to the Jacobian (or stiffness matrix) of the governing equations. This makes it possible to extract information from the stiffness matrix during the computations that can give some insight into the physical nature of the solutions obtained. Some of this information is directly available, while other can be obtained at little additional costs. We will illustrate this in the following discussion.

Suppose the equations of equilibrium are given by

$$f(\mathbf{d}, \lambda) = 0 \tag{2.1}$$

where $\mathbf{f} \in \mathbb{R}_N$ are nonlinear relations of the deformation variables $\mathbf{d} \in \mathbb{R}_N$ and the parameter λ denotes the load intensity. These equations are solved by considering the augmented form :

$$g(\mathbf{x}, \eta) = 0 \tag{2.2}$$

with g:

$$g(\mathbf{x}, \eta) = g(\mathbf{d}, \lambda, \eta) = \begin{pmatrix} f(\mathbf{d}, \lambda) \\ h(\mathbf{d}, \lambda) - \eta \end{pmatrix} \tag{2.3}$$

where the extra equation is introduced to define the parametrization of the solution path:

$$x(\eta) = \begin{pmatrix} d(\eta) \\ \lambda(\eta) \end{pmatrix}$$ (2.4)

It will thus be assumed that the F.E. continuation procedures that compute the static response of the shell structures have the formulation (2.2) + (2.3) as a basis [6].

2.1 Solution Corrector Equations

Suppose we have obtained a solution point of the curve (2.4) (see Figure 1) which we denote by $x(\eta_k)$. Consider now the corrections $\Delta\sigma^i$ to the prediction σ^0, that are calculated to find the solution point $x(\eta_{k+1})$ in the step $\eta_k \to \eta_{k+1}$. They are generally obtained by the repeated solution of a system of equations of the form:

$$A^i(k+1)\Delta\sigma^i + g(\sigma^i, \eta_{k+1}) = 0$$ (2.5)

where $A^i(k+1)$ is an approximation to the Jacobian

$$J = g_x = \begin{pmatrix} \partial f_i/\partial x_j \\ \partial h/\partial x_j \end{pmatrix} = \begin{Bmatrix} f_d & f_\lambda \\ h_d & h_\lambda \end{Bmatrix}$$ (2.6)

Note that with this notation f_d denotes the Jacobian (or stiffness matrix) of equations (2.1) and f_λ corresponds to the current load direction.

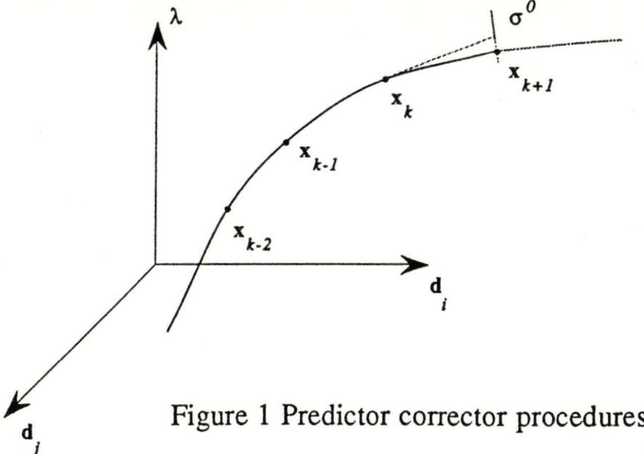

Figure 1 Predictor corrector procedures

In the pure Newton form, $A^i = J(\sigma^i)$ and A^i will approach the value that corresponds to $J(x_k)$ when the correction process (2.5) is convergent. In the modified Newton form, A^i is defined at some configuration σ in the neighborhood of the sought solution and is kept unaltered during the iteration process. In some applications, this value is the previous solution point x_k but in others it will be the prediction σ^0. This is dependent on the design of the particular algorithm in question. It is important to realize, however, that it is to some extent necessary to be able to compute J in the neighborhood of the points that are *critical* in the sense of the stability theory. We take the possibility to compute J for granted in what follows.

The value of the determinant $J = \det\{J(x(\eta_k))\}$ will change sign when passing a bifurcation point of uneven multiplicity. It will become zero at a bifurcation point with even multiplicity but not change its sign in passing. The value of the determinant $A = \mathrm{Det}\{A\}$ which is supposedly close to the value of J could thus serve as a rough indicator of the occurrence of critical points. If we monitor J together with the path derivative $\lambda' = d\lambda/d\eta$ (which can always be obtained during the computations), it is also possible to distinguish between bifurcation points and limit points [5,6].

2.2 <u>Solution Corrector Equations</u>

For reasons of efficiency, the system (2.5) is hardly ever solved in a straightforward manner because it is advantageous to make use of the solver for symmetric systems that is available in the code. It can be accomplished by a special implementation of the solution scheme for (2.5). A useful technique in this respect is block Gaussian elimination.[1]. The variant of this technique that we consider in this paper has some advantages that outweigh the slightly increased complexity of the scheme.

To simplify the notation we introduce the following shorthand:

$$K = f_d \; ; \; l = f_\lambda \; ; \; n^t = h_d \; ; \; n_0 = h_\lambda \tag{2.7}$$

Consider a partition of the Jacobian J in the following manner:

---

[1] Block Gaussian elimination as a stabilizing measure in the solution of the corrector process is well known in the Applied Mathematics literature, see for instance the application in [5].

$$J = \begin{bmatrix} K & l \\ \underline{n}^t & n_0 \end{bmatrix} = \begin{bmatrix} K & C & l \\ \underline{C}^t & C_1 & l_1 \\ \underline{n}^t & n^t_1 & n_0 \end{bmatrix} \tag{2.8}$$

where the matrices $\underline{C}$, $\underline{C}^T$ are $M \otimes (N-M)$, $(N-M) \otimes M$ respectively, and $C_1 = M \otimes M$. Similarly, l; $\underline{n}^t$ and l_1; n^t_1 are $(N-M)$ and (M) column and row vectors. The partition corresponds to a sorting out of a certain selection of M rows and columns. By what selection process it is effected will be explained later. At this moment, the most important thing to remember is that the particular choice of the partition must insure that $\underline{K}$ be nonsingular. In that case we can show with little difficulty that the (non-symmetric) matrix J^*, defined below, contains all the information we need to verify the stability or instability of the equilibrium solutions.

$$J^* = \begin{pmatrix} K^* & l^* \\ n^{*t} & n^*_0 \end{pmatrix} \tag{2.9a}$$

$$K^* = C_1 - \underline{C}^t (\underline{K})^{-1} \underline{C} \qquad ; l^* = l_1 - \underline{C}^t (\underline{K})^{-1} l$$

$$n^{*t} = n^t_1 - \underline{n}^t (\underline{K})^{-1} \underline{C} \qquad ; n^*_0 = n_0 - \underline{n}^t (\underline{K})^{-1} l_1 \tag{2.9b}$$

With the partitioning given above, equations (2.5) are written in the form:

$$\begin{pmatrix} K & C & l \\ \underline{C}^t & C_1 & l_1 \\ \underline{n}^t & n^t_1 & n_0 \end{pmatrix} \begin{pmatrix} \underline{t} \\ t_1 \\ \rho \end{pmatrix} = - \begin{pmatrix} \underline{F} \\ F_1 \\ F_0 \end{pmatrix} ; \begin{pmatrix} \underline{t} \\ t_1 \\ \rho \end{pmatrix} = \Delta \sigma \tag{2.10}$$

and the solution is carried out in terms of the following steps:

step 1

Factor $\underline{K}$ and set :

$$\underline{v} = - (\underline{K})^{-1} \underline{C} \; ; \qquad \underline{w} = - (\underline{K})^{-1} l \; ; \qquad \underline{u} = - (\underline{K})^{-1} \underline{F}$$

$$K^* = C_1 + \underline{C}^t \underline{v} \; ; l^* = l_1 + \underline{C}^t \underline{w} \; ; \quad F^*_1 = F_1 + \underline{C}^t \underline{u} \tag{2.11}$$

$$n^{*t} = n^t_1 + \underline{n}^t \underline{v} \; ; n^*_0 = n_0 + n^{*t} \underline{w} \; ; \quad F^*_0 = F_0 + \underline{n}^t \underline{u}$$

step 2

Solve :

$$\begin{Bmatrix} \mathbf{K}^* & \mathbf{l}^* \\ \mathbf{n}^{*t} & n^*_0 \end{Bmatrix} \begin{Bmatrix} t_1 \\ \rho \end{Bmatrix} = - \begin{Bmatrix} \mathbf{F}^*_1 \\ F^*_0 \end{Bmatrix} \tag{2.12}$$

for t_1, ρ and construct:

$$\Delta\sigma = \begin{pmatrix} \mathbf{t} \\ t_1 \\ \rho \end{pmatrix} \quad ; \quad \mathbf{t} = \underline{\mathbf{u}} + \underline{\mathbf{v}}t_1 + \rho\underline{\mathbf{w}} \tag{2.13}$$

Note that when M>1, $\underline{\mathbf{v}}$ is a (N-M)×(M) matrix of M column vectors and t_1 is a set of M parameters.

2.3 The selection of $\underline{\mathbf{K}}$

We will now discuss the selection of the minor $\underline{\mathbf{K}}$. In the STAGSC-1 code [7], we apply the scheme (2.10 & 2.11) with a default value M = 1. Analysis shows that it turns out to be advantageous to choose $\underline{\mathbf{K}}$ such that the row and column that are removed from $\mathbf{K}$ correspond to the coordinate direction of the fastest growing component of $\mathbf{d}$ at the particular step in question [8]. In formal terms the choice amounts to :

$$i^* = \text{Index}\{\ \text{column, row}\ \} = \text{ARG}\left\{ \text{Max}(\ |\ e_i^t \mathbf{d}'\ |\ ) \right\}$$
$$\forall\ i$$

$$(\ e_i\ [\text{for } i=1,....N]\ \text{is the set of base vectors that}$$
$$\text{span the space of deformation variables } \mathbf{d}, \text{ and } \mathbf{d}' \text{ is the tangent to } \mathbf{d}(\eta))$$

In that case we can show that at limit points $\underline{\mathbf{K}}$ is nonsingular so that the solution of (2.10) and (2.12) cannot be influenced by the nearly singular (and possibly badly conditioned) K in the neighborhood of these points.

At bifurcation points, the default setting (STAGSC-1) as described above, may be deficient because it can happen that the buckling mode **a** associated with this point does not have a component in the direction that was chosen for the selection of $\underline{\mathbf{K}}$. If that is the case, $\underline{\mathbf{K}}$ itself is singular and its null vector is

$$a = \begin{pmatrix} a \\ 0 \end{pmatrix} \in \mathbb{R}_N \qquad\qquad (2.14)$$

This observation only holds for simple [1] bifurcation points. In the event of a bifurcation point with multiplicity larger than one, say M^*, $\underline{K}$ will be nonsingular only if the index $M = M^*$, and none of the corresponding buckling modes a_i, has a component equal to zero in the directions e_k that determine the definition of $\underline{K}$. Consequently, to preserve a partition into a (large) nonsingular part and a (small) singular complementary part of the equations, it is necessary to choose $\underline{K}$ differently. This choice could for instance be based on the directions that correspond to the largest components of a_i. In that case it can be shown that $\underline{K}$ will be nonsingular at the bifurcation point.

As far it is known, there is no simple method available that produces information about the buckling mode(s) before the bifurcation points are reached. It is true that it is possible to compute the lowest eigenvalue and corresponding eigenvector of K with little additional cost during the correction process. This can be done by integration the corrector iteration loop with a procedure based on the inverse power method [6]. Of course, we can go even further, and use the factored stiffness matrix directly in a subspace iteration method, but this then will be at the cost of more computation. It is, in general, more practical to postpone calculation of the desired information after the critical point has been reached. This is, for instance, the approach that is used in the STAGSC-1 code where the bifurcation points are analyzed *after* they have been detected by one of the simple detection methods described in section 3.

2.2 Stability Coefficients

As was already mentioned the characteristic conditions for bifurcation and limit points can be cast in terms of the characteristics of the matrix J^* and K^*. This is so because the following relations exist between K, J, in one hand, and $\underline{K}$, $\underline{J}$ in the other [8]:

$$\text{Det}\{K\} = \text{Det}\{\underline{K}\}\text{Det}\{K^*\}$$

$$(2.15)$$

$$\text{Det}\{J\} = \text{Det}\{\underline{K}\}\text{Det}\{J^*\}$$

---

[1] See section 3.1 for a definition.

Thus, characteristic conditions for limit points and bifurcation points:

$$J = \text{Det}\{J\} \neq 0 \;\;;\; K = \text{Det}\{K\} = 0 \qquad (\textit{Limit point})$$

$$J = \text{Det}\{J\} = 0 \;\;;\; K = \text{Det}\{K\} = 0 \qquad (\textit{Bifurcation point})$$

$$(2.16)$$

can be replaced by:

$$J^* = \text{Det}\{J^*\} \neq 0 \;\;;\; K^* = \text{Det}\{K^*\} = 0 \qquad (\textit{Limit point})$$

$$J^* = \text{Det}\{J^*\} = 0 \;\;;\; K^* = \text{Det}\{K^*\} = 0 \qquad (\textit{Bifurcation point})$$

$$(2.17)$$

The condition numbers based on the complements J^* and K^* are in general considerably more manageable than the determinant values (2.15). But it should be noted, however, that the definition of K^* and J^* varies along the path because the index i* determining the values of (2.17) is changed, in principle, at every continuation step. Thus the numbers, $J^* = \text{Det}\{J^*(\eta; i^*)\}$ and $K^* = \text{Det}\{K^*(\eta; i^*)\}$ *are not* produced by functions that vary continuously along the path unless we keep (i*) unaltered. Incidentally, in the STAGSC-1 code , the determinant values J, K, J^*, K^* and that of $\underline{K}$ are monitored as well as the sign of the entries on the main diagonal of the matrix decomposition of $\underline{K}$. This is more than adequate for the detection and identification of critical points.

The conditions numbers (2.17) are useful as we will see later but there is yet another advantage of the block elimination scheme. When J is singular, *but $\underline{K}$ is not,* the null vector(s) of J are contained in the set of vectors $\underline{v}$, while the term $\underline{w}$ in (2.11) constitutes a particular solution of the system

$$g_x x' = \binom{0}{1} \qquad ; \qquad g_x = \frac{dg}{dx} \qquad x' = \frac{dx}{d\eta} \qquad\qquad (2.18)$$

that defines the tangent to the path at the bifurcation point. This means that $\underline{v}$, $\underline{w}$ span a subspace in which the tangents to the bifurcating branches are defined. It is a property that could be useful in the development of bifurcation procedures.

2.3 <u>Critical Points</u>

Continuation procedures usually do not fail to step over a critical point if this point is simple and if it is sufficiently separated from other critical points. Difficulties may occur if a certain stretch of the equilibrium path contains a cluster of bifurcation points as in the case of multi-mode buckling phenomena or in the case of the occurrence of spurious modes in connection with a defective finite element discretization of the given model.

If the procedure passes a bifurcation point or a limit point, the condition numbers that are monitored (2.16) and (2.17), will clearly record this event. In that case it is a simple matter to find an estimate of the critical point by means of a bisection procedure using the solution points that cradle the (yet unknown) critical point. A more accurate method can be based on a step length procedure that makes use of the values of the function $K^*(\eta) = \mathrm{Det}\{ K^*(\eta)\}$ during the approach to the critical state. In other words we replace the default step length procedure that determines the continuation step in terms of η, by

$$\Delta\eta_{k+1} = -\frac{K^*_k}{(K^*_k - K^*_{k-1})} (\eta_k - \eta_{k-1}) \tag{2.19}$$

where

$$K^*_k = \mathrm{Det}\{K^*(x(\eta_k))\} \qquad \text{for } k = 1,2,3,4,....$$

Bifurcation points can be computed in a different way using a nonsingular formulation in terms of an extended set of $2N +1$ equations. These methods as it were compute these points along an iteration path that is not necessarily close to the equilibrium path (2.4). Methods like these can be useful for the computation of critical points (limit points or bifurcation points) under variation of some design parameter, but we will not discuss these techniques here. For a thorough exposition of these possibilities we refer to [9]. A recent discussion of an engineering application of this technique in the context of an finite element formulation is given in [10].

3. BIFURCATION PROCEDURES

3.1 The Objective

By a bifurcation processor is meant an algorithm which is capable of computing bifurcation points and its branches. Although bifurcation phenomena do not as such occur in the physical world, -as a result of the presence of unavoidable imperfections-, they frequently appear in the world of our analytical models. Algorithms that are capable to analyze bifurcations are very useful in helping to understand the proper nature of many shell buckling problems.

It is conventional to make a distinction between simple, (or *distinct*), and non-simple bifurcation points. By the first is meant a bifurcation point that is characterized by two solution branches crossing or touching in one and the same point, while the latter definition means *more* than two solution branches crossing. In practice, however, the notion of a non-simple bifurcation point is not really meaningful. In our opinion, it is more appropriate to make a distinction between simple bifurcation and clustered bifurcations. This is so because, in the actual situation of the analysis of the discretized model of a given problem, bifurcation points with many branches do not really occur unless we tune the model by artificial means. Moreover, any change in the model, however small, may already affect the geometrical branching diagram considerably. Thus bifurcation points with more than two branches occur actually only by chance.

3.1 Linear Pre-Buckling States

In shell stability analysis [11] it makes sense to exploit the fact that many bifurcation problems occurring in this field are characterized by a trivial pre-buckling state. By trivial is meant here that the initial response of the structure is a linear (or an almost linear) function of the intensity of the applied loading. In these particular, but frequently occurring cases, it is possible to make use of a formulation whereby the bifurcation condition can explicitly be defined in terms of a linear eigenvalue problem.

The equations of equilibrium were defined in (2.1). Linearization at the undeformed state gives

$$\mathbf{K}(0)\mathbf{d}_I + \lambda\{f_\lambda(0,0)\} = 0 \tag{3.1}$$

The solution of this system is the pre-buckling state

$$d_I = d_I(\lambda) = \lambda d^0 \tag{3.2}$$

The bifurcation condition for this state is given by

$$\{K(0,0) - \lambda K'(0,0)\}a = 0 \tag{3.3}$$

where the geometric stiffness matrix K' is defined by

$$K'(0,0) = \left\{ \frac{dK}{d\lambda} \right\}_{d\,=\,0\,;\,\lambda\,=\,0} \tag{3.4}$$

In many codes the matrix K' is determined on the element level by taking into account the pre-stress s_I that is induced by the pre-buckling state (3.1). But it is also possible and quite convenient, to determine this matrix on the basis of the definition (3.4), taking the difference expression

$$K'(0,0) = \frac{K(0,\Delta\lambda) - K(0,0)}{\Delta\lambda} \tag{3.5}$$

for a small perturbation $\Delta\lambda$. It can be shown that the latter operation is justified if the initial assumption regarding the linear pre-buckling state is fulfilled and if $\Delta\lambda$ is chosen below a certain threshold value.

The solution of the symmetric system can be carried out by the standard eigenvalue solver available in the code. It will produce:

The eigenvalue $\leftrightarrow$ eigenvector pairs

$$(\lambda_i, a_i) : \text{for } i = 1,2,3,... \tag{3.6}$$

where we assume that the eigenvalues λ_i are all positive and ordered by

$$\lambda_1 \leq \lambda_2 \leq ... \leq \lambda_k \tag{3.7}$$

and a_i satisfy the ortho-normalization:

$$a_i.a_j = \delta_{ij} \tag{3.8}$$

If λ_1 is the smallest eigenvalue of (3.7), and $\lambda_1 < \lambda_2$, the bifurcation point determined by λ_1 is simple. In that case we can explore the post buckling branch of equilibrium states by starting the standard continuation procedure from a suitable prediction of a solution of the branch. But before showing the simplicity of such a scheme, we will first address a related topic which is concerned with the computation of the Taylor representation of the second solution branch.

3.2 <u>Perturbation Solution</u>

The equations of equilibrium can be written as

$$f(x) = 0 \; ; \; x = \begin{pmatrix} d \\ \lambda \end{pmatrix} \tag{3.9}$$

If a solution $x_1 = x(\eta_1)$ is known it is possible to seek approximate solutions in the neighborhood of this point in the form:

$$x(\eta) = x(\eta_1) + x'(\eta_1)\eta + \frac{1}{2}x''(\eta_1)\eta^2 + ... \tag{3.10}$$

We notice now, that it is this proposition on which the classical post-buckling analyses are based [11]. The form (3.10) is then used to represent the solution of the post buckling branch $x_{II}(\eta)$ with the critical state $^c x = x(\eta_c)$ determined by (3.3) as starting point of the development. Under some circumstances it is very useful to have the capability to compute this expansion explicitly following the eigenvalue analysis (3.3). The question is: Is this possible, and how can it be accomplished? We will now briefly look into this question.

The path derivatives to the solutions of (2.1) are defined by the following equations:

$$f_x x' = 0 \; ; \; f_x x'' = -f_{xx}x'x' \; ; \; f_x x''' = -\{3f_{xx}x'x''+ f_{xxx}x'x'x'\} \; ; \; etc. \tag{3.11}$$

with the normalizing conditions:

$$x'^t x' = 1; \; x'^t x'' = 0; \; x''^t x'' + x'^t x''' = 0; \; etc. \tag{3.12}$$

The sequence of equations (3.11) is obtained by successive differentiation with respect to η. At the bifurcation point these equations are based on the singular stiffness matrix

$$(f_d)^c = K^c = K(0,0) - \lambda_c K'(0,0) \tag{3.13}$$

and the solutions to these sets of equations are multiple valued. In that case, compatibility is preserved only if the right sides of these equations are in the range of K^c. If g is defined as the contraction $a^t f(x)$

$$g = a^t f(x) \tag{3.14}$$

we can express the compatibility conditions for the equation sets (3.11) as

$$g_{xx}x'x' = 0 \ ; \ 3g_{xx}x'x'' + g_{xxx}x'x'x' = 0 \ ; \ \text{etc.} \tag{3.15}$$

In the following we denote the tangent and curvature of the branch II at the critical point $^c x$ by:

$$t = x'_{II} \ ; \ k = x''_{II} \ ; \ t, k \in \mathbb{R}_{N+1} \tag{3.16}$$

The equations that determine these vector quantities are:

$$f_x t = 0 \ ; \ t^t t = 1; \ g_{xx}tt = 0 \quad (i)$$
$$\tag{3.17}$$
$$f_x k + f_{xx}tt = 0 \ ; \ t^t k = 0 \ ; \ 3g_{xx}tk + g_{xxx}ttt = 0 \quad (ii)$$

The result of the solution of (i) can be given in the form:

$$t = x'_{II} = \alpha\{b + \mu x'_I\} \ ; \ b = \begin{pmatrix} a \\ 0 \end{pmatrix} ; a \in \mathbb{R}_N \tag{3.18}$$

where the parameters α and μ are determined by

$$\mu = -\frac{A_3}{2A'_2} \qquad ; \quad \alpha = \sqrt{\{b+\mu x'_I\}^t\{b+\mu x'_I\}}$$
$$\tag{3.19}$$
$$A_3 = g_{xx}bb \ ; \ A'_2 = g_{xx}bx'_I$$

To present the solution for the curvature in a compact form, we first introduce some shorthand notation.

$$r = f_{xx}tt \; ; \; K = f_d \; ; \; l = f_\lambda \; ; \; k = \begin{pmatrix} \kappa \\ \rho \end{pmatrix} \; ; \; \kappa \in \mathbb{R}_N \; ; \; \rho \in \mathbb{R}_1 \tag{3.20}$$

where it is understood that these quantities are evaluated at the critical state. Equations (3.17) are then transformed to:

$$K\kappa = -r - \rho l \;\; (i) \; ; \; t^t k = 0 \quad (ii)$$

$$3g_{xx}tk + g_{xxx}ttt = 0 \quad (iii) \tag{3.21}$$

We can partition the first set (i) of these equations as follows:

$$\begin{pmatrix} K & \underline{c} \\ \underline{c}^T & c_1 \end{pmatrix}\begin{pmatrix} \underline{\kappa} \\ \kappa_1 \end{pmatrix} = -\begin{pmatrix} \underline{r} \\ r_1 \end{pmatrix} - \rho\begin{pmatrix} \underline{l} \\ l_1 \end{pmatrix} \tag{3.22}$$

where the column and row that are masked out correspond to the index of the largest component of the mode a. As was mentioned, this move is advantageous because it has the effect that the minor $\underline{K}$ (of the singular stiffness K^c) is nonsingular and in fact positive. The solution of the system (3.22) is then carried out with

$$\underline{\kappa} = \underline{u} + \kappa_1\underline{v} + \rho\underline{w} \tag{3.23}$$

where $\underline{u},\underline{v},\underline{w}$ like $\underline{\kappa}$ are vectors in $\mathbb{R}_N$ that do not have a component in the direction e_{i*}. It follows that these vectors are determined by

$$\underline{u} = -K^{-1}\underline{r} \; ; \; \underline{v} = -K^{-1}\underline{c} \; ; \; \underline{w} = -K^{-1}\underline{l} \tag{3.24}$$

and because we are at the bifurcation point, the last equation of the set (3.22) is automatically satisfied by (3.24). The remaining conditions to be satisfied (ii), (iii) determine then ρ and κ_1. After elimination of κ_1, the final solution can be presented as follows:

$$u = \begin{pmatrix} \underline{u} \\ 0 \\ 0 \end{pmatrix} \; ; \; v = \begin{pmatrix} \underline{v} \\ 1 \\ 0 \end{pmatrix} = b! \; ; \; w = \begin{pmatrix} \underline{w} \\ 0 \\ 1 \end{pmatrix} \tag{3.25}$$

$$k_1 = u + \alpha b \; ; \; k_2 = w + \beta b \tag{3.25}$$

$$\alpha = -\frac{t^t u}{t^t b} \; ; \; \beta = -\frac{t^t w}{t^t b}$$

$$\rho = \frac{A_4^*}{A_2^*} \; ; \; A_4^* = -\{3g_{xx}tk_1 + g_{xxx}ttt\} \; ; \; A_2^* = 3\,g_{xx}tk_2 \tag{3.27}$$

$$k = k_1 + \rho k_2 \quad ^{1)} \tag{3.28}$$

This completes the formal description of the solution for x'_{II} and x''_{II} at the buckling point. What is still left to be clarified is the way some of the vector and scalar quantities should be computed that are built on the derivatives of f. The terms that we have in mind are:

$$r = f_{xx}tt \; ; \; A_3 = g_{xx}bb \; ; \; A'_2 = g_{xx}bx'_I \; ; \tag{3.29}$$

$$A^*_4 = -\{3g_{xx}t\kappa_1 + g_{xxx}ttt\} \; ; \; A^*_2 = g_{xx}t\kappa_2$$

The derivatives f_{xx}, f_{xxx} are usually not routinely computed in a finite element code. The natural way to determine them is then to use some form of numerical differentiation on the residual functions f, or on the stiffness K. A proposal to this effect is given in the appendix where we use some general results concerning implicit function differentiations as presented in [13].

3.3 Continuation from the Buckling Point

If the quadratic expansion for the branch II is obtained in the way just described we can use this expression directly for the initialization of a continuation process along the branch. If it is not available as it will now be assumed, we must use another approach. A very simple method is described by the following steps [6,12]. Set:

$$\sigma^0 = {}^c x + \Delta \eta b^* \tag{3.30}$$

where the shifter b^* is defined by

$^{1)}$ The derivation given in [12] for the curvature k is incorrect and should be ignored.

$$b^* = vx'_I + \mu a_1$$

$$^cx_I = \begin{pmatrix} \lambda_1 d^0 \\ \lambda_1 \end{pmatrix} \; ; \; ^cx'_I = \begin{pmatrix} \lambda_1' d^0 \\ \lambda_1' \end{pmatrix} \tag{3.31}$$

$$(\lambda_1')^2 \{(d^0)^t(d^0) + 1\} = 1$$

with v and μ determined by

$$(^cx'_I)^t b^* = 0 \; ; \; b^{*t} b^* = 1 \tag{3.32}$$

If the arc-length continuation procedure is used, the vector b^* is directly inserted into the equations (2.3 ii), i.e.:

$$h(x) - \Delta\eta = 0$$
$$\tag{3.33}$$
$$h(x) = n^t(x - \sigma^0) \; ; \; n = b^*$$

The stepsize $\Delta\eta$ is here left to the choice of the user and belongs to the set of input data for the start of the computations of the branch $x_{II}(\eta)$. It may happen that the user is not able to find a suitable step $\Delta\eta^*$ and his attempts to start the branch computations fail. This can be caused by the circumstance that the convergence properties of the correction process are rather weak, even if the decomposition of J is carried out on the prediction σ^0 away from ^cx. If that occurs, there is a general approach available that can save the situation.

Consider again the prediction σ^0 as given by (3.30) but let us now not put particular restrictions on the step size $\Delta\eta^*$. We then define the set of equations

$$f(x\;;\mu) = 0 \tag{3.34}$$

with

$$f(x\;;\mu) = f(x) - (1-\mu)r_0 \tag{3.35}$$

$$r_0 = f(\sigma^0)$$

The new factor (μ) can be seen as a loading parameter because the residual r_0 is equivalent to a load system. We now seek the solution of the trajectory $\sigma(\mu)$, for $\mu = 0 \rightarrow 1$, defined by the set of equations

$$f(x\,;\,\mu) = 0$$

$$b^{*t}(x - \sigma^0) - \Delta\eta^* = 0$$

(3.36)

The solution arc $\sigma(\rho)$ is thus defined in the plane given by (3.36^b). For $\mu = 1$ it should coincide with a solution of the branch x_{II} if it has an intersection with this plane. Clearly this may not be the case if we take too large a step $\Delta\eta^*$, but for convenience we will assume that there is an intersection for our choice of $\Delta\eta^*$.

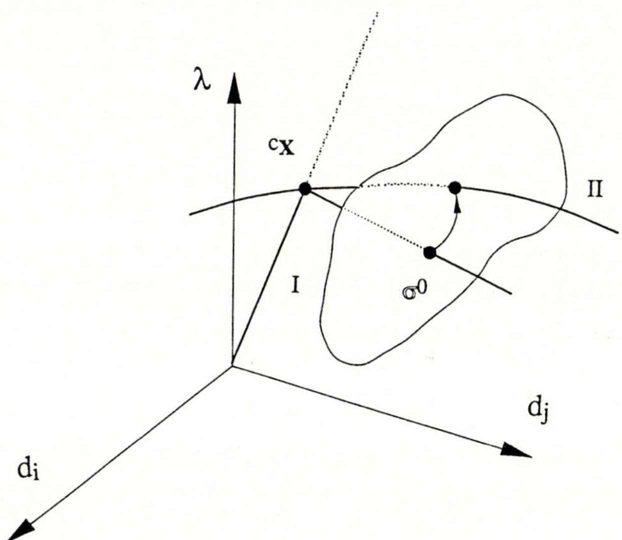

Figure 2   The Switch to II

The set of (N+1) dimensional equations can now be solved by applying the standard continuation procedure on equations (3.35). This means we view the solution points of (3.35) as parts of the curve

142

$$\tau = \begin{pmatrix} x \\ \mu \end{pmatrix} = \begin{pmatrix} x(\xi) \\ \mu(\xi) \end{pmatrix} \in \mathbb{R}_{N+2} \qquad (3.37)$$

where ξ is the new path parameter. In that case equations (3.36) are augmented by

$$h(x , \mu) - \xi = 0$$
with
$$h(x , \mu) = n^t(x - x_k) - n_0(\mu - \mu_k) - \Delta\xi \qquad (3.38)$$

and, we have to solve:

$$f(x ; \mu) = 0 \qquad (3.39a)$$

$$b^{*t}(x - \sigma^0) - \Delta\eta^* = 0 \qquad (3.39b)$$

$$h(x , \mu) - \xi = 0 \qquad (3.39c)$$

Apart from the starting point $(\tau^0)^t = [(\sigma^0)^t , 0]$, the initiation of this process requires an approximate or exact value for the tangent to the path $\tau(\xi)$ at τ^0 [1] . Note now that this information is contained in the set of equations:

$$f_x(\sigma^0)x' + f_\mu(\sigma^0) = 0 \qquad (3.40)$$

which is the linearization of (3.34) at σ^0. It follows then that the corrector equations for this particular continuation process are of the form:

$$A\Delta\tau = R$$

with :

$$A = \left\{ \begin{matrix} K & l & r_0 \\ b_2^t & b_1 & 0 \\ n_2^t & n_1 & n_0 \end{matrix} \right\} \;;\; \Delta\tau = \begin{pmatrix} \Delta d \\ \Delta\lambda \\ \Delta\mu \end{pmatrix} \;;\; R = \begin{pmatrix} f \\ g \\ h \end{pmatrix} \;;\; b^* = \begin{pmatrix} b_2 \\ b_1 \end{pmatrix} \;;\; \begin{pmatrix} n \\ n_0 \end{pmatrix} = \begin{pmatrix} n_2 \\ n_1 \\ n_0 \end{pmatrix} \qquad (3.41)$$

$$g = b^t(x - \sigma^0) - \Delta\eta^* \;;\; h = n^t(x - x_k) - n_0(\mu - \mu_k) - \Delta\xi$$

---

[1] This tangent is denoted as $\begin{pmatrix} n \\ n_0 \end{pmatrix}$ in (3.41).

It is clear that the block elimination procedure will be useful for the solution of these equations but we will not go into the details of this scheme here.

3.3 General Procedures

In the general case, the solution path that is computed by the continuation procedure is nonlinear. In that case, we cannot formulate the bifurcation condition explicitly as in the case of a linear pre-buckling state. It is then necessary to cradle the bifurcation point first, for instance, by a bisection method or other means. During these computations it is helpful if we get as much information about the nature of the bifurcation point as possible. By this is meant that one should get estimates of the buckling modes associated with the bifurcation point and use these in the construction of the initialization of the switch procedure. As far is known, only in one case has a buckling processor of the general type been implemented in a general finite element code. This is the so-called Thurston Processor described in [12] and [14], which is part of the STAGSC -1 code [7]. In principle, the STAGS bifurcation processor is capable of dealing with compound phenomena, but experience gained with its use shows that one runs quickly up against the limitations of perception of the user to interpret the results. In other words, in general, compound bifurcations are extremely difficult to deal with. We will not discuss the Thurston Processor here for lack of space ; descriptions of it can be found in [12 & 14].

4. STABLE POST BUCKLING SOLUTIONS

It is sometimes of interest to compute the (further a field), stable post buckling or post snapping solutions. The question is then how can this in general achieved? Questions like these are *important* in for instance shell buckling phenomena where the post buckling states are reached in a series of jumps from one mode shape to another.

In the cases indicated above, the "far field" solutions may or may not be connected with the undeformed state of the shell. In other words, it is possible that in the space of static solutions isolated but stable post buckling states exist which cannot be reached following the principle of continuation with a one parameter loading program. That such states exist can be deduced from the studies reported in [15,16,17] which deal with the equi-

exist can be deduced from the studies reported in [15,16,17] which deal with the equilibrium states of shell structures under multi-parameter loading systems. One of the basic conclusions of these studies can be phrased as follows. The equilibrium state of a structure that is reached at a certain combination of loads is not unique, but is dependent on the sequence by which the loading is applied even if the loads are conservative.

In the literature, methods have been reported which are able to make jumps in the static solution space by means of which geometrically isolated solutions can be reached [18,19,20]. However, these methods do not tell the user, whether the solutions obtained correspond to the states at which the actual structure will come to rest, after it has experienced the snap from the unstable critical point, a split moment before. It will therefore be necessary to find out afterwards whether the far field equilibrium states computed are physically meaningful. It is not clear, however, by which criterion this should be ascertained.

Observations of this kind seem to imply that the most reliable way to solve these particular problems is to simulate the dynamical behavior of a structure as soon it has reached an unstable critical state (limit point or bifurcation point). Consequently, one should integrate the equations of motion taking into account the characteristics of the problem as faithfully as possible. The idea to solve the equations of motion rather than the equations of statics as a means to analyze buckling problems is of course not new, see for instance [21]. What we propose in this paper, however, is not the replacement of one method against the other but an unification of both techniques. In the following we will illustrate this idea by means of a simple example.

5 EXAMPLE AND CONCLUSION

5.1 Example problem

A cylindrical shell segment of dimensions presented in figure 3 is loaded in compression in the axial direction. If simple support conditions are enforced along all four edges the cylinder will exhibit a behavior as pictured in figure 4. Bifurcation buckling does not take place in this case because the simple support conditions disturb the development of a uniform state of stress at the onset of the loading. The problem was modeled with the standard STAGSC-1 elements (code: 410) [7]. We took only half of the panel into account assuming symmetry around the generator $\alpha = 4.5^0$. The mesh was 12×7 in an even distribution.

The static response is pictured in figure 4. This part of the calculations was carried out by the standard continuation procedure of STAGSC -1 as described in [22]. In a second run we attempted to start the calculations from the limit point at the load $\lambda_c = 1.87$ (with a slightly increased , but fixed, load) using one of the time integrators of STAGS.[1] It was done using a small amount of damping[2] with the other initial conditions furnished by the static solution at $\lambda = \lambda_c$.

$R = 586$ mm

$L = 92$ mm

$t = 1.$ mm

$\alpha = 9^0$

$N_x = \lambda N_0$

$N_0 = 73.4285$ N/mm

$E = 70000$ N/mm^2

$v = 0.3$

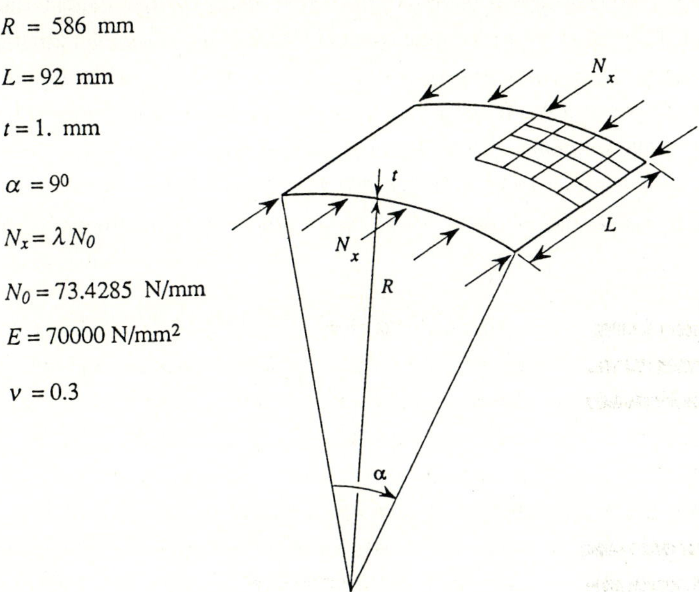

Figure 3 Cylindrical Panel

5.2 Calculations

The aim was to reach the post snapping (stable) state II directly and indeed this attempt was successful although not without some difficulty. The plot of the variation of the velocity in normal direction of the center of the panel versus time during the snapping process (as given in Figure 5) gives the illustration of this difficulty. We observe that the snapping sequence can roughly be divided into three parts: (i) an initiation period

---

[1] The results shown here were obtained by the method of K. C. Park [7].

[2] We used for the damping matrix in this example $D = .3M + .3K$ (M = mass matrix) which is a completely arbitrary choice.

whereby the structure moves in a relatively "slow" motion away from the limit point, (ii) the actual snap which takes place in a time interval of roughly one third of the total time, and (iii) a consolidation period whereby the structure moves with damped oscillations towards the stable post snapping state. The difficulties occured at the onset of the second period whereby the procedure could not adapt itself to the sudden acceleration of the motion. A restart was necessary with a considerably scaled down time step. In fact, it seemed that the integrator could not properly calculate the velocities in this case (the peak value is unreasonably high i.e. the computed kinetic energy seemed to be in conflict with the energy conservation law). All four available integrators in STAGS suffer from this deficiency and its remedy is still a matter of further investigation. It is clear, however, that qualitatively, the observed behaviour does comply with our intuitive notion of the buckling process. Incidentally, it was very easy to restart the static continuation procedure along the stable post buckling state using the last dynamical solution as predictor.

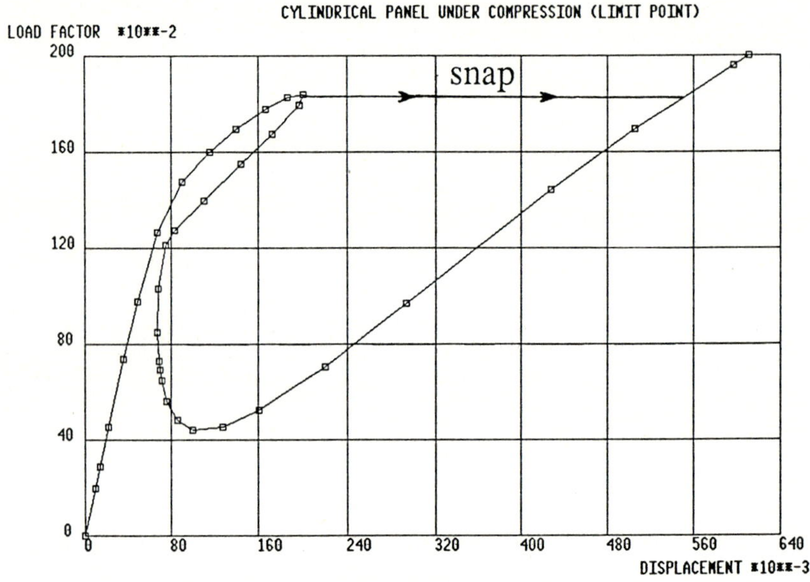

Figure 4 Static Solution

Our experimentations were not confined to this example alone. Several other versions of the problem were investigated with different boundary conditions and loading conditions i.e. : prescribed displacements versus dead weight loads. In all these cases we obtained similar results in a qualitative sense. In other words, the plot in figure 5 is quite characteristic of the snapping process for this particular type of problem.

In all cases considered, the snap through motion returned to a motion of damped oscillations as soon as a certain neighborhood of the stable post buckling state II was reached. For this particular model, there are no *other* stable post buckling states and therefore the observed behavior is predictable. It is conjectured, however, that for more complicated models where more than one stable post buckling state exists, the snapping process will strongly depend on the characteristics of the problem and the precise

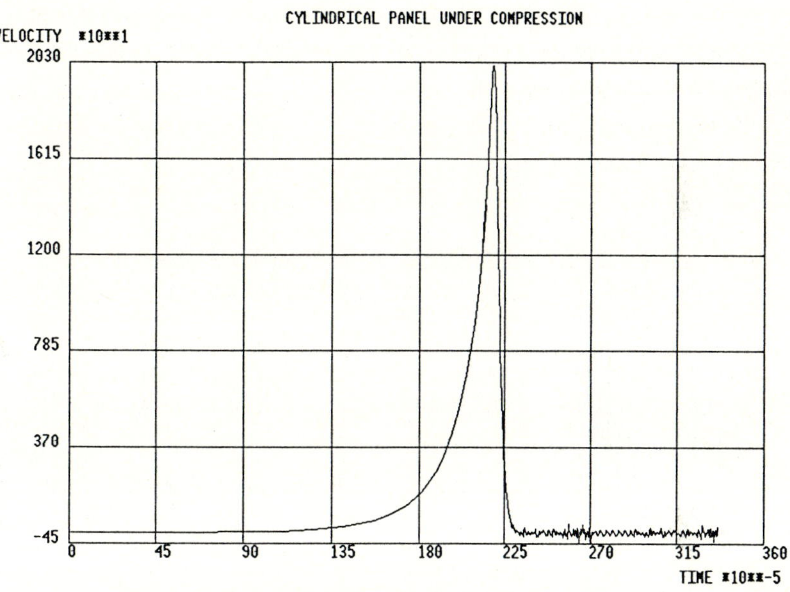

Figure 5 . Velocity in Radial Direction at Center

conditions under which the snap will be initiated. In such cases it will not longer be possible to know beforehand to "which" state the structure will move.

What we omitted to mention so far was the expense of the computations in the time domain. As expected, the dynamical part of the solutions shown here required more computation (about 2 times of the computational effort that was spent to obtain the static solution diagram). On the other hand we did not attempt to speed up the computations in any way as this was not part of the objective of this exercise. "Tuning " of the integrators to the specific needs of these violent motions is still a task that lays ahead.

5.3 <u>Conclusion</u>

In many cases of the analysis of the load carrying capacity of shell structures the methods of statics in terms of continuation methods will be able to produce useful results. We reviewed some particular aspects of this approach in the first part of this paper. Sometimes, however, it will be necessary to compute the "far field"solutions. We claim that in these cases continuation procedures by themselves will not always be suitable to compute the stable post buckling states. In such cases, a possible way out is to use time integration in conjunction with the methods of continuation. Although in no way infallible (because of our ignorance of the precise conditions that should be applied), this approach seems to give the best possible assurances to bring these special buckling problems to a satisfactory solution.

<u>References</u>

1   Felippa C.A.: Solution of Nonlinear Static Equations. Handbook on Computational Mechanics", Volume on Large Deflection and Stability of Structures, (Bathe K.J. ed.), to appear.

2   Riks E.; den Reijer P.J.: A Finite Element Analysis of Cracks in a Thin Walled Pressurized Cylinder. National Aerospace Lab., NLR , NLR TR 87021 U,The Netherlands, Jan. 1987.

3   Riks E.: Bulging Cracks in Pressurized Fuselages: A Numerical Study, NLR MP 87058U, National Aerospace Lab., NLR, The Netherlands, Sept. 1987.

4   Waszcyszyn, X.: Numerical Problems of Nonlinear Stability Analysis of Elastic Structures. Computers & Structures 17, no.1 (1981), (13-24).

5   Rheinboldt W.C.: Numerical Analysis of Continuation Methods for Nonlinear Structural Problems, Computers & Structures Vol.13, (1981) (103-113).

6   Riks E.: Some Computational Aspects of the Stability Analysis of Nonlinear Structures , Comp. Meth. in Appl. Mech. in Engrg. 47 (1984) 219-259.

7   Almroth B.O., Brogan F.A.: Structural Analysis of General Shells, Volume II, User's Instructions for STAGSC-1". LMSC-D633873, Lockheed Palo Alto Research Lab., Palo Alto, CA. Dec. (1982)

8   Riks E.; Rankin C.C.: Bordered Equations in Continuation Methods: An Improved Solution Technique, NLR MP 87057U, National Aerospace Lab., NLR, The Netherlands, Jan. 1987.

9   Rheinboldt W.C. : The Computation of Critical Boundaries on Equilibrium
        Manifolds, SIAM J. Numer. Anal. Vol. 19, no. 3, June 1982.

10   Wriggers P., Wagner W., Miehe C.: A Quadratically Convergent Procedure
        For the Calculation of Stability Points in Finite Analysis. Comp. Metds., in Appl.
        Mech. and Engrng., Vol. 70, 1988, 329-347.

11   Budiansky B. : Theory of Buckling and Postbuckling of Elastic Structures,.In:
        Advances in Applied Mechanics 14, C.S. Yih (ed.), Academic Press, New York.
        (1974).

12   Riks. E.: Bifurcation and Stability, A Numerical Approach. National Aerospace
        Lab., The Netherlands, NLR MP 84078 U, 1984. Also in: Innovative Methods
        for Nonlinear Problems, (W.K. Liu, T. Belytschko, K.C. Park eds.) Pineridge
        Press, 0-906674-41-7, (1984)

13   Mackens W.: Numerical Differentiation of Implicitly Defined Space Curves,
        Bericht Nr. 45 Institut fur Geometrie und Praktische Mathematik, RWTH
        Aachen, March 1987.

14   Thurston G.A.; Brogan F.A.; Stehlin P.: Postbuckling Analysis Using a General
        Purpose Code . AIAA Paper No. 85-0719-CP. Presented at the AIAA/ ASME/
        AHS 26th Structures, Structural Dynamics, and Materials Conference,
        ORLANDO, FLORIDA, April 15-17, 1985.

15   Shilkrut D.: Investigation of Axisymmetric Deformation of Geometrically
        Nonlinear, Rotationally, Orthotropic, Circular Plates. Int. J. Non-Linear
        Mechanics, Vol. 18, No.2, pp 95-118, 1983.

16   Shilkrut D.:   Stability and Vibration of Geometrically Nonlinear Cylindrically
        Orthotropic Circular Plates. J. of Applied Mechanics, 345-360, Vol. 51, June
        1984.

17   Shilkrut D.: The Influence of the Paths of Multiparametrical Conservative
        Loading on the Behaviour of a Geometrically Nonlinear Deformable Elastic
        Body. In: "Buckling of Structures ; Theory and Experiment". The Joseph Singer
        Anniversary Volume, (I. Elishakoff, J. Arbocz. C.D. Babcock, jr., A. Libai,
        eds.), Elsevier 1988.

18   Kroplin B.H.: A Viscous Approach to Post-Buckling Analysis. Eng. Struct.
        Vol., July 1981.

19   Kroplin B.H., Dinkler D.: Eine Methode zur Directen Berechnung von
        Gleichgewichtslagen im Nachbeulbereich. Ingenieur Archiv, 51 (1982), pp.
        415-420.

20   Petiau C., Cornuault C.: Efficient Algorithms for Post Buckling Computations,
        in: Computing Methods in Applied Science and Engineering. (R. Glowinsky and
        J.L. Lions eds.) Elseviers Science Publishers B.V. (North-Holland) @ 1984

21   Feodos'ev V.I.: On a Method of Solution of the Nonlinear Problems of Stability of Deformable Systems. PPM Vol. 27, No 2, 1963, pp. 265-274.

22   Riks E.: Progress in Collapse Analysis. Journal of Pressure Vessel Technology, Vol. 109, 35, February 1987.

Appendix

For the construction of the perturbation solution the following vector and scalar quantities need to be calculated (eqs. (3.19) & (3.27)):

$$\mathbf{r} = f_{xx}\mathbf{tt}$$

$$A_3 = g_{xx}\mathbf{bb} \;;\; A'_2 = g_{xx}\mathbf{bx'}_I \tag{A1}$$

$$A^*_4 = -\{g_{xx}\mathbf{tk}_1 + g_{xxx}\mathbf{ttt}\} \;;\; A^*_2 = g_{xx}\mathbf{tk}_2$$

The question is now how can this be accomplished? First of all we note that:

$$A'_2 = g_{xx}\mathbf{bx'}_I = \mathbf{a}^t f_x(^c\mathbf{x})\mathbf{x'}_I\mathbf{a} = \mathbf{a}^t \mathbf{K'}(^c\mathbf{x})\mathbf{a} = \mathbf{a}^t \mathbf{K'}(0)\mathbf{a} \tag{A2}$$

The path derivative of $\mathbf{K}$ along the basic state is identical to $\mathbf{K'}(0)$ as given by (3.5).

Let us now look at the vector $\mathbf{r}$. At the equilibrium state $\mathbf{x}$, where $f(\mathbf{x}) = 0$, the vector function $f(\mathbf{x} + \mu\mathbf{b})$ can be expanded as:

$$f(\mathbf{x}) = f_x(\mathbf{x})\mathbf{b}\mu + \frac{1}{2}f_{xx}(\mathbf{x})\mathbf{bb}\mu^2 + O(\mu^3) \tag{A3}$$

and we find that:

$$\mathbf{r} = f_{xx}(\mathbf{x})\mathbf{bb} = \frac{d^2}{d\mu^2}\{f(^c\mathbf{x} + \mu\mathbf{t})\} \,\Big|_{\mu=0} \tag{A4}$$

Similarly:

$$A_3 = \frac{d^2}{d\mu^2}\{g(^c\mathbf{x} + \mu\mathbf{b})\}\Big|_{\mu=0} \tag{A5}$$

$$A'_2 = \frac{1}{2} \frac{d^2}{d\mu^2} \{ g(^cx + \mu t + \mu^2 k_1) - g(^cx + \mu t - \mu^2 k_1) \} \big|_{\mu=0}$$

(A6)

$$A^*_4 = - \frac{d^3}{d\mu^3} \{ g(^cx + \mu t + \mu^2 k_1) \} \big|_{\mu=0} \tag{A7}$$

Mackens [13] derived formula's for these particular cases where we have

$$f(0) = 0 \; ; \; f^\circ(0) = 0 \; ; \; f^{\circ\circ}(0) \neq 0 \; ; \; f^{\circ\circ\circ}(0) \neq 0 \;\; [1]$$

where f = either $f\,(^cx + \mu b)$ or $g(^cx + \mu b)$, $g(^cx + \mu t + \frac{1}{2} \mu^2 k_1)$ etc..

The following two expressions should give $O(\mu^4)$ accuracy:

$$f^{\circ\circ}(0) \;\; = \mu^{-2} \left\{ \frac{4}{3} \{ f(\mu) + f(-\mu) \} - \frac{1}{12} \{ f(2\mu) + f(-2\mu) \} \right\} + O(\mu^4)$$

$$f^{\circ\circ\circ}(0) = \mu^{-3} \left\{ 4 \{ f(\mu) - f(-\mu) \} - \frac{1}{8} \{ f(2\mu) - f(-2\mu) \} \right\} + O(\mu^4) \tag{A8}$$

They correspond to the formulas designated by $S^{2,2,4}$ and $S^{3,2,4}$ respectively. For further details and recommendations for the size of the step μ we refer to [13].

---

[1] $(\;)^\circ = \dfrac{d}{d\mu}$

Dynamic Stability Analysis of Shell Structures

A. Burmeister, E. Ramm

Institut für Baustatik, University of Stuttgart, FR Germany

Summary

Time dependent pressure loads may cause dynamic buckling instabilities in the response of thin-walled structures. A typical example is the dynamic snap-through of a spherical shell under sudden external pressure. Usually, the critical load is obtained performing several time step analyses under different load magnitudes to localize the critical step load. In this paper a more efficient alternative is described allowing to determine the critical load directly or with at least considerably less time history analyses. The dynamic stability is evaluated in the sense of Liapunov's first method. The geometrically and materially nonlinear finite element method is applied.

In the above mentioned simplified procedure stability criteria are evaluated parallel to one nonlinear time response analysis. Starting with the eigenproblem of the system matrix the applicability of the kinetic stability criterion is shown. The procedure can be extended to the eigenproblem for the load parameter in order to approximate the critical step load to some extent directly. Numerical examples demonstrate the quality of the proposed method.

1.0 Introduction

Structures under time dependent loading may be exposed to different kinds of dynamic instabilities, such as parametric resonance or dynamic buckling. As an example shallow shells under extreme loading conditions are mentioned where important applications in practice caused several research activities in that area /8/, /13/, /1/ in the last years. The present study is restricted to dynamic buckling phenomena. These instabilities are characterized by a sudden change in structural behaviour which occurs as soon as the loading reaches a critical value. Similarities to instabilities under static loading exist because both bifurcation and snap-through may occur. However, load levels as well as buckling modes are in general different between the static and dynamic load case.

An often employed method to solve dynamic buckling problems is to monitor the structural response under increasing load levels applying the criterion of Budiansky and Roth /4/ in which a sudden increase of a characteristic displacement defines the critical load step. The main objective of the present study is to replace this time consuming procedure by a more efficient method. Using the finite element method in connection with Liapunov's definition of stability, criteria for dynamic buckling are derived. Based on these criteria a method is proposed in order to find the dynamic buckling load with one or only a few time history analyses. This so-called scaling method is based on an eigenproblem for the load parameter which is solved parallel to nonlinear time history analyses.

2.0 Equations of Motion

Based on a total Lagrangian description the principle of virtual work supplemented by the inertia forces at time t leads to:

$$\int_{0_v} {}^0\varrho \; {}^t\ddot{u}_k \; \delta u_k \; d({}^0v) = {}^t\delta W_{(ext)} - \int_{0_v} {}^tS_{ij} \; \delta \; {}^t\epsilon_{ij} \; d({}^0v) \tag{1}$$

with the Green – Lagrange strains ${}^t\epsilon_{ij}$, the 2nd Kirchhoff – Piola stresses ${}^tS_{ij}$ and the virtual displacements δu_k. A finite element displacement discretization of equation (1) leads to the equation of motion:

$$\mathbf{M} \; {}^t\ddot{\mathbf{u}} = {}^t\mathbf{R} - {}^t\mathbf{F} \tag{2}$$

with the constant, positive definite mass matrix **M**, the vectors of nodal accelerations ${}^t\ddot{\mathbf{u}}$, external nodal forces ${}^t\mathbf{R}$ and internal nodal forces ${}^t\mathbf{F}$, respectively. ${}^t\mathbf{F}$ is in general a nonlinear function of the nodal displacements ${}^t\mathbf{u}$. Based on the usual series expansion at time $t + \Delta t$, equation (2) is linearized with respect to the incremental displacements:

$$ {}^{t+\Delta t}\mathbf{F} \; ({}^t\mathbf{u} + \mathbf{u}) \simeq {}^t\mathbf{F} \; ({}^t\mathbf{u}) + {}^t\mathbf{F} \; ({}^t\mathbf{u}),_{{}^t\mathbf{u}} \, \mathbf{u} \tag{3}$$

leading to the incremental equation of motion:

$$\mathbf{M} \; {}^{t+\Delta t}\ddot{\mathbf{u}} + {}^t\mathbf{K} \; \mathbf{u} = {}^{t+\Delta t}\mathbf{R} - {}^t\mathbf{F} \tag{4}$$

with the tangent stiffness matrix:

$$ {}^t\mathbf{K} = {}^t\mathbf{K}_e + {}^t\mathbf{K}_g = {}^t\mathbf{F} \; ({}^t\mathbf{u}),_{{}^t\mathbf{u}} \tag{5}$$

The consistent linearization of degenerated elements exhibiting large rotations is described in /11/. Usually, the physical damping is added directly to equations (2) or (4):

$$\mathbf{M} \; {}^t\ddot{\mathbf{u}} + \mathbf{D} \; {}^t\dot{\mathbf{u}} = {}^t\mathbf{R} - {}^t\mathbf{F} \tag{2 a}$$

$$\mathbf{M} \; {}^{t+\Delta t}\ddot{\mathbf{u}} + \mathbf{D} \; {}^{t+\Delta t}\dot{\mathbf{u}} + {}^t\mathbf{K} \; \mathbf{u} = {}^{t+\Delta t}\mathbf{R} - {}^t\mathbf{F} \tag{4 a}$$

Explicit time integration methods are based on (2 a). Implicit schemes like Newmark's method resort to equation (4 a) and require additional equilibrium iterations. Considering the sensitivity of structures against perturbations, especially in the range of the stability limit /5/, it is obvious that time integration schemes have to meet high stability and accuracy requirements. Usually, these can be satisfied using explicit time integration schemes limited by a critical time step Δt_{crit}. Because dynamic instability phenomena are part of the global structural response which can be obtained using $\Delta t \gg \Delta t_{crit}$ usually implicit algorithms are more efficient. Problems concerning the numerical stability of implicit time stepping algorithms /10/, especially if the tangent stiffness matrix is indefinite, can be overcome. An appropriate procedure is described in /5/. Due to its accuracy, stability and efficiency characteristics Newmark's method ($\beta = 1/4$, $\delta = 1/2$) as well as the related α – methods (Hilber, Hughes, Taylor /7/, Wood, Bossak, Zienkiewicz /15/)

are well suited for dynamic buckling problems. A fourth order very accurate implicit method using Padé approximations /6/, /14/ is recommended only for extremely sensitive dynamic buckling problems and for parametric resonance instabilities.

With regard to stability criteria the equation of motion including nonlinear behaviour is cast into a first order form. Equation (4 a) is augmented on both sides by $^tK\,^tu$. Using the identity $^{t+\Delta t}\dot{u} = {}^{t+\Delta t}\dot{u}$ results in /5/:

$$^{t+\Delta t}\dot{z} = {}^tA\; {}^{t+\Delta t}z + {}^{t+\Delta t}f \tag{6}$$

with the state vector:

$$^{t+\Delta t}z^T = [{}^{t+\Delta t}u, \quad {}^{t+\Delta t}\dot{u}] \tag{7}$$

generalized exitation vector:

$$^{t+\Delta t}f^T = [O, \; M^{-1}\,({}^{t+\Delta t}R - {}^tF + {}^tK\,{}^tu)] \tag{8}$$

and :

$$^tA = \begin{bmatrix} O & I \\ -M^{-1}\,{}^tK & -M^{-1}\,D \end{bmatrix} \tag{9}$$

3.0 Dynamic Buckling

Apart from the question whether a system under a defined loading is in stable or in unstable motion, the determination of the dynamic buckling load is a major task in dynamic buckling analysis. In the next chapter it will be shown that the first question can be answered applying stability criteria parallel to a nonlinear time response analysis. In the following chapter 3.2 these criteria are used to formulate an eigenproblem for the load parameter. The solution of this eigenproblem allows to estimate the critical step load directly. Considering arbitrary time functions of the load, the knowledge of the critical step load is advantageous in so far as the step load represents the worst time function of the load. Thus the critical step load is a lower limit of the dynamic buckling load.

3.1 Dynamic Buckling Criteria

Nonlinear static and dynamic analyses usually require an incremental solution process. By this incremental solution process the real time-variant system is approximated in a step by step way assuming time-invariance within a step. This also holds for stability criteria. In order to take into account the nonlinear structural behaviour the stability check is done at each increment, that is at discrete times t, t + Δt ...

To derive the stability criteria according to Liapunov's definition of stability an unperturbed (fundamental) motion ^tu (equation (2)) and a perturbed (neighboured) motion ^tu_s are considered:

$$M\,{}^t\ddot{u}_s + D\,{}^t\dot{u}_s = {}^tR - {}^tF\,({}^tu_s) \tag{10}$$

where:

$$^{t}\mathbf{u}_{s} = {}^{t}\mathbf{u} + {}^{t}\tilde{\mathbf{u}} \tag{11}$$

with the difference motion $^{t}\tilde{\mathbf{u}}$. Furthermore, identical load histories are assumed for the perturbed and unperturbed motion. Linearizing the internal forces with respect to $^{t}\tilde{\mathbf{u}}$:

$$^{t}\mathbf{F}\,(^{t}\mathbf{u}_{s}) = {}^{t}\mathbf{F}\,(^{t}\mathbf{u} + {}^{t}\tilde{\mathbf{u}}) = {}^{t}\mathbf{F}\,(^{t}\mathbf{u}) + {}^{t}\mathbf{F}\,(^{t}\mathbf{u}),_{^{t}\mathbf{u}}\,{}^{t}\tilde{\mathbf{u}} \tag{12}$$

considering (5):

$$^{t}\mathbf{F}\,(^{t}\mathbf{u}),_{^{t}\mathbf{u}} = {}^{t}\mathbf{K} \tag{13}$$

and using (2 a) the linearized incremental equation for the difference motion is:

$$\mathbf{M}\,{}^{t}\ddot{\tilde{\mathbf{u}}} + \mathbf{D}\,{}^{t}\dot{\tilde{\mathbf{u}}} + {}^{t}\mathbf{K}\,{}^{t}\tilde{\mathbf{u}} = \mathbf{O} \tag{14}$$

Corresponding to the requirement of time – invariance a relation for the increment of the difference motion is necessary.

Similar to the fundamental solution, equation (3), the neighboured solution at time $t + \Delta t$ is expanded in a Taylor series and linearized with respect to $\mathbf{u}_{s}$. Assuming that $^{t}\mathbf{K}_{s} = {}^{t}\mathbf{K}$ and using (4 a) results in:

$$\mathbf{M}\,{}^{t+\Delta t}\ddot{\tilde{\mathbf{u}}} + \mathbf{D}\,{}^{t+\Delta t}\dot{\tilde{\mathbf{u}}} + {}^{t}\mathbf{K}\,{}^{t+\Delta t}\tilde{\mathbf{u}} = \mathbf{O} \tag{15}$$

Equations (15) and (14) define the incremental difference solution which reads in a first order form /5/:

$$\dot{\tilde{\mathbf{z}}} = {}^{t}\mathbf{A}\;\tilde{\mathbf{z}} \tag{16}$$

with the solution:

$$\tilde{\mathbf{z}} = ({}^{t}\underline{\Phi}\,(\Delta t) - \mathbf{I})\;{}^{t}\tilde{\mathbf{z}} \tag{17}$$

If the vector norm at the beginning is limited:

$$|\,{}^{0}\tilde{\mathbf{z}}\,| < \delta \tag{18}$$

then stability of the motion follows from /9/:

$$|\,{}^{t}\underline{\Phi}\,(\Delta t)\,| \leq c\,, \quad 0 < c < \infty\,, \quad t \geq 0 \tag{19}$$

A consistent matrix norm for the fundamental matrix $|{}^{t}\underline{\Phi}\,(\Delta t)|$ is used.

In order to obtain a more convenient criterion for stability the difference motion is represented by /9/:

$$\tilde{\mathbf{z}} = e^{^{t}\kappa \Delta t}\;\bar{\mathbf{z}} \tag{20}$$

Introducing (20) into (16) gives:

$$({}^{t}\mathbf{A} - {}^{t}\kappa\,\mathbf{I})\;\bar{\mathbf{z}} = \mathbf{O} \tag{21}$$

156

For the following considerations multiple eigenvalues are excluded without loss of generality. Using matrix notation the eigenvalues:

$$^t\underline{K} = \begin{bmatrix} ^t\kappa_1 & & & \\ & ^t\kappa_2 & & \\ & & \ddots & \\ & & & ^t\kappa_n \end{bmatrix} \tag{22}$$

the eigenvectors:

$$\overline{Z} = [\ \overline{z}_1 \ \ \overline{z}_2 \ \ ... \ \ \overline{z}_n] \tag{23}$$

and:

$$e^{t\underline{K}\Delta t} = \begin{bmatrix} e^{t\kappa_1\Delta t} & & & \\ & e^{t\kappa_2\Delta t} & & \\ & & \ddots & \\ & & & e^{t\kappa_n\Delta t} \end{bmatrix} \tag{24}$$

lead to the solution of equation (16):

$$\tilde{z} = \left(\overline{Z}\ e^{t\underline{K}\Delta t}\ \overline{Z}^{-1} - I \right)\ ^t\tilde{z} \tag{25}$$

Comparing equation (25) to (17) $^t\underline{\Phi}$ is defined as:

$$^t\underline{\Phi}\ (\Delta t) = \overline{Z}\ e^{t\underline{K}\Delta t}\ \overline{Z}^{-1} \tag{26}$$

Because $\overline{Z}$ is constant the limitation of $^t\underline{\Phi}\ (\Delta t)$ can be replaced considering the eigenvalues $^t\kappa$. $^t\underline{\Phi}\ (\Delta t)$ is limited and therefore the structure is in stable motion if the real parts of all $^t\kappa_i$ are:

$$\text{Re}\ ^t\kappa_i\ < \ 0 \tag{27}$$

The critical load level is defined:

$$^t\kappa_i\ =\ 0 \tag{28}$$

Alternatively, stability can be judged using the circular frequency $^t\omega$ instead of $^t\kappa$. Decomposing equation (21) results in:

$$(\ ^t\text{K} + {^t\kappa}^2\ \text{M}\ )\ \overline{u}\ =\ O \tag{29}$$

which can be compared to:

$$(\ ^t\text{K} - {^t\omega}^2\ \text{M}\ )\ \overline{u}\ =\ O \tag{30}$$

so that:

$$^t\omega^2\ =\ -\ {^t\kappa}^2 \tag{31}$$

Therefore, using equation (28) the critical load level is also identified by /8/:

$$^t\omega_1 = 0 \tag{32}$$

if:

$$^t\omega_1 \leq {}^t\omega_2 \leq {}^t\omega_3 \ldots \tag{33}$$

The judgement of dynamic buckling problems using the criteria for time-invariant systems is straightforward if the structure is in the stable region. In the region of instability a problem arises since the dynamic buckling load obtained by the condition $^t\omega_1^2 \leq 0$ at only one time t in general does not cause instability of the structure. Usually, the instability phenomena are observed at a slightly higher load level where the condition $^t\omega_1^2 \leq 0$ holds for several time steps /5/ ($^t\omega_1^2 \leq 0$, $^{t+\Delta t}\omega_1^2 \leq 0, \ldots {}^{t+n\Delta t}\omega_1^2 \leq 0$). This effect may be explained by the assumptions introduced above (linearization of internal forces $^t\mathbf{K}_s = {}^t\mathbf{K}$). In order to decide to what extent it is also effected by the nonlinear structural behaviour further studies are necessary.

Summarizing $^t\omega_1^2 \leq 0$ can be used as a conservative condition for the dynamic buckling load. It can be shown /5/ that the procedure corresponds to a freezing at that specific time when $^t\omega_1^2 \leq 0$ occurs.

3.2 Scaling Method

It is intended to perform a nonlinear time response analysis for a step load lower than the critical one and to calculate a multiplier $^t\lambda = {}^t\mathbf{R}_{crit}/{}^t\mathbf{R}$ that scales the actual load level $^t\mathbf{R}$ to the critical one $^t\mathbf{R}_{crit}$.

If $^t\mathbf{R}$ is not too far away from the critical value, stresses may be assumed linearly proportional to the load:

$$^t\mathbf{R} \;\rightarrow\; {}^t\lambda \; {}^t\mathbf{R} \tag{34}$$

$$^t\mathbf{K}_e + {}^t\mathbf{K}_g \;\rightarrow\; {}^t\mathbf{K}_e + {}^t\lambda \; {}^t\mathbf{K}_g \tag{35}$$

or:

$$({}^t\mathbf{K}_e + {}^t\mathbf{K}_g - {}^t\omega^2 \, \mathbf{M}) \, \bar{\mathbf{u}} = \mathbf{O} \rightarrow ({}^t\mathbf{K}_e + {}^t\lambda \, {}^t\mathbf{K}_g - {}^t\omega^2 \, \mathbf{M}) \, \bar{\mathbf{u}} = \mathbf{O} \tag{36}$$

In the critical case of buckling $^t\omega = 0$ and:

$$^t\mathbf{R}_{crit} = {}^t\lambda \; {}^t\mathbf{R} \tag{37}$$

Therefore, equation (36) with $^t\omega = 0$:

$$({}^t\mathbf{K}_e + {}^t\lambda \; {}^t\mathbf{K}_g) \, \bar{\mathbf{u}} = \mathbf{O} \tag{38}$$

defines an eigenproblem for the scaling factor $^t\lambda$ and the buckling modes $\bar{\mathbf{u}}$. Solving equation (38) supplementary to a nonlinear time history analysis renders the eigenvalues $^t\lambda$ and the corresponding dynamic buckling modes $\bar{\mathbf{u}}$ as functions in time. They are periodic if the structure is in stable vibration ($^t\lambda > 1$). The minimum value of the first eigenvalue (min $^t\lambda_1$) is of special interest. Obviously dynamic buckling occurs if min $^t\lambda_1 = 1$. Suppose a load $^t\mathbf{R} < {}^t\mathbf{R}_{crit}$ leads to min $^t\lambda_1 > 1$ then the estimated critical load can be obtained by $^tR_{crit} = \min {}^t\lambda_1 \, {}^tR$. The assumption of linear proportionality between load and stresses does not hold if min $^t\lambda_1$ is too big, i.e. differs too much from the critical value 1. In this case ^tR is augmented and the calculation is repeated, hopefully leading to a better approximation of $^tR_{crit}$.

The entire procedure can be summarized as follows:

1. Perform a nonlinear time response analysis for $^tR < {}^tR_{crit}$. Because $^tR_{crit}$ is unknown, $^tR = 0.7 \; R_{crit}^{static}$ (static analysis) can be used as a starting value.

2. Solve the eigenproblem equation (38) for t, t + Δt ...

3. Find min $^t\lambda_1$. Usually, min $^t\lambda_1$ is obtained at that time when a characteristic displacement reaches a maximum. Therefore, the eigenproblem has to be solved only once.

4. If min $^t\lambda_1$ turns out to be 1.0, the actual load level is already the critical one $^tR_{crit} = {}^tR$ (min $^t\lambda_1 = 1.0$). If $0.9 \leq \min {}^t\lambda_1 \leq 1.5$ the critical load level $^tR_{crit} = \min {}^t\lambda_1 \, {}^tR$ is a reasonable estimate and the problem is solved.

5. If min $^t\lambda_1 > 1.5$ or min $^t\lambda_1 < 0.9$ repeat steps 1 to 4 with a better choice of ^tR.

It should be mentioned that this procedure is based on the same assumption as the criterion $^t\omega_1 = 0$. Therefore, the dynamic buckling load obtained by the scaling method will be conservative.

An eigenproblem of the form of equation (38) has also been used in static stability analysis /3/. The derivation presented above also proves that this static stability criterion can be derived from Ziegler's kinetic stability criterion /16/.

4. Examples

For illustration the scaling method as well as the stability criterion are applied to a simple beam under axial step loading (Fig. 1). Bifurcation occurs for a step load of p = 1.72 kN which is in this simple case identical to the static buckling load.

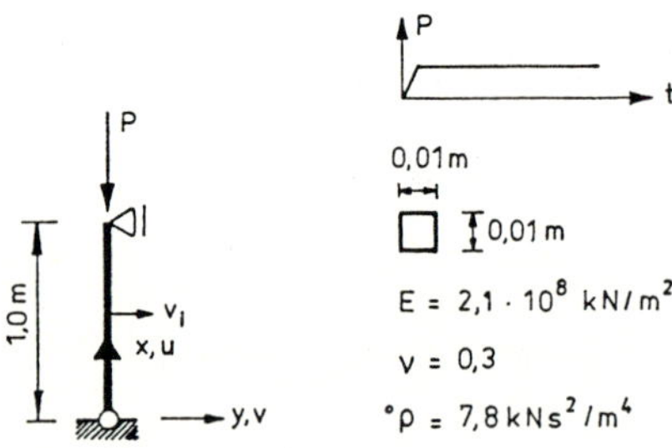

Figure 1: Axially loaded beam

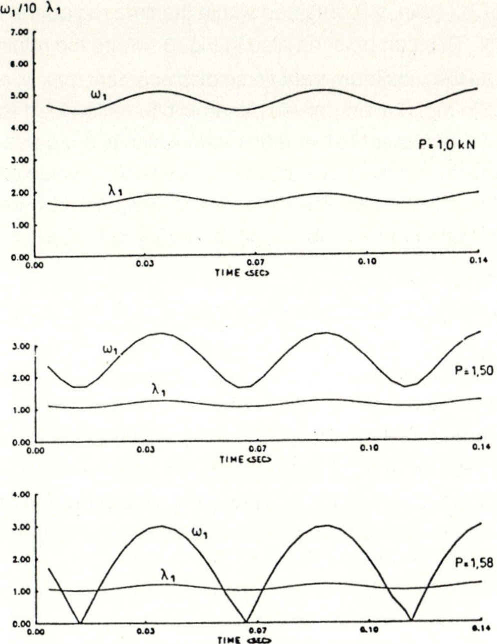

Figure 2: Time history of ${}^t\omega_1$, ${}^t\lambda_1$ ($\Delta t = 0.002$ sec)

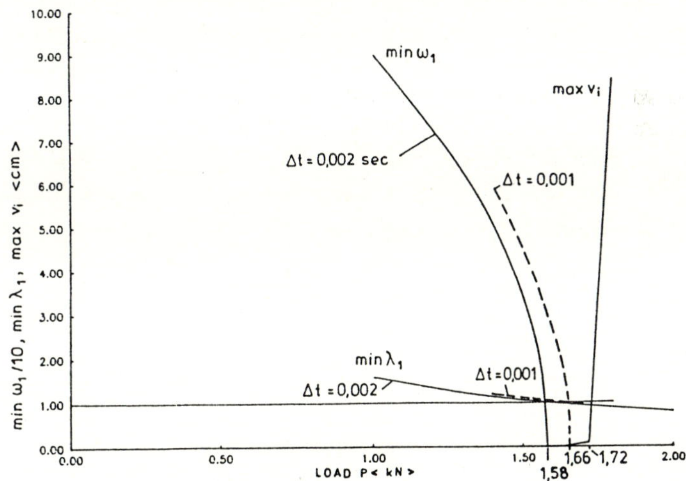

Figure 3: Application of different methods

In Fig. 2 the eigenvalues ${}^t\omega_1$, ${}^t\lambda_1$ are plotted versus time for different load levels. For $p = 1.0$ kN, $p = 1.5$ kN stability follows from ${}^t\omega_1 > 0$ and ${}^t\lambda_1 > 1$, respectively. Using the scaling method from the minimum value of $^t\lambda_1$ (min $^t\lambda_1$) obtained within the time region the limit of stability is reached for $p_{crit} = 1.58$ kN. This can be seen also in Fig. 3 where the minimum value of $^t\omega_1$ (min $^t\omega_1$), min $^t\lambda_1$ as well as the maximum transverse displacement max ^tv_i are plotted versus the load level. As indicated in Fig. 3 an improved dynamic bifurcation load ($p_{crit} = 1.66$ kN) is obtained with a smaller time increment ($\Delta t = 0.001$ sec) which is 3.5 percent lower than the exact solution 1.72 kN. Further studies showed that the exact critical value cannot be reached even if the time step is further decreased. The difference is due to the assumption introduced in chapter 3.1. This shows that the definition of instability ($^t\omega_1^2 \leq 0$ at any instant–freezing in time) is conservative.

Next the scaling method is applied to the symmetrical arch (Fig. 4). In Fig. 5 the corresponding eigenvalues min $^t\omega_1$, min $^t\lambda_1$ as well as the maximum vertical displacement of the center node 1 max w_1 are plotted versus the actual load level ^tR. It can be seen that the eigenvalue $^t\lambda_1$ leads already to a sufficient estimate of the critical load at a lower load level (Fig. 5). Furthermore, the critical step load obtained from equation (32) or (38) ($p_{crit} = 380$ lb/in^2) is less than the value at which max w_1 is rapidly increasing ($p_{crit} = 447$ lb/in^2). This observation can be generalized /5/. Again it is a result of different assumptions introduced with the derivation of the criterion (32) (freezing in time). The maximum difference was found to be less than 15 percent.

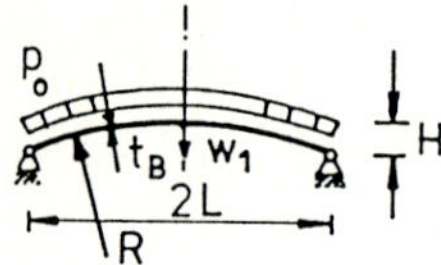

R = 67,115 in
L = 17,371 in
H = 2,287 in
t_B = 1,0 in
E = 10^7 lb/in^2
v = 0,2
ρ = $2,44 \times 10^{-6}$ lb sec^2/in^4

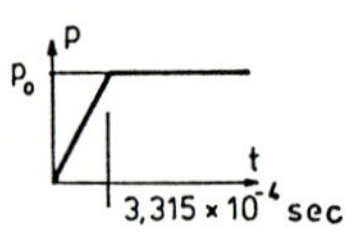

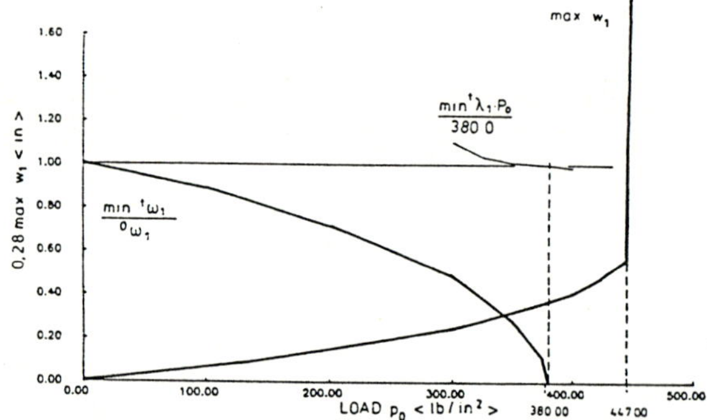

Figure 4:
Symmetrical arch

Figure 5:
Dynamic buckling analysis
($\Delta t = 0.000166$ sec)

Stimulated by practical problems the following spherical cap (Fig. 6) under a ring load is analysed considering only geometrical nonlinearities. Further studies of nozzle reinforced spherical caps including also material nonlinearities can be found in /12/.

These dynamic buckling loads are 70 and 76.1 percent, respectively of the static buckling load /5/, /2/.

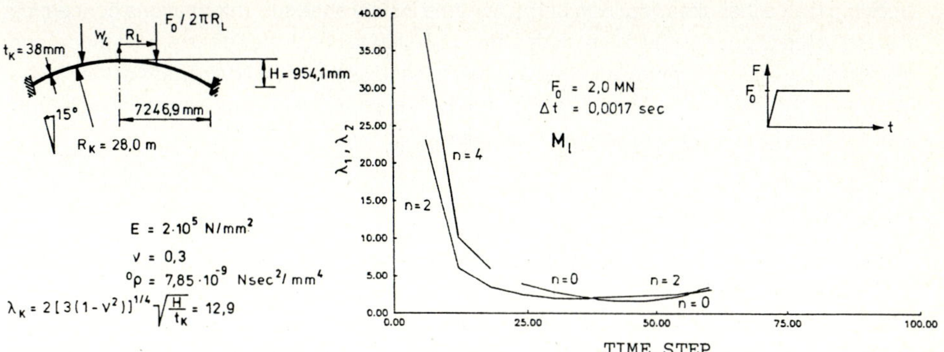

Figure 6:
Spherical cap

Figure 7:
Scaling method, $R_L = 1611.8$ mm

In order to determine the dynamic buckling mode one quarter of the shell was idealized using six biquadratic isoparametric degenerated serendipity shell elements in meridional and eight elements in hoop direction together with symmetry constraints. Applying the scaling method axisymmetric dynamic buckling was found under $F_{crit} = 2.74$ MN (Fig. 7). Therefore, an axisymmetric finite element idealization could be used for additional detailed analyses.

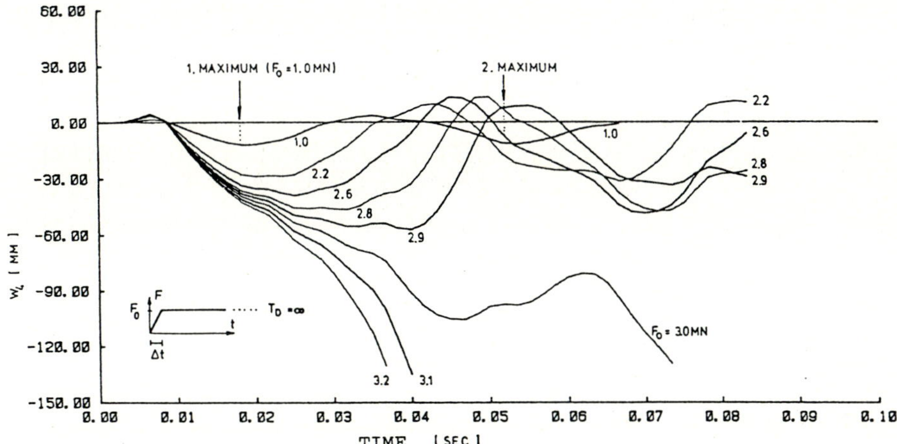

Figure 8:   Dynamic buckling under ring load $R_L = 1611.8$ mm

From several time response analyses (Fig. 8) (Newmark's method – $\beta = 1/4$, $\delta = 1/2$, $\Delta t = 0.0017$ sec $\simeq T_1/20$) again a slightly higher dynamic buckling load ($F_{crit} = 3.0$ MN) is obtained.

162

5. Conclusions

It is demonstrated that the proposed scaling method allows to estimate the critical dynamic buckling load either directly or with only a few time history analyses. Informations concerning the sequence of possible buckling modes are worthwhile to check the finite element model and the sensitivity against imperfections. As a further advantage this method operates at load levels lower than the critical one. Therefore, the demands to be made on time integration schemes are less than following the procedure of Budiansky/Roth /4/.

6. Acknowledgement

The financial support of the Deutsche Forschungsgemeinschaft under grant Ra 218/4–2 is gratefully acknowledged.

References

/1/   Ahmadian, M.; Inman, D.J.: On the Stability of General Dynamic Systems Using a Liapunov's Direct Method Approach. Computers & Structures 1–3 (1985) 287–292.

/2/   Akkas, N.; Bauld, N.R.: Buckling and Postbuckling Behavior of Clamped Spherical Shells under Axisymmetric Ring Loads. J. Applied Mechanics 9 (1971) 996–1002.

/3/   Brendel, B.: Geometrisch nichtlineare Elastostabilität. Dissertation. Bericht Nr. 79–1, Institut für Baustatik, Universität Stuttgart, 1979.

/4/   Budiansky, B.; Roth, R.S.: Axisymmetric Dynamic Buckling of Clamped Shallow Spherical Shells. NASA TND–1510 (1962) 597–609.

/5/   Burmeister, A.: Dynamische Stabilität nach der Methode der finiten Elemente mit Anwendungen auf Kugelschalen. Dissertation. Bericht Nr. 6 (1987) Institut für Baustatik, Universität Stuttgart.

/6/   Burmeister, A.; Ramm, E.: Dynamic Buckling FE Analysis – Conventional Versus Direct Approach. To be published.

/7/   Hughes, T.J.R.: The Finite Element Method – Linear Static and Dynamic Finite Element Analysis. Prentice Hall Inc., Englewood Cliffs, New Jersey, 1987.

/8/   Kleiber, M.; Kotula, W.; Saran, M.: Dynamic Quasi–Bifurcations in Structures Subjected to Step Loadings. Europe – US Symposium 'FE Methods for Nonlinear Problems', Trondheim, August 1985, Springer–Verlag.

/9/   Müller, P.C.; Schiehlen, W.O.: Lineare Schwingungen. Akademische Verlagsgesellschaft, Wiesbaden, 1976.

/10/   Ortiz, M.: A Note on Energy Conservation and Stability of Nonlinear Time – Stepping Algorithms. Computers & Structures 1 (1986) 167–168.

/11/   Ramm, E.; Matzenmiller, A.: Large Deformation Shell Analyses Based on the Degeneration Concept. Proceedings 'Finite Element Methods for Plate and Shell Structures' (eds. F.J.R. Hughes, E. Hinton), Pineridge, Swansea, UK, 1986, 365–393.

/12/   Ramm, E.; Burmeister, A.: Dynamic Buckling of Spherical Shells Using a Direct Method. To be published.

/13/   Rammerstorfer, F.G.; Auli, W.: Computation of the Stability of Dynamically Loaded Non-Linear Structures. 3rd Int. Conference on Num. Meth. Nonlin. Prob., Dubrovnik, Sept. 1986.

/14/   Trujillo, D.M.: The Direct Numerical Integration of Linear Matrix Differential Equations Using Padé Approximation. J. Numerical Methods in Engineering 9 (1975) 259-270.

/15/   Wood, W.L.; Bossak, M.; Zienkiewicz, O.C.: An Alpha Modification of Newmark's Method. J. Numerical Methods in Engineering 15 (1980) 1562-1566.

/16/   Ziegler, H.: Die Stabilitätskriterien der Elastomechanik. Ingenieur – Archiv 20 (1952) 49-56.

Free Formulation Elements with Drilling Freedoms for Stability Analysis of Shells

P.G. Bergan, M.K. Nygård and R.O. Bjærum

A.S Veritas Research, P.B. 300, 1322 Høvik, Norway

Abstract

The paper describes a class of shell elements based on the free formulation by Bergan and Nygård. The elements include the "drilling" freedom that gives a significantly improved membrane action as compared to traditional elements. The elements may undertake arbitrarily large displacements and rotations and material nonlinearities are accounted for. The "hyperplane displacement control" method is applied for solution of the static instability problems including singularities. A linearized eigenvalue analysis may be carried out at any stage during the incremental solution in order to find the limit point or bifurcation paths. The examples given cover verification-type problems like snap-through analysis of shells and buckling of plates. Further, an extensive buckling analysis of a corrugated aluminum shear panel is described. The examples verify that the shell elements and the solution algorithms are accurate, reliable and efficient.

1. Introduction

A major problem in analysis of shells is to find sufficiently reliable and accurate shell elements. For more than twenty years a vast amount of different types of shell elements have been proposed. The most used elements types are displacement based nonconforming elements, degenerated 3-D elements with various types of modifications, assumed stress or strain hybrid formulations, and, finally, new types of element formulations like the free formulation elements [1-6]. This last group of elements allows for nonconforming modes as well as the so-called "drilling" freedoms [2-6] and display highly improved membrane action. Such elements have proven to be superior to other elements for a wide range of applications, including linear and nonlinear shell analysis.

In addition to finding the best possible elements, an equally challenging problem is to predict the buckling load and to follow the post-buckling behaviour of shells. Shell structures, particularly stiffened shells, are often optimized such that several critical buckling patterns appear at nearly the same load level. In practice, it is very difficult to construct a reliable solution algorithm that always picks up the lowest possible buckling mode. The buckling shape will often change during the postbuckling phase, and structures may snap from one mode of buckling to another. These problems require sophisticated quasi-static solution algorithms with automatic computation of load steps. The capacity of solving the eigenvalue problem at any stage during the incremental solution has proven to be very useful. This can be utilized to pin-point bifurcation points, to guide the solution along new paths as well as to find limit points.

2. Shell Analysis with Free Formulation Elements

Free Formulation in brief

The "Patch Test" [7] has been established as a standard test for proving convergence of finite elements [8,9]. The Free Formulation proposed by Bergan and Nygård [1] is based on the so-called "Individual Element" (IE) test by Bergan and Hanssen [10]. This test gives explicit mathematical constraint equations which correspond to the patch test imposed on a single element. A main advantage by this form is that it may be built directly into the finite element formulation, as in the free formulation, and it opens the possibility for using non-conforming trial functions.

In mathematical terms the IE-test applied to an element stiffness matrix $\mathbf{k}$ may be written as

$$\mathbf{k}\mathbf{G}_{rc} = \mathbf{P}_{rc} \tag{1}$$

Let it be assumed that the element has n freedoms and that the type of problem has n_r rigid body modes and n_c constant strain modes. These displacement patterns may be written in terms of generalized displacement modes $\mathbf{N}$

$$\mathbf{u} = \begin{bmatrix} \mathbf{N}_r & \mathbf{N}_c \end{bmatrix} \begin{bmatrix} \mathbf{q}_r \\ \mathbf{q}_c \end{bmatrix} = \mathbf{N}_{rc}\mathbf{q}_{rc} \tag{2}$$

where $\mathbf{q}$ are the associated generalized coordinates. The corresponding nodal displacements for the element are obtained from

$$\mathbf{v} = \begin{bmatrix} \mathbf{G}_r & \mathbf{G}_c \end{bmatrix} \begin{bmatrix} \mathbf{q}_r \\ \mathbf{q}_c \end{bmatrix} = \mathbf{G}_{rc}\mathbf{q}_{rc} \tag{3}$$

Note that the n_{rc} nodal displacement patterns form the basis for the IE-test of equation (1).

It is obvious that pure rigid body modes produce no nodal forces, $\mathbf{P}$. The constant strain patterns $\mathbf{N}_c$, on the other hand, give non-zero nodal force patterns $\mathbf{P}_c$.

The constant strain modes corresponding to equation (2) may be written as

$$\epsilon = \Delta\mathbf{u} = \Delta\mathbf{N}_{rc}\mathbf{q}_{rc} = \Delta\mathbf{N}_c\mathbf{q}_c = \mathbf{B}_c\mathbf{q}_c \tag{4}$$

166

where Δ is the strain producing differential operator for the type of problem considered. It can be shown, see e.g [1], that the nodal force patterns for constant strain modes may be written

$$\mathbf{P}_c = \mathbf{LCB}_c \tag{5}$$

where $\mathbf{L}$ is the "lumping matrix" that transfers edge tractions to nodal forces and $\mathbf{C}$ is the constitutive matrix of the material. Explicit expressions for the lumping matrix of membrane elements may be found in [5], [6] while $\mathbf{L}$ for bending elements are given in [1], [3].

A stiffness formulation that implicitly satisfies the individual element test is

$$\mathbf{k} = \int_V \mathbf{B}^T \mathbf{CB} dV \tag{6}$$

provided a special form of the strain-displacement matrix $\mathbf{B}$ [1],[3] is applied in order to meet the constraints set by the IE-test;

$$\mathbf{B} = \frac{1}{V}\mathbf{L}^T + \tilde{\mathbf{B}}_h \mathbf{H}_h \tag{7}$$

It is assumed that n_h displacement patterns $\mathbf{N}_h$ are chosen to make out a complete set of n linearly independent modes. Subscript h refers to these higher order modes. Let $\mathbf{H}$ be the inverse to the complete $\mathbf{G}$ matrix for the n, compare equation (3). The contribution from the higher order modes in $\mathbf{B}$ is

$$\tilde{\mathbf{B}}_h = \mathbf{B}_h - \frac{1}{V}\int_V \mathbf{B}_h dV \tag{8}$$

This form implies that $\tilde{\mathbf{B}}_h$ is energy orthogonal [1] to the rc-modes, hence

$$\int_V \tilde{\mathbf{B}}_h dV \equiv 0 \tag{9}$$

V is the volume of the element. The strains may now be found in a straight-forward manner as

$$\epsilon = \mathbf{Bv} \tag{10}$$

The traditional form of the stiffness matrix as given in equation (6) corresponds to a potential energy formulation or a virtual work formulation. It can be shown that an extended, nonlinear strain form of equation (1) together with a virtual work formulation may be used to arrive at a general, large displacement version of the free formulation. Extensive discussion of this may be found in Bergan and Nygård [2] and Nygård [3]. According to the incremental virtual work equations the tangential element stiffness may be separated into two contributions, the material and geometric stiffness matrix, respectively. The material stiffness is directly the stiffness as derived in equation (6).

$$k_{mat} = \int_{V_o} B^T C_T B \, dV \tag{11}$$

C_T is the tangential constitutive tensor obtained from the material law, while **B** is the strain-displacement matrix of equation (7). A Corotated Lagrange (CL) formulation is applied for the description of motion, implying that the integration is performed over the initial volume.

The geometric stiffness matrix, k_{geom}, depends directly on the current state of stress. Applying the incremental Green strains in the incremental virtual work expression, k_{geom} may be obtained from

$$\delta v^T k_{geom} \Delta v = \int_{V_o} S_{ij} \frac{1}{2} (\delta u_{m,i} \Delta u_{m,j} + \Delta u_{m,i} \delta u_{m,j}) \, dV_o \tag{12}$$

S_{ij} is here the component form of the symmetric 2nd Piola-Kirchhoff stress tensor. The tangential or incremental element stiffness matrix is now found to be

$$k_T = k_{mat} + k_{geom} \tag{13}$$

The internal nodal reaction forces for an element, S_{int}, are found from the virtual work expressions on a total form.

$$S_{int} = \int_{V_o} B^T S \, dV_o \tag{14}$$

Shell elements

The Kirchhoff thin shell theory which is the most common theory for thin shell or plate elements is adopted here for three node and four node shell elements. This theory neglects the transverse shear strains, and the kinematics of the shell may be described as

$$u = u_o - z w_{,y}$$

$$v = v_o - z w_{,x} \tag{15}$$

$$w = w_o$$

Quantities with subscript o indicate values at the middle reference plane.

Each shell element has a reference base vector system attached to the co-rotated configuration of the element. All the kinematic assumptions that follow are made entirely with reference to this flat ghost reference element [2],[3].

The elements have six degrees of freedom at each node, three translational and three rotational freedoms, all oriented in accordance with the base vector system of the reference element. These freedoms may be separated into two groups, $\mathbf{v}_m$ is connected to the in-plane or membrane behaviour of the element, and $\mathbf{v}_b$ with the bending action of the element.

For a single node the membrane freedoms are given as

$$\mathbf{v}_{mi} = \begin{bmatrix} u_o \\ v_o \\ \dfrac{1}{2}(v,_x - u,_y) \end{bmatrix}_i = \begin{bmatrix} u_o \\ v_o \\ \Theta_z \end{bmatrix}_i \tag{16}$$

The rotational freedom Θ_z about the z-axis ("drilling freedom") is defined as the mechanical rotation derived from the in-plane displacement gradients. This quantity is invariant with respect to the reference coordinates. However, there is no unique relationship between the freedom Θ_z and the rotations of the adjacent element sides, see discussion by Bergan and Felippa [4]. This kind of freedom can therefore not be used in connection with the usual displacement formulations which requires C^o interelement continuity for the membrane action. However, C^o continuity is no requirement for the free formulation elements. This "drilling freedom" in the context of free formulation elements was first utilized for linear triangular membrane elements by Bergan and Felippa [4,5], and followed by extensions to quadrilateral membrane elements [3,6] and nonlinear shells [2,3] by Bergan and Nygård. This freedom gives highly improved membrane action of the elements [2-6], and overcomes the normally stiff behaviour of lower order shell elements. Considering membrane action only, the three node elements with drilling freedoms typically display a reduction of error with a factor of ten [4,5] as compared with elements without this freedom. This improved membrane behaviour is essential for predicting the stability of shells since a major source to the instability is associated with the action of membrane forces.

The bending freedoms for a node are the same as those of a traditional thin shell formulation,

$$\mathbf{v}_{bi} = \begin{bmatrix} w \\ \Theta_x \\ \Theta_y \end{bmatrix}_i = \begin{bmatrix} w_o \\ w,_y \\ -w,_x \end{bmatrix}_i \tag{17}$$

The total set of displacement functions of the element are

$$\mathbf{u} = \begin{bmatrix} \mathbf{u}_m \\ \mathbf{u}_b \end{bmatrix} = \begin{bmatrix} \mathbf{N}_m & 0 \\ 0 & \mathbf{N}_b \end{bmatrix} \begin{bmatrix} \mathbf{v}_m \\ \mathbf{v}_b \end{bmatrix} \tag{18}$$

A further detailing of the expressions for the membrane displacements gives

$$\mathbf{u}_m = \begin{bmatrix} u_o \\ v_o \end{bmatrix} = \begin{bmatrix} \mathbf{N}_u \\ \mathbf{N}_v \end{bmatrix} \mathbf{v}_m \tag{19}$$

Note that the drilling freedoms are coupled to both u and v displacements. The bending field is defined by another set of interpolation functions

$$\mathbf{u}_b = w = \mathbf{N}_b \mathbf{v}_b \tag{20}$$

3 node element

The displacement functions for the membrane part of the three node element are given in a non-dimensional coordinate system with origin at the centroid of the element (x_c, y_c)

$$\xi = \lambda(x - x_c) \ , \qquad \eta = \lambda(y - y_c) \tag{21}$$

where the length scaling factor is

$$\lambda = \frac{1}{\sqrt{A}} \tag{22}$$

x and y are the local element coordinates, and A is the area of the element.

The complete set of six rigid-body and linear strain modes may be written

$$\begin{bmatrix} u \\ v \end{bmatrix} = \mathbf{N}_{rcm} \mathbf{q}_{rcm} = \begin{bmatrix} 1 & 0 & -\eta & \xi & 0 & \eta \\ 0 & 1 & \xi & 0 & \eta & \xi \end{bmatrix} \mathbf{q}_{rcm} \tag{23}$$

There are a total of nine freedoms controlling the membrane displacements; three additional higher order modes are thus required. Following [4], the best choice is the in-plane "bending" modes oriented along the median lines of the triangle, see Fig. 1,

$$\mathbf{N}_{hm,i} = \begin{bmatrix} \bar{\xi}_i \bar{\eta}_i \\ -\dfrac{1}{2} \bar{\xi}_i^2 \end{bmatrix} \ , \quad i = 1,2,3 \tag{24}$$

These natural modes are objective with respect to the element reference system.

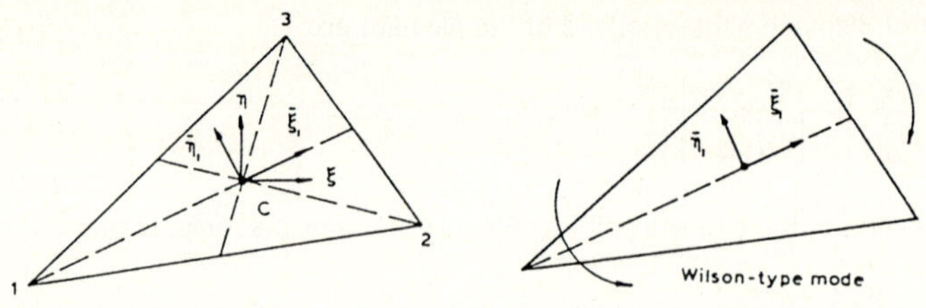

Figure 1. Local coordinate system used for "Wilson-type" modes for the triangular shell element.

The plate bending modes are expressed most conveniently in terms of area coordinates ς_i. The complete set of rigid-body and constant curvature modes may be written

$$\mathbf{N}_{rcb} = \begin{bmatrix} \varsigma_1 & \varsigma_2 & \varsigma_3 & \varsigma_1\varsigma_2 & \varsigma_2\varsigma_3 & \varsigma_3\varsigma_1 \end{bmatrix} \tag{25}$$

The three higher order modes suggested in [22] are used,

$$\mathbf{N}_{hb} = \begin{bmatrix} \varsigma_1\varsigma_2(\varsigma_1-\varsigma_2) & \varsigma_2\varsigma_3(\varsigma_2-\varsigma_3) & \varsigma_3\varsigma_1(\varsigma_3-\varsigma_1) \end{bmatrix} \tag{26}$$

The element satisfy the condition of correct rank for the stiffness matrix both for the membrane and the plate bending action when applying a three-point integration in the element plane.

4 node element

The same basic rc-modes as for the triangle in equation (23) are used. A fundamental requirement for selection of the six additional higher order modes is that the **G**-matrix must be regular. The final form of the six h-modes then becomes [3],[6]

$$\mathbf{N}_{hm} = \begin{bmatrix} \xi\eta & \eta^2 & 0 & 0 & \eta(\xi-\eta)^2+c_{11}\xi+c_{12}\eta & -\eta(\xi+\eta)^2+c_{21}\xi+c_{22}\eta \\ 0 & 0 & \xi\eta & \xi^2 & -\xi(\eta-\xi)^2+c_{13}\xi+c_{14}\eta & \xi(\eta+\xi)^2+c_{23}\xi+c_{24}\eta \end{bmatrix} \tag{27}$$

where

$$\begin{aligned}
c_{11} &= 2(-p_2+p_3) & c_{21} &= 2(p_2+p_3) \\
c_{12} &= -p_1+4p_2-3p_3 & c_{22} &= p_1+4p_2+3p_3 \\
c_{13} &= 3p_1-4p_2+p_3 & c_{23} &= -3p_1-4p_2-p_3 \\
c_{14} &= 2(-p_1+p_2) & c_{24} &= -2(p_1+p_2)
\end{aligned} \tag{28}$$

and

$$p_1 = \frac{1}{A}\int_A \xi^2 dA \quad , \quad p_2 = \frac{1}{A}\int_A \xi\eta\, dA \quad , \quad p_3 = \frac{1}{A}\int_A \eta^2 dA \qquad (29)$$

p_1, p_2 and p_3 are easily calculated analytically since the ξ,η-system is a Cartesian system. The linear terms in the two last h-modes are especially constructed to give energy-orthogonal modes which in turn make the expression for the **B** matrix of equations (7-9) simpler.

The rc-modes for the out-of-plane displacements are expressed as the six terms of a complete second degree polynomial by use of the same coordinate system as the membrane part

$$\mathbf{N}_{rcb} = \begin{bmatrix} 1 & \xi & \eta & \xi^2 & \xi\eta & \eta^2 \end{bmatrix} \qquad (30)$$

The choice of the remaining six high order modes are discussed in [1-3]. The four first terms form a complete and, thus, an objective third degree polynomial. Here, the expression in [3] is applied to satisfy energy orthogonality.

$$\mathbf{N}_{hb} = \begin{bmatrix} \xi^3 & \xi^2\eta & \xi\eta^2 & \eta^3 & \xi^3\eta - c_2\xi^2 - c_1\xi\eta & \xi\eta^3 - c_3\xi\eta - c_2\eta^2 \end{bmatrix} \qquad (31)$$

c_1, c_2 and c_3 are constants determined from the energy orthogonality requirement.

Use of a 3x3 Gauss integration scheme in the element plane gives a correct rank of the stiffness matrix. However, a 2x2 reduced integration scheme will give proper rank for the plate bending action, and one spurious zero energy mode for the membrane action. This mode, however, can not be shared with other elements and cannot propagate throughout a mesh to give a singular system. This is thus still a sound element which has been used for a wide range of problems without getting any problems with the zero energy mode. However, some care should always be taken if the 2x2 integration is applied.

The present triangular three node elements and quadrilateral four node elements are denoted FFTR and FFQ, respectively. The elements are assumed to be initially plane, but may undergo arbitrarily large displacements and rotations. The concept of a co-rotating "ghost" reference element [11,12] is utilized. This ghost element to which all stresses and strains are referred, represent a close fit of the initial element to the current, deformed element. A totally objective procedure is used to establish the ghost element configuration [2,3,12].

Finite rotations

Finite rotations in space are here dealt with by attaching an orthogonal triad of unit base vectors $\mathbf{i}_i$ to each node. The rotation of the node is thus uniquely defined by

$$\mathbf{i}_i = T_{ij}\mathbf{l}_j \tag{32}$$

where $\mathbf{l}_j$ represents the global base vector system. The transformation matrix is updated step by step during the solution process maintaining its orthonormality properties, see also [3,11,12].

3. Solution Algorithms

Incremental analysis

Detection of instabilities and following instable branches of a solution path requires a robust solution algorithm. Some problems are by nature such that prescribed displacements may be used rather than applying external loads, and this may simplify the solution in the way that unloading becomes unnecessary and that singular modes do not appear. Such problems can be solved with a Euler forward incrementation combined with true or modified Newton-Raphson iterations.

Other problems are more complex by nature and require solution algorithms that can handle singularities and instable branches. An algorithm of this kind is the so-called "Hyperplane Displacement Control Method" [13]. A displacement measure is defined from the combination of one or more displacement components to form a hyperplane in the multidimensional displacement space. The load application is controlled by keeping the increase of this displacement measure constant for each load increment. Trial choices of load levels are corrected during the iterations to give the correct displacement measure. An advantage by this procedure over the arc-length method [14-16] is that the characteristic equation that determines the step length is linear (no problem with double roots).

Eigenvalue analysis

In some situations it may be desirable to estimate the locations of critical points along the solution path with a minimum of computational effort. This may be achieved by an eigenvalue analysis based on a linearization of the stiffness reduction as the applied load increases.

Subtracting the tangential stiffness matrices of two following equilibrium states yields

$$\Delta\mathbf{K} = \mathbf{K}(\lambda_o) - \mathbf{K}(\lambda_o + \Delta\lambda) \tag{33}$$

where λ_o may be any load level within the prebuckling range, and $\Delta\lambda$ is the difference in the load parameter between the two states. ΔK may incorporate stiffness changes due to geometrical as well as material nonlinearities. The linearization lies in the assumption that the total stiffness changes proportionally to the load parameter λ, see Fig. 2.

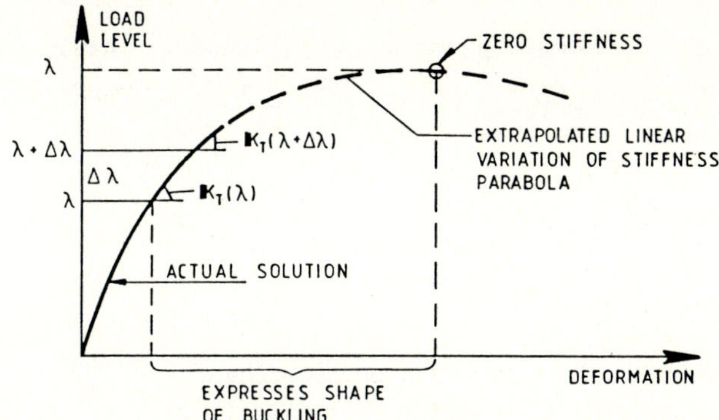

Figure 2. Concept for finding critical load from linearized buckling analysis.

At critical points along the solution path the tangential stiffness matrix becomes singular whether it is a limit point or a bifurcation point. The critical load level is thus found by determining the load parameter λ_{cr} for which the stiffness matrix has a zero determinant,

$$\det\left[K(\lambda_o) - \frac{\Delta\lambda_{cr}}{\Delta\lambda}\Delta K \right] = 0 \tag{34}$$

This is equivalent to solving the generalized eigenproblem

$$\left[K(\lambda_o) - \xi_{cr}\Delta K \right] r = 0 \tag{35}$$

The smallest eigenvalue ξ_{cr} corresponds to the first critical state. The critical load level is then

$$\lambda_{cr} = \lambda + \xi_{cr}\Delta\lambda \tag{36}$$

The associated eigenvector r represents the corresponding buckling or bifurcation mode. Bifurcation points and limit points are distinguished by projecting the eigenmode on the current direction of the incremental displacements. In the case of a limit point the buckling mode will continue along the current displacement path, while for bifurcations the eigenmode points in direction of a new branch orthogonal to the current displacement path.

4. Examples

Large rotations of cantilever beam

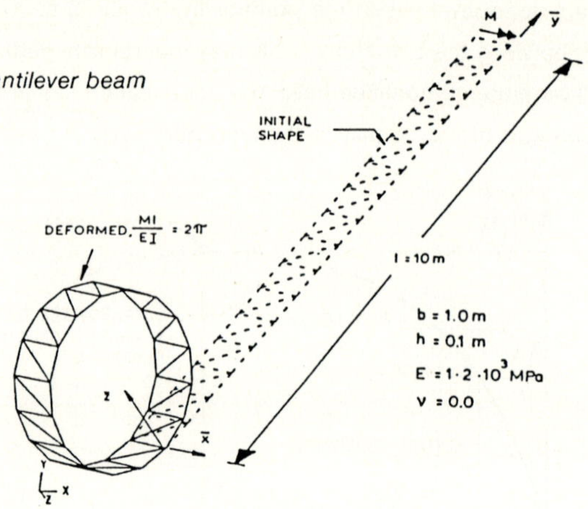

Figure 3. Bending of cantilever oriented "arbitrarily" in space.

The procedure for treating the finite rotations is validated through the example shown in Fig. 3, where a cantilever plate oriented "arbitrarily" in space is bent 360 degrees around by a concentrated end moment. The results obtained with the triangular, as shown in Fig. 3, as well as the quadrilateral free formulation elements are almost identical to the analytical solution of this problem. Note that this example is a "complete" test on the finite rotation capabilities only when the orientation of the beam and the bending rotations are not parallel to the global axes.

The position of the cantilever in the figure is defined by using an auxiliary $(\bar{x},\bar{y},\bar{z})$-system given by the transformation

$$\begin{bmatrix} \bar{x} \\ \bar{y} \\ \bar{z} \end{bmatrix} = \frac{1}{7} \begin{bmatrix} 6 & -2 & -3 \\ 3 & 6 & 2 \\ 2 & -3 & 6 \end{bmatrix} \begin{bmatrix} x \\ y \\ z \end{bmatrix} \tag{36}$$

One side of the cantilever follows the $\bar{y}$-axis. This auxiliary system is used for modelling purposes only, all actual computations are performed in the global (x,y,z)-system. The six nodal degrees of freedom are free except at the clamped edge where all freedoms are fixed. The moment is also applied in the global coordinated system with components $M_x = -\frac{6}{7}M$, $M_y = \frac{2}{7}M$ and $M_z = \frac{3}{7}M$.

The problem is solved using prescribed increments of the end moment and a true Newton-Raphson iteration scheme.

Postbuckling of a plate

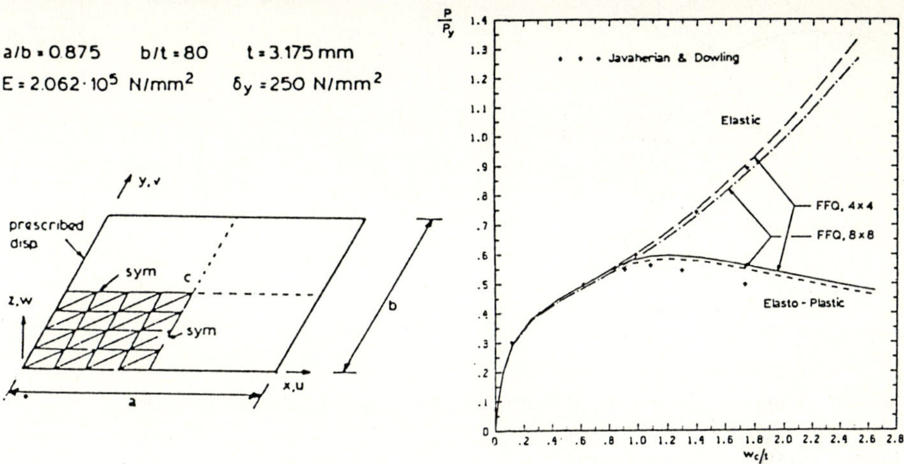

Figure 4. Buckling of rectangular plate

A rectangular plate with geometry given in Fig. 4 is subjected to end shortening along the side $x=0$. All edges are transversally supported. Two cases are analyzed, one with linear elastic material and one with elasto-plastic von-Mises material. A radial return algorithm is used for the elasto-plastic iterations. The end shortening of the plate is given as prescribed increments of displacements, giving a stable postbuckling behaviour. The plate has initially been given a small geometric imperfection to trigger the anticipated buckling behaviour. The imperfection constitutes a single sine wave in the x- and y-directions, with a maximum amplitude of 0.001b at the center of the plate.

The sharpness of the critical load increases with decreasing initial imperfection. The results obtained are found very close to those reported by Javaherian and Dowling [17] for the elastic case. However, the results differ somewhat for the elasto-plastic case; this is probably mainly due to use of different plasticity formulations in the two studies.

Buckling of a cylindrical panel

The snap-back behaviour of the cylindrical panel shown in Fig. 5a is solved using the hyperplane displacement control method. The results obtained for the FFQ element are given in Fig. 5b. Comparison with results by Oliver and Onate [18] show very good agreement. The FFQ element shows rapid convergence, and even the rather coarse 4x4 mesh gives very accurate results.

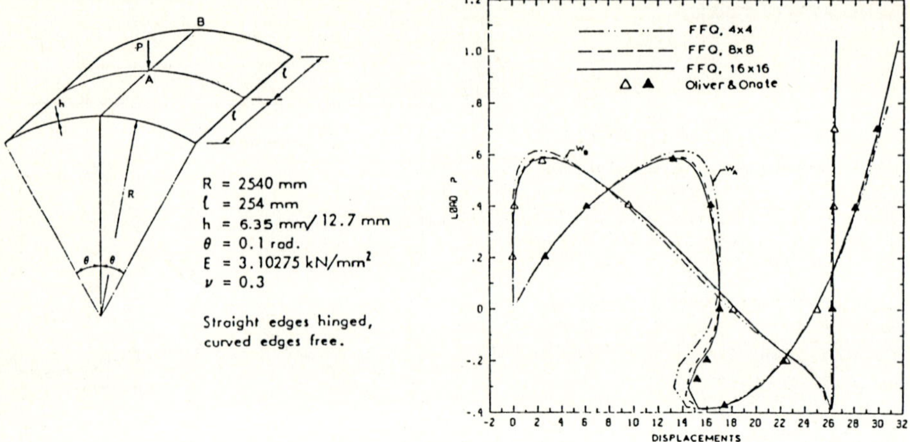

Figure 5. Snap-through of cylindrical panel; a) geometry, b) load-deflection curve.

Shear buckling of corrugated panel

The shear loaded corrugated aluminium shear panel in Fig. 6 is analyzed to determine the overall structural behaviour, the ultimate capacity and the post-buckling behaviour. This analysis constitutes the second phase in a project of designing corrugated panels with a ductile post-buckling behaviour. The first phase consisted of laboratory tests, which were also used for the initial verification of the computer model. A close agreement between the computed and the measured results was obtained. The second phase included a redesign if the panel based entirely on new computer simulations. Details of these analyses are described in [19].

The final version of the corrugated panel is fixed at one end, it has box stiffeners along the two horizontal edges, and it has a plate stiffener along the free vertical edge. The panel is subjected to uniformly distributed shear loads acting along the upper edge, the lower edge and the vertical end stiffener, see Fig. 6.

The computational model has 18 equally spaced elements along the height of the panel. There are two elements in the horizontal direction in the bottom of each corrugation channel and there are three elements between each corrugation. The inclined walls in the corrugation channels are modelled with a single element in the horizontal direction. This means that the mesh is too coarse to model local buckling of the corrugation channels, however, the mesh is sufficiently accurate to model the global buckling properly. It should also be noted that local buckling is more likely to happen in the plate sections between the channels, and that the three element wide model of these fields is capable of reproducing such buckling.

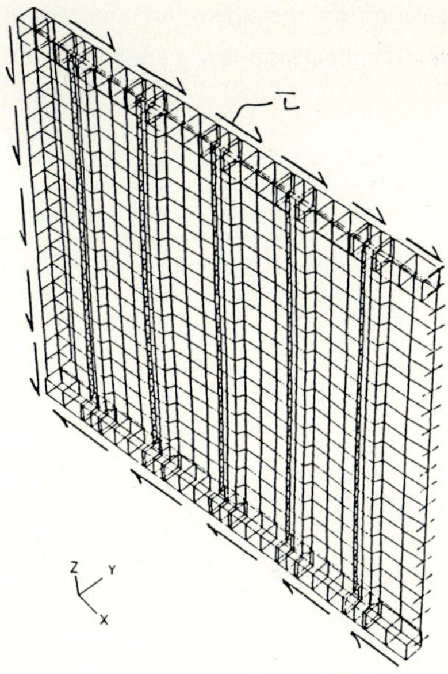

Figure 6. Corrugated aluminum shear panel subjected to shear loading.

As mentioned earlier, the drilling freedoms of the free formulation shell element are of great importance when modelling the stiffness effects in folded plates such as in this case. Of particular importance is that the relatively low bending stiffness associated with the out-of-plane rotations gets directly coupled with the high membrane rotational stiffness of the drilling freedoms at the edges of the channels. In this way the effect of the locking of the bending perpendicular to edges gets represented correctly by the present type of element.

A global imperfection is introduced in the model in order to trigger the shear buckling mode. The shape of the imperfection is based on analytical formulas for the buckling shape of corrugated panels, given in [20]

Two different analyses are performed using the same computer model. First, a linearized buckling analysis is carried out to obtain an estimate of the critical load level. Then a complete nonlinear analysis is performed to study the buckling behaviour of the structure. For this purpose the hyperplane displacement control method is used to guide the automatic solution algorithm.

The first two buckling modes, resulting from the eigenvalue analysis, are shown in Figs. 7 and 8. The corresponding normalized critical load-levels are 5.94 and 7.11, respectively.

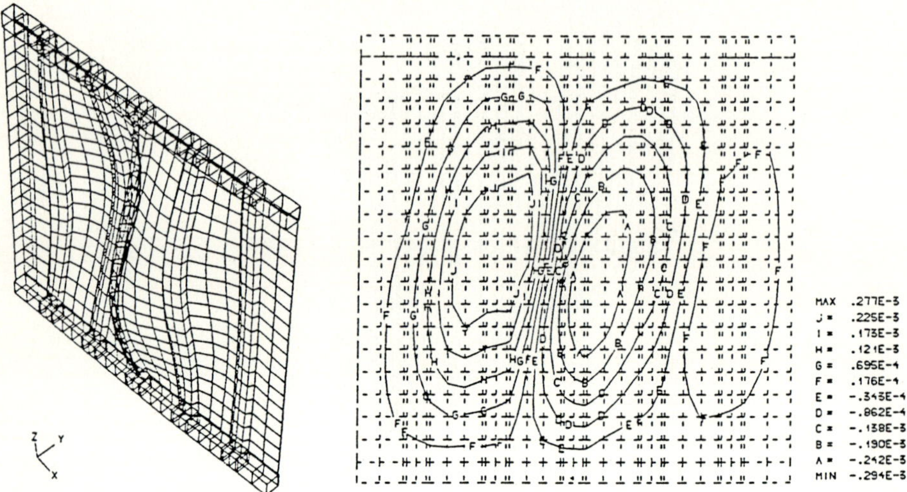

Figure 7. First linearized buckling mode corresponding to load level 5.94. a) plot of mode shape, b) contours plot of y-displacements.

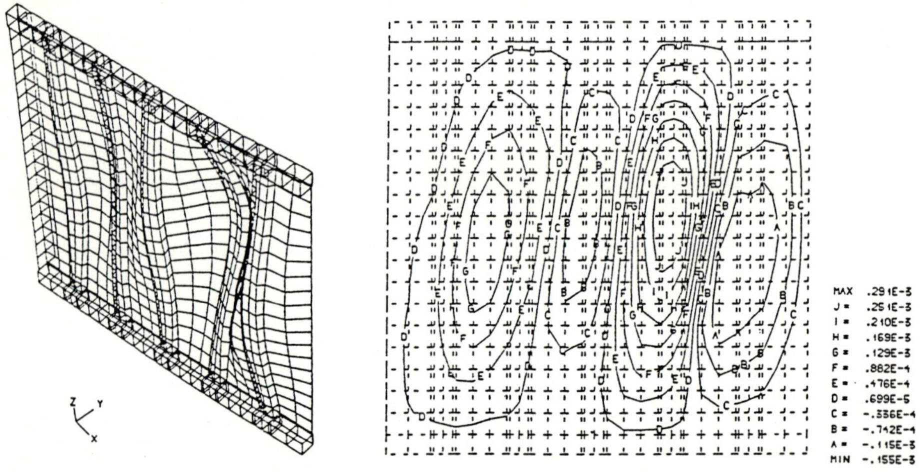

Figure 8. Second linearized buckling mode corresponding to load level 7.11. a) Plot of mode shape, b) Contours plot of y-displacements.

A complete nonlinear analysis gives an ultimate capacity of the corrugated panel of 6.91, which is 16% higher than that of the linearized buckling analysis. The deformed shape of the computational model for the ultimate load level is very similar to that of the first buckling mode from the eigenvalue analysis, as can be seen by comparing Fig. 7 with Fig. 9.

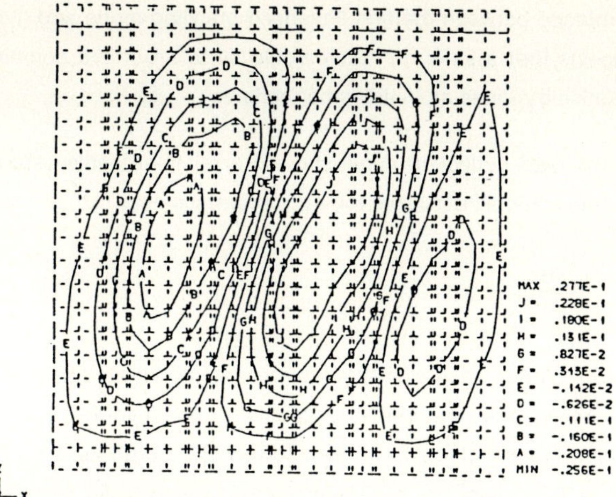

Figure 9. Contour plot of y-displacements at buckling load level 6.91.

The difference in the results of the two analyses may be explained by the increasing stabilizing effect of tensile stress bands acting in a skew direction. Note that, as opposed to a plane panel, there are in fact several stress bands developing simultaneously.

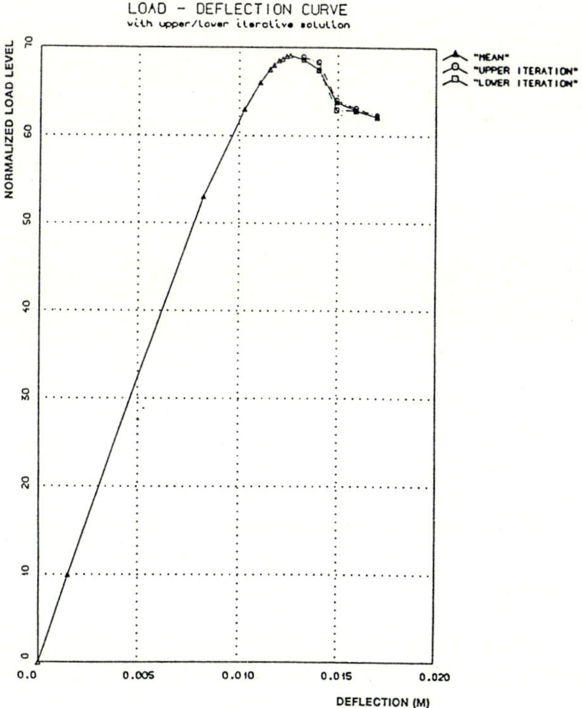

Figure 10. Load deflection curve from nonlinear analysis.

The close resemblance between the first linearized buckling mode and the deformed shape at buckling suggests that a safe approach would be to assume a combination of the first few linearized buckling modes as initial imperfections.

In this analysis the load-deflection curve showed a relative ductile behaviour in the post-buckling range; this is highly desirable for a structural member like this, see Fig. 10.

5. Conclusions

The use of free formulation shell elements has been demonstrated for plate and shell problems with large deflections, material and geometric nonlinearities, instability and post-buckling behaviour. The problems have been solved using both the triangular and the quadrilateral elements giving very similar results. Results obtained for buckling load prediction and postbuckling behaviour for a corrugated shear panel compare also very well with test results. The new elements have an especially efficient membrane action due to the utilization of drilling freedoms. These normal rotations also eliminate the traditional problem of shell analysis regarding how to deal with an incomplete set of rotational freedoms for the elements. The elements are recommended for a wide range of shell analysis including instability problems.

6. Acknowledgement

The authors wish to acknowledge the participation of colleagues in Veritas Offshore Technology and Services A/S in modelling and analyzing the corrugated panel described herein.

7. References

1   Bergan, P.G. and Nygård, M.K.: Finite elements with increased freedom in choosing shape functions. *Int. J. for Numerical Methods in Engineering*, Vol. 20, 643-664, (1984).

2   Bergan, P.G. and Nygård, M.K.: Nonlinear shell analysis using free formulation finite elements. In P.G. Bergan, K.J. Bathe and W. Wunderlich (Eds.), *Finite Element Methods for Nonlinear Problems.* Springer Verlag 1985.

3   Nygård, M.K.: The Free Formulation for nonlinear elements with applications to shells, Dr. ing. thesis, Report no. 86-2, Division of Structural Mechanics, The Norwegian Institute of Technology, Trondheim, Norway, 1986.

4   Bergan, P.G. and Felippa, C.A.: A triangular membrane element with rotational degrees of freedom, *Computer Methods in Applied Mechanics and Engineering,* Vol. 50, 25-69, (1985).

5   Bergan, P.G. and Felippa, C.A.: Efficient implementation of a triangular membrane element with drilling freedoms. In T.J. Hughes and E. Hinton (Eds.): *Finite Element Methods for Plate and Shell Structures, Vol. 1, Element Technology,* pp.128-152, Pineridge Press, Swansea, 1986.

6   Bergan, P.G. and Nygård, M.K.: A Quadrilateral Membrane Element with Rotational Freedoms. In G. Yagawa and S. N. Atluri: *Computational Mechanics'86, Proc. Int. Conf. on Computational Mechanics,* Tokyo, Springer Verlag, 1986.

7   Irons, B.M. and Razzaque, A.: Experience with the Patch Test for convergence of finite elements. In K. Aziz (Ed.): *Mathematical Foundations of the Finite Element Method with Applications to Partial Differential Equations. 557-587,* Academic Press, 1972.

8   Taylor, R., Zienkiewicz, O.C., Simon, J. and Chan, A.H.C.: The Patch Test - a condition for assessing f.e.m. convergence. In K.J. Forsberg and H.H. Fong: *Finite Element Standards Forum, Book 2 of 2,* Proceedings of AIAA/ASME/ASCE/AMS 26th Structures, Structural Dynamics, and Material Conference, Orlando, USA, 1985.

9   Strang, G. and Fix, G.J.: *An Analysis of the Finite Element Method,* Prentice-Hall, Englewood Cliffs, N.J. (1973).

10   Bergan, P.G. and Hanssen, L.: A new approach for deriving "good" finite elements. In J.R. Whiteman (ed.), *The Mathematics of Finite Elements and Applications,* Vol. II. 483-498, London: Academic Press (1976).

11   Bergan, P.G. et al.: FENRIS Manuals, Theory - Program Outline - Data Input, NTH, SINTEF, and A.S. Veritec, Høvik, Norway, 1984.

12   Nygård, M.K. and Bergan, P.G.: Advances in treating large rotations for nonlinear problems. In A.K. Noor (Ed.), *State-of-the-art Surways on Computational Mechanics,* To be published, ASME, 1989.

13   Simons, J.F., Bergan, P.G. and Nygård, M.K.: Hyperplane displacement control methods in nonlinear analysis. In W.K. Liu, T. Belytschko and K.C. Park (Eds.), *Innovative Methods for Nonlinear Problems,* 345-364, Swansea, Pineridge Press, 1984.

14   Riks, E.: The application of Newton's method to the problem of elastic stability, *Journal of Applied Mechanics,* Vol. 39, 1060-1066 (1972).

15   Crisfield, M.A.: Incremental/iterative solution procedures for non-linear problems. In C. Taylor, E. Hinton and D.R.J. Owen (Eds.), *Numerical Methods for Nonlinear Problems.* Vol. 1, 261-290, Swansea, Pineridge Press, 1984.

16 Ramm, E.: Strategies for tracing non-linear response near limit points, In W. Wunderlich, E. Stein and K.J. Bathe (Eds.): *Non-Linear Finite Element Analysis in Structural Mechanics*, 68-89, Spinger-Verlag, New York, 1981.

17 Javaherian, H. and Dowling, P.J.: Large deflection elasto-plastic analysis of thin shells, *Engng. Structures*, Vol. 7, pp.154-163 (1985).

18 Oliver, J and Onate, E.: A total Lagrangian formulation for the geometrically nonlinear analysis of structures using finite elements. Part I: Two-dimensional problems: Shell and plate structures, *Int. J. for Numerical Methods in Engineering*, Vol. 20, 2253-2281 (1984).

19 Veritas Offshore Technology and Services A/S 1988, *Nonlinear Analysis of Corrugated Shear Panel, Snorre Living Quarter,* VT-report 88-3285.

20 Easley, J.T.: Buckling Formulas for Corrugated Metal Shear Diaphragms, *Journal of Structural Division,* ASCE, (1975).

21 Mazzolani, F.: *Aluminium Alloy Structures,* Pitman Publishing Inc., 1985.

22 Hanssen, L., Syvertsen, T.G. and Bergan, P.G.: Stiffness derivation based on element convergence requirements. In J.R. Whiteman (Ed.): *The Mathematics of Finite Elements and Applications. Vol. III.* 83-96, London, Academic Press, 1979.

Stability of Dynamically Loaded Structures

Dieter Dinkler and Bernd Kröplin

Institut für Statik und Dynamik - Universität Stuttgart

Summary

A method is suggested, which simplifies the investigations to the safety of buckling structures under time depending actions. The static analysis of the structure is used in order to fix the critical deformation state and hence the critical strain energy. Representing the deformation behaviour and the loading by some energetic measures an approximation for the development of the motion during the loading may be computed with relatively little effort by using Galerkin's procedure. If the effect of the external actions is expressed by the energy which is induced to the structure during the loading, the proof of the stability may be given by comparing this external energy to the critical strain energy.

1 Introduction

In engineering it is often necessary to design structures not only for static but also for time dependend loads. For structures with linear behaviour, 'load factors' are well known in order to estimate the load–time dependence. Nonlinear structures require, especially when instabilities as buckling are concerned, a more detailed treatment in order to obtain reliable safety estimates. In extreme cases the equations of motions have to be integrated in time including all nonlinearities. Because of its effort such investigations seem to be often inadequate and besides this hard to interpret because of the numerous modes of motions which appear.

This paper gives a method to calculate estimates for the instationary nonlinear buckling behaviour in a simplified manner without direct integration of the equations of motion. The method consists of three steps:

- Definition and calculation of critical states of motion as limit states in the prebuckling range. These states describe the limit between prebuckling and postbuckling. Calculation of the energy related to the limit states.

- Calculation of the external energy, which describes the time dependent load influence.

- Comparison of the critical energy from step one and the load energy.

The assumptions are based on elastic material behaviour. The load functions are short-time loadlevel changes or impulse loads.

2   Fundamentals

Fundamental works considering the theory of stability of elastic structures are e.g. Koiter/1/, Leipholz/2/, Huseyin/3/, Thompson/Hunt/4/ and Ziegler/5/. These papers discuss mostly questions of stability of equilibrium states, while the stability of nonlinear dynamic systems is discussed e.g. in Hayashi/6/, Thompson/Stewart/7/ and Haken/8/.

The proof of stability of an equilibrium state does not meet the task which is discussed here since the considered structures are assumed to be in a stable equilibrium position. The further questions which have to be asked are:

- How stable is the considered equilibrium state?

- What is the size of additional loads which cause loss of stability?

The basis of the theory of stability of elastic continuous systems is – neglecting temperature effects /9/ – the energy balance in rateform(1). It contains the rates of the potential energy, the kinetic energy and the external loads

$$\dot{V}_v = \int \{\dot{\gamma}_{ij} e_{ijkl} \gamma_{kl} + \rho \dot{v}_i \ddot{v}_i\} dR^3 - \int \dot{v}_i \bar{s}_i dO = 0 \tag{1}$$

with the initial condition $V_{v_0} = c$ at time $t = t_0$.

$\dot{()}$: derivative with respect to time
e_{ijkl} : elasticity of the material
ρ : material density
v_i : displacements in direction of the space coordinates x_i , $i = 1, 2, 3$
$\bar{s}_i$: loads

The Green strains γ_{ij} are defined as

$$\gamma_{ij} = \frac{1}{2}(v_{i,j} + v_{j,i} + v_{k,i} v_{k,j}). \tag{2}$$

Alternatively the potential energy can be formulated in stresses and strains as a mixed formulation. Equation (3) contains the nonlinearity one order lower as (1), which leads to some advantages in numerical investigations.

$$\dot{V}_{\sigma,v} = \int \{\dot{\gamma}_{ij}\sigma_{ij} + \gamma_{ij}\dot{\sigma}_{ij} - \dot{\sigma}_{ij}f_{ijkl}\sigma_{kl} + \rho \dot{v}_i \ddot{v}_i\} dR^3$$
$$- \int \dot{v}_i \bar{s}_i dO_\sigma - \int s_i(v_i - \bar{v}_i) dO_v = 0. \tag{3}$$

f_{ijkl} : flexibility of the material
σ_{ij} : 2. Piola-Kirchhoff stress–tensor
s_i : stress–vector

The common treatment of thin–walled shell structures is to reduce the continuum by means of the Kirchhoff–Love–Hypothesis in order to simplify the calculation by introducing a reference surface. The mixed formulation with moderately large rotations describes the rate of the potential dependent on the displacements u_α, u^3, the velocities $\dot{u}_\alpha, \dot{u}^3$, the stress resultants axial forces $n^{\alpha\beta}$ and bending moments $m^{\alpha\beta}$ and the loads p_α, p^3 of the reference surface /11/

$$\dot{V} = \dot{V}(u_\alpha, u^3, \dot{u}_\alpha, \dot{u}^3, n^{\alpha\beta}, m^{\alpha,\beta}, p_\alpha, p^3). \tag{4}$$

The discretisation of the reference surface with finite elements leads after the integration over the reference surface to the matrix notation (5)

$$\dot{V}_z = \dot{z}^t \{ \mathbf{A}^L(z_0)z + \mathbf{A}^N(\tfrac{1}{2}z)z + \mathbf{M}\ddot{z} - p \} = 0. \tag{5}$$

z : vector of unknowns $z^t = \{u, n, m\}^t$
$\mathbf{A}^L(z_0)$: linear matrix, dependent on the initial state z_0
$\mathbf{A}^N(\tfrac{1}{2}z)$: geometric nonlinear matrix
$\mathbf{M}$: mass matrix
p : vector of loads

3   Stability of motions

The fundamental investigations concerning stability of motions with perturbed initial conditions date back to Liapunov /10/. A motion is 'stable', if equation (6) holds. Otherwise the motion is said to be 'unstable'

$$\begin{aligned}
]v_{t_0,a_S} - v_{t_0,a_0}] &\leq \delta_1(\epsilon) \quad , \quad t = t_0, \\
]v_{t\ ,a_S} - v_{t\ ,a_0}] &< \epsilon \qquad\quad , \quad t > t_0.
\end{aligned} \tag{6}$$

'Asymptotic stability' is given if equation (7) holds

$$\lim_{t \to \infty}]v_{t,a_S} - v_{t,a_0}] = 0. \tag{7}$$

In case of a permanent perturbation R_s, which is not only a perturbation of the initial conditions, the motion is said to be 'totally stable' if also (8) is satisfied

$$]R_{s(t,v)}] \leq \delta_2(\epsilon) \quad , \quad t > t_0. \tag{8}$$

For nonlinear structures 'local stability' is given if motions with the same energy level are possible in different regions, e.g. in the pre– and postbuckling range. In nonlinear case the wavelength depends on the amplitude of the deformations and the stability in the sense of (6) is not satisfied. In this case the motion is called 'orbitally stable' if the trajectories of the unperturbed and the perturbed motion in the phase plane are close to each other. Criteria, which allow the judgement of stability without direct solution of the equations of the perturbed motions Liapunov gives in his second method. If a functional V exists for the perturbed motion, which satisfies conditions (9) and (10) the motion is stable. Such a Liapunov–Functional is given for the here considered case with (1) or (5). The first condition requires the positive definitness of the functional

$$V[z_s] > 0. \tag{9}$$

The second condition is a 'growth condition' (10). For undamped motions this condition is identical to the energy balance (1)

$$\dot{V}[z_s] \cdot sign(V[z_s]) \leq 0. \tag{10}$$

Both conditions are a priori satisfied. Hence the considered motions are 'orbitally stable'. In case of a finite, not arbitrary small perturbation of the equilibrium state (6) in the sense of 'orbital stability' is necessary to consider. This can be shown with the structure given in figure 1 /2/.

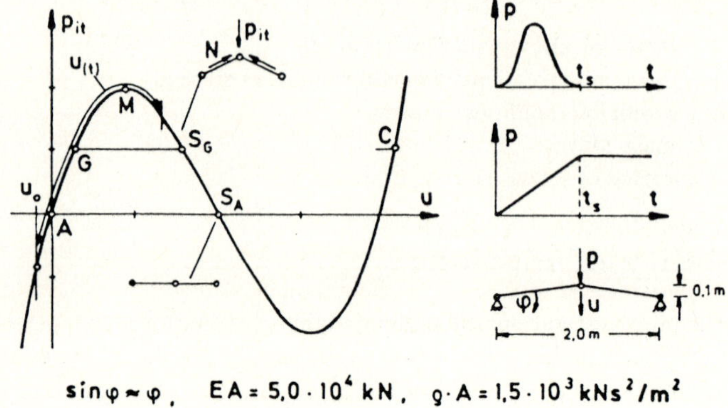

$$\sin\varphi \approx \varphi, \quad EA = 5{,}0 \cdot 10^4 \, kN, \quad \varrho \cdot A = 1{,}5 \cdot 10^3 \, kNs^2/m^2$$

Fig. 1 : Internal reactions of a flat truss

The reaction forces p_{iT} of the structure in direction of the displacements of the middle node are considered as control parameter and are plotted with respect to the related displacement u.

First, an initial displacement u_0 is given. The system moves around the unloaded state A. If u_0 is large enough, the system may pass the maximum of the inner reaction forces. It returns as long as the flat position S_A is not passed. Hence the flat position is the critical deformation state for the motion around A. If S_A is passed, condition (6) is not satisfied. With respect to the state A the critical deformations are characterized by a vanishing 'secant–stiffness' $A - S_A$. The same is true for a preloaded structure, e.g. the equilibrium state G.

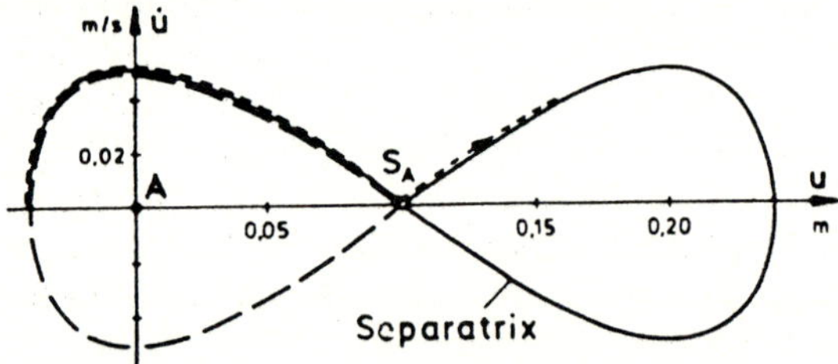

Fig. 2 : Motion of the truss of figure 1

The motion describes a trajectory in the phase-plane of displacements and velocities, figure 2. The distance of the trajectory from the static equilibrium state is a measure for the induced energy by the initial conditions. The critical deformations, see figure 1, are here represented on the static axis ($\dot{u} = 0$) as saddle-point S_A. For investigations to stability the energylevel ΔV_{crit} is of interest. This holds for all states of motion on the limit-curve, which is called 'critical trajectory' or 'separatrix'. Therefore this curve is the limit-curve for all 'local stable' motions in the pre- and post-buckling range.

A judgement of the stability of a motion against finite perturbations, which is defined by it's initial conditons needs the solution of the following two tasks:

- The critical energy ΔV_{crit} has to be calculated.

- The energy V_0 which is induced by the initial conditions and the energy T_e which is a function of perturbations have to be quantified.

The proof of stability is then possible with (11).

$$\Delta V_{crit} - T_e - V_0 \quad \begin{matrix} > 0 : & stable \\ = 0 : & indifferent \\ < 0 : & unstable \end{matrix} \tag{11}$$

In a normalized version (11) may be written as

$$\eta = \frac{\Delta V_{crit} - T_e - V_0}{\Delta V_{crit}}. \tag{12}$$

η is called the 'degree of stability'.

4   Calculation of the critical energy

As lined out in paragraph 3 the critical deformations can be defined as difference between stable and unstable equilibrium state. They can be calculated step by step with the well-known arc–length–methods for a given distribution of the perturbation forces.

More simple is the investigation of the energy surface of the structure. Figure 3 gives an energy profile orthogonal to the phase–plane for the special case of $\dot{u} = 0$.

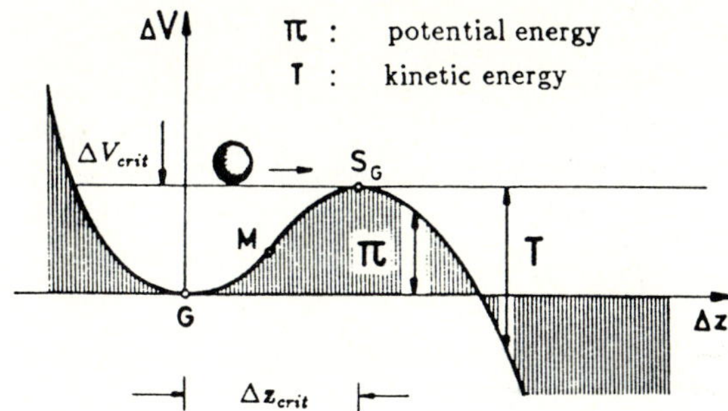

Fig. 3 : Energy profile for $\dot{u} = 0$

The stable equilibrium state is given by the minimum G of the potential energy. A perturbation of this equilibrium state leads to a 'local stable' motion of the structure with the energy level ΔV. An increase from ΔV to ΔV_{crit} shifts the structure to the saddle–point S_G. S_G gives the critical deformation state of the structure, since two 'local stable' motions in the pre– and the postbuckling range meet here. A further increase of the energy leads to a motion in the whole pre– and postbuckling state. Hence the decisive factor for the stability of the equilibrium state G is the energy–level, which is given by the maximum of the potential energy. The maximum is related to the unstable equilibrium state.

Therefore the first variation of the potential energy has to vanish in the case of the critical deformations.

$$\delta\Pi = 0. \tag{13}$$

(13) leads to a nonlinear algebraic system of equations for the critical state vector Δz_{crit}, which non–trivial solution describes the unstable equilibrium state.

$$\bar{z}_i^t\{\mathbf{A}^L(z_G) + \mathbf{A}^N(\tfrac{1}{2}\Delta z_{crit})\}\Delta z_{crit} = 0. \tag{14}$$

Some possibilities for an approximate calculation of the critical vector Δz_{crit} are given in $/12/$.

The critical energy, which is necessary for the stability criteria can be calculated now for the equilibrium state G by means of integration of (5) for the static case.

$$\Delta V_{crit} = \Delta z_{crit}^t\{\frac{1}{2}\mathbf{A}^L(z_G) + \frac{1}{6}\mathbf{A}^N(\Delta z_{crit})\}\Delta z_{crit}. \tag{15}$$

5   Calculation of the load energy

The aim of the investigations is to compare the critical energy with the load energy which is induced by perturbations, see (11). The external energy depends on the function in time of the perturbation and the related state vector development. Hence the energy has to be integrated over the time of perturbation t_s

$$V_{t_s} = \int_{t_0}^{t_s} \dot{z}^t\{\mathbf{A}^L(z_0)z + \mathbf{A}^N(\tfrac{1}{2}z)z + \mathbf{M}\ddot{z} - \mathbf{p}\}dt = 0. \tag{16}$$

The load is given by

$$\mathbf{p} = \Delta p_i \phi_{pi} \quad , \quad t_0 \leq t \leq t_s \tag{17}$$

Since the exact function is not known a Ritz–like method is chosen. The spatial distribution of the unknown state–vector is estimated by a static analysis of a reference load of the same shape as the perturbation. In case of different critical modes every mode has to be investigated with a special perturbation. The approximation of the space–time function of the state–vectors

$$\mathbf{z} = \alpha_{ij}\Delta z_{i(\Delta p_i)}\phi_{j(t)} \tag{18}$$

consists of the spatial pattern $\Delta z_{i(\Delta p_i)}$ and a number of free chosen shape–functions $\phi_{j(t)}$ with unknown scalers α_{ij} for the time–dependent part. The shape functions have to satisfy the initial conditions of the perturbed motion for the deformations and the velocities. In our case this is

$$\phi_{j(t)} = t^{j+1} \quad , \quad j \geq 1. \tag{19}$$

Experience has shown that polynomials are well suited since the initial motion oscillates very little.

With the approximation (18) the energy balance (16) is first calculated for spatial coordinates. The calculation of the quadratic forms can be performed with standard finite element programs under static loading. The second step is the integration of the energy rates for the time of perturbation. This is shown in (20).

$$\alpha_i \ \{ \ \int_{t_0}^{t_*} \dot{\phi}_i \Delta \mathbf{z}^t \{ \mathbf{A}^L(\mathbf{z}_0) \phi_j + \mathbf{A}^N(\tfrac{1}{2}\Delta \mathbf{z}\phi_k \alpha_k)\phi_j + \mathbf{M}\ddot{\phi}_j \} \Delta \mathbf{z} dt \ \alpha_j$$

$$- \int_{t_0}^{t_*} \dot{\phi}_i \Delta \mathbf{z}^t \Delta \mathbf{p} \phi_p \} dt \ \} \ = 0 \tag{20}$$

The integration in time can be performed analytically or numerically for example with Gauss. After integration a nonlinear algebraic system of equations has to be solved for the unknown factors α_j.

$$\{V_{AL} + V_{AN}(\alpha_k) + V_M\}_{ij}\alpha_j - V_{pi} = 0 \tag{21}$$

With little effort (21) can be solved iteratively. With the calculated α_j the function of the motion is known and the induced energy T_e follows

$$T_e = \alpha_i V_{pi}. \tag{22}$$

With less effort the energy V_0 can be calculated. This is the energy, which is induced by the initial conditions $\mathbf{z}_0, \dot{\mathbf{z}}_0$ related to the equilibrium state $\mathbf{z}_G$. With

$$\begin{aligned} \Delta \mathbf{z} &= \mathbf{z}_0 - \mathbf{z}_G \\ \Delta \dot{\mathbf{z}} &= \dot{\mathbf{z}}_0 \end{aligned} \tag{23}$$

follows V_0 using (5) and (15)

$$V_0 = \Delta \mathbf{z}^t \{ \frac{1}{2}\mathbf{A}^L(\mathbf{z}_G) + \frac{1}{6}\mathbf{A}^N(\Delta \mathbf{z})\} \Delta \mathbf{z} + \frac{1}{2}\Delta \dot{\mathbf{z}}^t \mathbf{M} \Delta \dot{\mathbf{z}}. \tag{24}$$

6 Applications to snap–through–problems

The aim of the applications is to show the method for different loads. For clarity the investigation is limited for two simple structures. For complex shell structures the method is totally analog if the discretisation leads to a form similar to (5). If the potential describes the nonlinearities exactly, the method leads to a illimit exact judgement of the stability.

6.1 Impact loading

The structure, see figure 1 and its initial equilibrium state G in the prebuckling range is given. For this state first the critical energy ΔV_{crit} is calculated after paragraph 4. A sufficiently large perturbation, e.g. with the time distribution after figure 4 accelerates the structure beyond the separatrix such that loss of stability occurs. For small perturbations the structure moves in the prebuckling range around the equilibrium state

190

G. The level of the initial load and the perturbation pattern in time are arbitrary and do not influence the accuracy of the result.

$$p_{(t)} = p_0 \;+\; \Delta p \,(1 - \cos t) \qquad , \quad t_0 \leq t \leq t_s = 2\pi$$
$$p_{(t)} = p_0 \qquad\qquad\qquad\qquad , \quad t \;>\; t_s$$

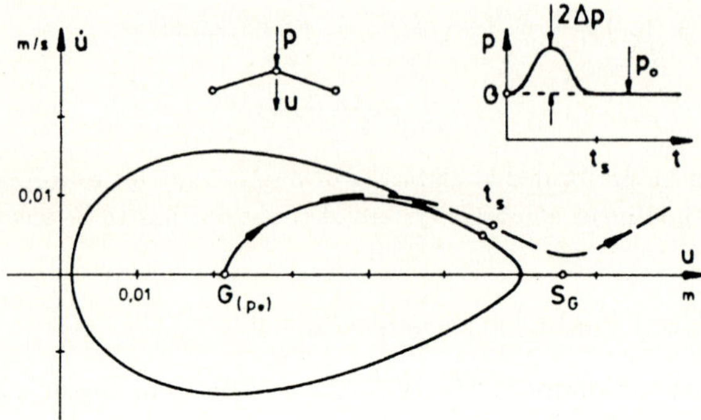

Fig. 4 : Motion of a flat truss under impact loading

Table 1

Δp	$10^7 \Delta V_{crit}$	$\leftrightarrow$	$10^7 T_e$	
—— 0,0031	1250	$<$	1313	$\Rightarrow$ unstable
—— 0,0030	1250	$>$	1224	$\Rightarrow$ stable

The approximation of the state–vector (18) is done with the first eigenmode of the equilibrium state. The motion in time caused by the perturbation and the related energy T_e are calculated after paragraph 5. The values in table 1 are calculated with a polynominal (19) with eight terms. The comparison of critical energy and external energy shows that the error can be kept arbitrarily small. The convergence of the external energy T_e for different approximations in time is given in /12/.

6.2   Transition loads

Abrupt changes of the load–level cause dynamic transitions. These have to be investigated in a different manner since not the stability with respect to the initial state A has to be proved but the stability of the motion around the new equilibrium state G. Dependent on the load–velocity the structure reaches at time $t = t_s$ the state of motion $z_0, \dot{z}_0$ which gives the initial condition for the following motion around the new equilibrium state G.

The calculation of this critical energy ΔV_{crit} can be performed after paragraph 4 with respect to the new equilibrium state G. The state of motion $z_0, \dot{z}_0$ is calculated after paragraph 5. Then the calculation of the energy V_0 with respect to G can be performed with (24). The values in table two show, that the loss of stability can be predicted also for a transition.

$$p_{(t)} = \Delta p \frac{t}{t_s} \quad , \quad t_0 \le t \le t_s = \frac{\Pi}{2}$$
$$p_{(t)} = \Delta p \quad , \quad t > t_s$$

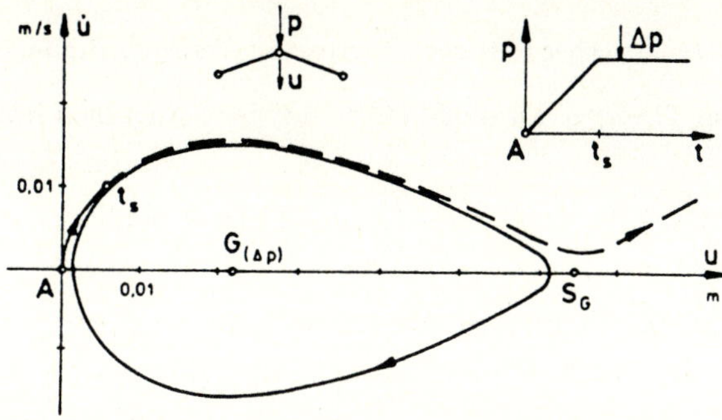

Fig. 5 : Motion under abrupt transition loads

Table 2

Δp	$10^7 \Delta V_{crit}$	$\leftrightarrow$	$10^7 T_e$	
$--$ 0,0152	1163	$<$	1198	$\Rightarrow$ unstable
$--$ 0,0151	1206	$>$	1189	$\Rightarrow$ stable

7 Conclusion

A method is given, which allows a simplified stability investigation of snapping and buckling structures under time–dependent loads. The degree of stability is calculated from the comparison of the load–induced energy and the critical energy stored in the structure. The deformation state related to the critical energy is calculated from the condition of vanishing secant–stiffness with respect to the initial load state of the structure. For different limit criteria it is possible to use other deformation state vectors without changing the method.

The results show, that the loss of stability is not related to the induced impulse but to the work of the perturbating loads, in this case the induced energy.

Acknowledgement : This paper was sponsored by Deutsche Forschungsgemeinschaft.

References

1. Koiter,W.T. : General Equations of Elastic Stability for thin Shells. Proc. Symp. in Honor of Lloyd H. Donnell, D.Muster Ed. . Houston : University of Houston Press 1966 187 - 228

2. Leipholz,H. : Stabilität elastischer Systeme. Karlsruhe : Verlag G.Braun 1980

3. Huseyin,K. : Nonlinear Theory of Elastic Stability. Leyden : Noordhoff Int. Publ. 1975

4. Thompson,J.M.T.; Hunt,G.W. : A General Theory of Elastic Stability. London : John Wiley & Sons 1973

5. Ziegler,H. : Die Stabilitätscriterien der Elastomechanik. Ingenieur Archiv 20 (1952) 49 - 56

6. Hayashi,C. : Nonlinear Oscillations in Physical Systems. New York : Mc Graw-Hill Book Company 1964

7. Thompson,J.M.T.; Stewart,H.B. : Nonlinear Dynamics and Chaos. Chichester : John Wiley & Sons 1986

8. Haken,H. : Order in Chaos. Computer Methods in Applied Mechanics and Engineering 52 (1984) 635 - 652

9. Koiter,W.T. : On the Thermodynamic Background of Elastic Stability Theory. Problems of Hydrodynamics and Continuum Mechanics. L.I.Sedov Anniversary Vol. SIAM . Philadelphia : 1969

10. Malkin,J.G. : Theorie der Stabilität einer Bewegung. München : Verlag R.Oldenbourg 1959

11. Dinkler,D. : Stabilität dünner Flächentragwerke bei zeitabhängigen Einwirkungen. Bericht Nr. 87 - 52, Institut für Statik - TU Braunschweig, Herausgeber Prof.Dr.-Ing.H.Duddeck. Braunschweig : Eigenverlag 1987

12. Dinkler,D. : Stabilität elastischer Tragwerke mit nichtlinearem Verformungsverhalten bei instationären Einwirkungen. Ingenieur-Archiv 59 (1989). Berlin : Springer Verlag

Part 3
New Finite Element Derivations for Nonlinear Shell Analysis

Analysis of Finitely-Deformed Shells Using Low-Order Mixed Models

A.F. Saleeb, T.Y. Chang, and S. Yingyeunyong

Department of Civil Engineering, University of Akron, Akron, OH 44325 USA

Summary

The general framework for the fully nonlinear analysis of shells is developed using two, low-order, mixed models of the quadrilateral (HMSH5) and triangular (HMSH3) type. An updated-Lagrange description is adopted, and the governing equations are derived based on a consistent linearization of an incremental modified Hellinger-Reissner mixed variational principle with independently-discretized displacement and strain fields. Effective solution algorithms are utilized for updating the element configuration in the presence of large nodal rotations in space, as well as for the "objective" time integration of its spatial stress (strain) fields when finite stretches occur. The results of a number of practical shell applications are also presented.

1. Introduction

The development of finite element models for the fully-nonlinear analysis of plates and arbitrary-curved shells, including the effects of both material (e.g. plasticity) and geometric nonlinearities (e.g. large displacement/rotations and finite stretches), has been a very active research area over the years, and especially in recent years [1,3-23,27]. This extensive work was prompted mainly by the many conceptual, theoretical, as well as computational difficulties and challenging problems involved in nonlinear shell formulations and solutions. In addition, there is an increased industrial need to perform large-scale computations utilizing these general shell models, such as sheet metal forming processes, progressive failure analysis (buckling, post-buckling and limit collapse states) of stiffened plates and shells, etc.

As evidenced by numerous recent works [1,3,7-12,14-22], the degenerated-shell approach has apparently continued to be the most popular in such general developments. In particular, important contributions here include many variants of the displacement-based formulations for nine-noded (biquadratic) elements [e.g., 10,12,14,16-20] and three-noded (linear) triangular models [e.g. 3,15], as well as a number of different hybrid/mixed developments in [21-27].

However, developing a fully nonlinear solution procedure for arbitrarily-curved shells is a very demanding task. Indeed, in such a general development, it is only a minimal requirement to use a *robust* linear shell element (i.e., rank-sufficiency, locking free, accurate stress predictions, insensitivity to geometric distortions, etc.) as a starting point. Additionally, several other fundamental issues must be carefully investigated. Confining attention to the quasistatic problem, the following are of utmost importance from a theoretical standpoint: (i) *consistent linearization* of the underlying weak, or variational, form of the governing equations [12,13,20,23,29]; (ii) treatment of *large rotations*, in both stiffness derivation as well as the configuration update procedure [16,20,27, 30-33]; (iii) use of objective *measures* of stress and strain, and their rates, suitable for the particular form of the nonlinear material model employed {e.g. hyperelasticity, hypoelasticity, plasticity, etc.) [34-40]; (iv) use of a valid method for stress integration (update) that maintains *incremental objectivity* in the presence of large rotations and stretches, when dealing with rate-type constitutive equations [37-43]; and (v) the ability to handle *nonconservative* loading [13,33], which often lead to nonsymmetric geometric stiffness contributions.

The objective in the present paper is to review some of our recent work dealing with the formulation of "simple", *low-order* nonlinear degenerated-shell models. To this end, we first summarize the development of a general theoretical framework for the analysis of finitely stretched and rotated shells using a hybrid/mixed method [27]. In particular, this is utilized here as a basis for deriving two representative nonlinear models of this type; i.e., HMSH3 (three-noded triangle) and HMSH5 (four-noded quadrilateral). Numerical results illustrating the performance of these elements in test cases are also reported in order to demonstrate the effectiveness of various proposed solution schemes.

We finally note that the present mixed elements HMSH3 and HMSH5 are "extended" nonlinear versions of their linear counterparts in [24-26]. The crucial point in their mixed formulations concerns the *judicious selection* of the parameters in the polynomial strain approximation. In particular, this was facilitated by the use of a set of *bubble functions* (i.e., kinematic degrees of freedom associated with an interior fifth node) in HMSH5

[24,25], or by imposing the so-called *relaxed* or *shear-edge* constraints for HMSH3 [26].

For conciseness, the scope of the present paper is limited to quasistatic problems with conservative-type loads. In addition, special instability phenomena and post-buckling responses are not addressed. Further, in all the applications presented, we utilize a "semi-linear" elastic isotropic model, thus enabling the comparison of our results with other independent solutions available in the literature.

2. Geometric and Kinematic Descriptions

2.1 Basic Hypotheses

The main assumptions underlying the present mixed-element developments are summarized as follows:

(i)   straight and inextensible shell "normals" (or "thickness" fibers).

(ii)   zero "through-thickness" normal stress (i.e., plane-stress assumption).

(iii) "small" *transverse* shear strains, but *large* membrane and bending strain components.

The first two simplifying hypotheses (i) and (ii) are typical in any degenerated shell model [e.g., 1,2,16], based on classical Mindlin/Reissner plate theories. However, note that, even though assumption (i) is utilized as a basis for deriving the element governing "stiffness" equations , fiber "thinning" effects due to large membrane strains can still be recovered in the computations; e.g., using the incremental "average-thickness" update procedure in [17].

On the other hand, certain important implications resulting from the additional assumption (iii) can be exploited to develop simplified and effective numerical algorithms [27]. From a practical standpoint, its justification is that "large" transverse shear strains are *not* typical in common thin or moderately-thick shell structures.

2.2 Coordinate Reference Frames

Three different types of Cartesian reference frames are defined here for convenience in subsequent derivations:

(1)   The *fixed* or global reference frame, with its orthonormal base vec-

tors $\underset{\sim}{e}_i (i=1,2,3)$, is used to define element geometry and its translational displacement DOF (degrees of freedom).

(2) A unique *local* fiber coordinate system is constructed at *each node*, with associated base vectors $\underset{\sim}{e}_i^f (i=1,2,3)$, where $\underset{\sim}{e}_3^f$ coincides with the nodal fiber. During the incremental analysis, this orthogonal fiber triad is continuously updated and used as a *moving basis* for defining the "finite" rotational DOF at the node.

(3) A *local* lamina system at *each integration point* in an element, $\underset{\sim}{e}_i^\ell (i=1,2,3)$, with $\underset{\sim}{e}_3^\ell$ taken to be normal to the lamina surface, and the other two in-plane lamina-tangent vectors $\underset{\sim}{e}_1^\ell$ and $\underset{\sim}{e}_2^\ell$ selected as in [27] (see Fig. 1).

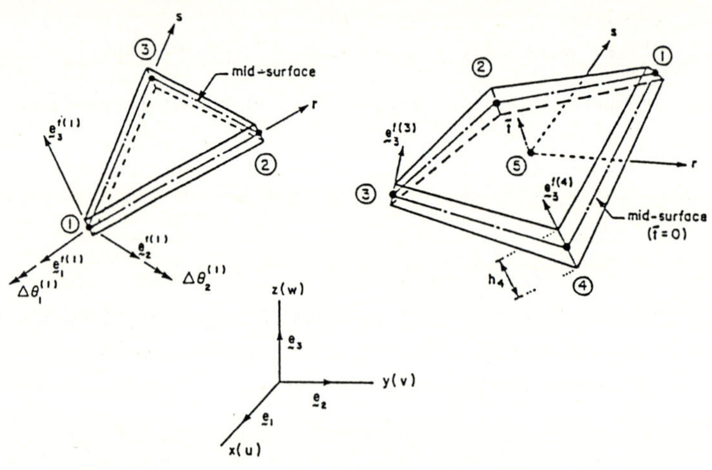

(a) Geometry, Displacement, and Fiber Basis

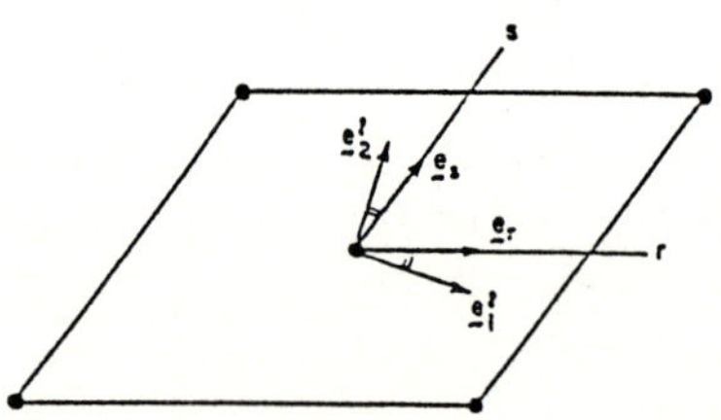

(b) Typical Lamina Coordinates and In-plane Skewness

Fig. 1. Typical Elements HMSH5 and HMSH3

This latter lamina basis rotates rigidly as the element deforms, and it is found to be most convenient for invoking the plane stress assumption in the e_3^{ℓ}-direction in all configurations as well as for ensuring the invariant property in the interpolation of the assumed strain field given later.

2.3 Geometry and Kinematics

With its midsurface taken as the reference surface, the position vector to an arbitrary point in the shell element at any time instant "t" is defined in terms of the natural coordinates (r,s,t) as follows:

$$
{}^{t}\underset{\sim}{x} = \sum_{k=1}^{n_e} N_k \; {}^{t}\underset{\sim}{x}_k + \frac{t}{2} \sum_{k=1}^{n_e} {}^{t}h_k \; N_k \; {}^{t}\underset{\sim}{e}_3^{f}(k)
\tag{2.1}
$$

where ${}^{t}\underset{\sim}{x}_k$, ${}^{t}\underset{\sim}{e}_3^{f}(k)$, and ${}^{t}h_k$ are the position vector, components of a unit pseudonormal (fiber) vector, and fiber-dimension (shell-thickness) parameter, respectively, at nodal point k on the reference surface. The $N_k(r,s)$ are the two-dimensional shape functions associated with node k [25,26], and n_e the number of nodes per element.

There are five DOF per node used to parameterize the element configuration in the shell space: three global translations (u,v,w), and two "fiber" rotations θ_1 and θ_2 about axes $\underset{\sim}{e}_1^{f}$ and $\underset{\sim}{e}_2^{f}$. Note that following [27], these latter nodal rotation freedoms are appropriately viewed here as generalized *finite* rotational coordinates of the Rodrigues-Euler type [30-33], thus providing a convenient, singularity-free (no drilling freedoms) parameterization for the director fields of the elements. In total, an HMSH5 (HMSH3) element thus contains 25 (15) DOF.

In the context of the present incremental analysis, three successive configurations, at time "0" (initial), "t" (current) and "(t+Δt)" (incremented or neighboring) are considered. Then, the total and incremental displacement fields of the element can be expressed as

$$
\underset{\sim}{u} = \sum_{k=1}^{n_e} N_k \; {}^{t}\underset{\sim}{u}_k + \frac{t}{2} \sum_{k=1}^{n_e} N_k \; ({}^{t}h_k \; {}^{t}\underset{\sim}{e}_3^{f}(k) - {}^{0}h_k \; {}^{0}\underset{\sim}{e}_3^{f}(k))
\tag{2.2}
$$

$$\Delta \underline{u} = \sum_{k=1}^{n_e} N_k \, \Delta\underline{u}_k + \frac{t}{2} \sum_{k=1}^{n_e} {}^t h_k \, N_k \, ({}^{t+\Delta t}\underline{e}_3^{f(k)} - {}^t\underline{e}_3^{f(k)}) \qquad (2.3)$$

where

$$\underline{u}_k = {}^t\underline{x}_k - {}^o\underline{x}_k \quad \text{and} \quad \Delta\underline{u}_k = {}^{t+\Delta t}\underline{x}_k - {}^t\underline{x}_k$$

The relation in Eq. (2.2) is directly employed to evaluate the total ele-ment displacements and their derivatives (i.e., total *geometric* "Almansi" strains in Eq. 3.3). In this, components of the "updated" director vec-tors, $\underline{e}_3^f$, in the configuration "t" are determined in terms of nodal rota-tions using the (geometrically-exact) update procedure of Sec. 4.4.

On the other hand, the relation in Eq. (2.3) provides the basis for de-riving the "linearized" governing equations of the element. To this end, the second term must first be expanded in terms of nodal rotation incre-ments, and here we utilize the following "linearized" kinematic approxima-tion:

$$\Delta \underline{u} = \sum_{k=1}^{n_e} N_k \, \Delta\underline{u}_k + \frac{t}{2} \sum_{k=1}^{n_e} {}^t h_k \, N_k \, (-\Delta\theta_1^k \, {}^t\underline{e}_2^{f(k)} + \Delta\theta \, {}^t\underline{e}_1^{f(k)}) \qquad (2.4)$$

for the *stiffness* evaluation of both elements considered. We particularly note that this leads to the well-known expression (i.e., similar to the "true" continuum case [1]) of a *single* geometric stiffness matrix (see Sec. 4.2).

Remark 2.1   Alternative forms for nonlinear kinematic approximations of the rotational term in Eq. (2.3) have been also employed in several recent studies [e.g. 20,23,31-33]. This led to the introduction of various *addi-tional* geometric stiffness contributions, which were found to be necessary in order to attain quadratic rate of (asymptotic) convergence in Newton-Raphson iterative schemes for solutions of large-rotation shell problems. However, as was demonstrated in [27], even with the single geometric stiffness based on (2.4) above, the same "good" convergence rate is also exhibited by the present mixed model HMSH5; the crucial point here is the "exact" updating of the orientations of nodal fiber triads (Sec. 4.4).

3. Variational Principle

A modified Hellinger-Reissner variational principle [e.g. 5,27,44] provides the starting point for the present incremental, step-by-step analysis. This takes the following form in *updated* Lagrangian (UL) description with "t" as reference; i.e., $\delta\Delta\tau_{HR} = 0$,

$$\Delta\tau_{HR} = \int_V \left[-\frac{1}{2}\Delta e^T c \Delta e + \sigma^T \Delta\hat{e} + \Delta e^T c \Delta\hat{e} - \Delta e^T c(e-\hat{e}) \right] dV - \Delta W \tag{3.1}$$

where Δe and $\Delta\hat{e} = \Delta\hat{e}$ (linear) + $\Delta\hat{\eta}$ (nonlinear) are, respectively, independently-assumed and geometric (from displacements) Green strain increments; e and $\hat{e}$ the corresponding total Alamansi strains; c the material stiffness; $\Delta\sigma = c\Delta e$ the Truesdell stress increment; σ the true (Cauchy) stress; and ΔW is the work of prescribed forces. The last term in the bracket of Eq. (3.1) is due to compatibility-mismatch [5,45].

Note that, for the purpose of implementing elements HMSH3 and HMSH5, all strain/stress vectors in the above will be defined with respect to the *lamina* basis at "t". In particular, this implies the use of a "reduced" (5x5) material matrix, in accordance with assumption (ii) of Sec. 2.1.

4. Finite Element Formulation

4.1 Strain-Field Discretization

In addition to displacement interpolation, a polynomial for Δe in the present mixed elements is also needed. To this end, we utilize the same specific *"least-order"* polynomial strain approximations for these elements proposed previously for linear analysis [24-26]. Thus, in tensor-component form, the incremental *lamina* strains $\Delta e'$ for *undistorted* geometry are:

For HMSH5 Element:

$$\Delta e'_{11} = \beta_1 + \beta_2 r + \beta_3 s + t(\beta_4 + \beta_5 r + \beta_6 s)$$

$$\Delta e'_{22} = \beta_7 + \beta_8 r + \beta_9 s + t(\beta_{10} + \beta_{11} r + \beta_{12} s)$$

$$\Delta e'_{12} = \beta_{13} + \beta_{14} t \tag{4.1}$$

$$\Delta e'_{23} = \beta_{15} + \beta_{16} r + \beta_{17} s$$

$$\Delta e'_{13} = \beta_{18} + \beta_{19} s + \beta_{17} r$$

202

For HMSH3 Element:

$$\Delta e'_{11} = \beta_1 + \beta_2 t$$
$$\Delta e'_{22} = \beta_3 + \beta_4 t$$
$$\Delta e'_{12} = \beta_5 + \beta_6 t \qquad (4.2)$$
$$\Delta e'_{23} = (2r-1)\beta_7 + (2r+2s-1)\beta_8$$
$$\Delta e'_{13} = (1-2s)\beta_9 + (1-2r-2s)\beta_8$$

Note that the above strain distribution needs to be further modified to account for geometric distortions; namely the "important" in-plane lamina (skewness) distortion. To this end, we use a "constant" Jacobian transformation as described in [24,27]. With this, we can finally write

$$\Delta\underline{e} = \underline{P}\ \Delta\underline{\beta} \qquad (4.3)$$

where $\Delta\underline{\beta}$ are generalized strain parameters, and $\underline{P}$ are "modified" strain-interpolation functions.

4.2 Element Stiffness Equations

In view of Eqs. (2.4) and (4.3), the appropriate "linearized" form of the variational principle in (3.1) is simply obtained as in the convectional "true" continuum case. This yields, after invoking the stationarity conditions with respect to $\Delta\underline{\beta}$, and then $\Delta\underline{q}$,

$$\Delta\underline{\beta} = \underline{H}^{-1}\left[\underline{G}\ \Delta\underline{q} + (\underline{Z}_1 - \underline{Z}_2)\right] \qquad (4.4)$$

where

$$\underline{H} = \int_V \underline{P}^T\ \underline{c}\ \underline{P}\ dV \ ; \quad \underline{G} = \int_V \underline{P}^T\ \underline{c}\ \underline{B}_L\ dV \qquad (4.5)$$

$$\underline{Z}_1 = \int_V \underline{P}^T\ \underline{c}\ \hat{\underline{e}}\ dV \ ; \quad \underline{Z}_2 = \int_V \underline{P}^T\ \underline{c}\ \underline{e}\ dV \qquad (4.6)$$

as well as the desired final stiffness relationships:

$$(\underset{\sim}{K}_L + \underset{\sim}{K}_{NL}) \; \Delta \underset{\sim}{q} = \underset{\sim}{Q} - (\underset{\sim}{Q}_1 + \underset{\sim}{Q}_2) \tag{4.7a}$$

$$\underset{\sim}{K}_L = \underset{\sim}{G}^T \; \underset{\sim}{H}^{-1} \; \underset{\sim}{G} \quad ; \quad \underset{\sim}{K}_{NL} = \int_V \underset{\sim}{B}_{NL}^T \; \bar{\sigma} \; \underset{\sim}{B}_{NL} \; dV \tag{4.7b}$$

$$\underset{\sim}{Q}_1 = \int_V \underset{\sim}{B}_L^T \; \sigma \; dV \quad ; \quad \underset{\sim}{Q}_2 = \underset{\sim}{G}^T \; \underset{\sim}{H}^{-1} \; (\underset{\sim}{Z}_1 - \underset{\sim}{Z}_2) \tag{4.7c}$$

see details in [27]. The $\underset{\sim}{K}_L$ is the element "linear" (or material) stiffness, $\underset{\sim}{K}_{NL}$ its geometric stiffness (shown in [27] to exhibit second-order "accuracy"), and the right-hand side of Eq. (4.7a) includes "correction" terms due to both equilibrium imbalance ($\underset{\sim}{Q}-\underset{\sim}{Q}_1$) as well as compatibility mismatch terms in $\underset{\sim}{Q}_2$.

4.3 Solution Procedure

Once assembled, the linearized equations above are utilized in the following *global* incremental-iterative full Newton-Raphson scheme:

$$({}^{t+\Delta t}\underset{\sim}{K}_L + {}^{t+\Delta t}\underset{\sim}{K}_{NL})^{(n)} \; \Delta \underset{\sim}{q}^{(n+1)} = {}^{t+\Delta t}\underset{\sim}{Q} - {}^{t+\Delta t}(\underset{\sim}{Q}_1 + \underset{\sim}{Q}_2)^{(n)} \tag{4.8}$$

for the solution for the incremental nodal displacements in the $(n+1)$th iteration within the time step "t"→"t+Δt". A displacement-type convergence criterion is adopted here (0.001 tolerance for the ratio of norms of incremental to total nodal displacement vectors).

4.4 Large-Rotation Configuration Update

Configuration update involves the calculations of (i) new nodal coordinates, as well as (ii) orientations of the associated fiber triads. Although (i) is trivial, (ii) is complicated by the non-vectorial character of finite space rotations. Here, we make use of the so-called *exponential mapping* algorithm for rotational transformation of vectors [30,33]; i.e. at the end of the $(n+1)^{th}$ iteration,

$$
\begin{bmatrix} e_1^{f(k)} \\[2mm] e_2^{f(k)} \\[2mm] e_3^{f(k)} \end{bmatrix}^{(n+1)}
=
\begin{bmatrix} 1-g_1\beta^2 & g_1\alpha\beta & -g_2\beta \\[2mm] g_1\alpha\beta & 1-g_1\alpha^2 & g_2\alpha \\[2mm] g_2\beta & -g_2\alpha & 1-g_1(\alpha^2+\beta^2) \end{bmatrix}
\begin{bmatrix} e_1^{f(k)} \\[2mm] e_2^{f(k)} \\[2mm] e_3^{f(k)} \end{bmatrix}^{(n)}
\tag{4.9}
$$

$$
g_1 = \frac{1 \cos \|\Delta\theta\|}{\|\Delta\theta\|^2} \quad ; \quad g_2 = \frac{\sin \|\Delta\theta\|}{\|\Delta\theta\|} \qquad \|\Delta\theta\| = (\alpha^2 + \beta^2)^{1/2}
\tag{4.10}
$$

where the *local* components of the rotation pseudovector $\Delta\theta_k^{(n+1)}$ along the fiber axes $e_i^{f(k)}$ (at configuration "n") are conveniently defined as $(\alpha,\beta,0)$.

4.5 Strain Update

The calculation of the updated *independent* (Almansi) strain field e at the quadrature points in the element is needed to evaluate the iterative compatibility mismatch "force" vector Q_2. A simple "approximate" procedure [27] is utilized here. In this, the updated e is given by a *push-forward* transformation [29] for the (covariant) tensor components of the total (incremented) strains using the *relative* deformation gradient for configurations (n+1) and (n). For convenience, the latter is written as follows, using the polar-decomposition theorem [29],

$$
F_{n+1} = {}_{n}^{n+1}F = \frac{\partial x(t_{n+1})}{\partial x(t_n)} = R_{n+1}\, U_{n+1}
\tag{4.11}
$$

where R and U are rigid-rotation and pure stretch tensors, respectively. With appropriate reference to lamina coordinate systems in different configurations, we may then show that

$$
e^{(n+1)} = U_{n+1}^{-1}(e^{(n)} + \Delta e^{(n+1)})\, U_{n+1}^{-1}
\tag{4.12}
$$

where now *all* components of tensors on the right-hand side are referred to lamina coordinates in (n), whereas $e^{(n+1)}$ are taken with reference to lamina axes in the new configuration (n+1).

<u>Remark 4.1</u> In keeping with the present mixed formulation, it is crucial to determine the relative stretches (as well as all other *pure* "strain-like" quantities) from the "true" strains. To this end, we use the following *approximation* [27]:

$$\underline{U}_{n+1} = (1 + \tfrac{1}{2}I_3^e)\underline{I} + (1 - \tfrac{1}{2}I_2^e)\Delta\underline{e}^{(n+1)} - \tfrac{1}{2}(1 - I_1^e)\Delta\underline{e}^{(n+1)}\Delta\underline{e}^{(n+1)} \qquad (4.13)$$

where $I_i^e = (i=1,2,3)$ are the invariants of $\Delta\underline{e}^{(n+1)}$.

This formula obviates the need for "costly" procedures to formally obtain the square-root of a positive-definite matrix, while it still maintains the condition of *same* principle axes for $\underline{U}_{n+1}$ and $\Delta\underline{e}^{(n+1)}$.

<u>Remark 4.2</u> However, when needed later (Sec. 5.1), $\underline{R}_{n+1}$ must be calculated from element displacement field. For "relatively" small strain increments, it can be approximated [27] by the coordinate-transformation matrix for lamina bases in configurations (n) and (n+1).

5. Stress Update

For a class of large-strain constitutive models, a general spatial *rate-form*, consistent with the variational statement in Eq. (3.1), is considered here (a superposed dot indicates a material time derivative)

$$\overset{o}{\sigma}_{ij} = c_{ijkl}\, d_{kl} \quad ; \quad \overset{o}{\underline{\sigma}} = \dot{\underline{\sigma}} - \underline{\ell}\,\underline{\sigma} - \underline{\sigma}\,\underline{\ell}^T + (d_{kk})\,\underline{\sigma} \qquad (5.1)$$

where $\overset{o}{\underline{\sigma}}$ is the (objective) Truesdell rate of Cauchy stress $\underline{\sigma}$, $\underline{d}$ the (spatial) deformation-rate tensor, $\underline{\ell}$ the velocity gradient, and $\underline{c}$ may generally depend on stress and/or deformation history (e.g. plasticity).

6. Sample Applications

We consider here several numerical simulations for HMSH3 and HMSH5. The former HMSH3 element is used in meshes of the cross-diagonal type [26] in all cases. In addition, all results are obtained assuming isotropic linear elastic material behavior, and *except* for some large-strain applications in Secs. 6.3 and 6.4, all other problems were solved using the small- strain/large-rotation assumption (Eq. 5.5).

6.1 Clamped Square Plate Under Uniform Load

This problem is taken from [16]. Figure 2 depicts the results of HMSH3 and HMSH5 utilizing the same 4x4 mesh as for the 4-noded (reduced-integration) model Q4-UI in [16]. The solutions from HMSH3 and Q4-UI are identical, and it is also interesting to note that the average number of iterations per load step (about 3) for all three elements was the same.

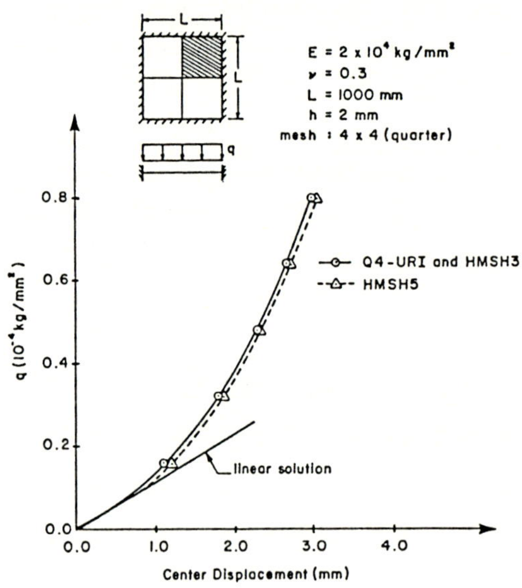

Fig. 2. A Clamped Square Plate Under Univorm Load

6.2 A Pinched Cylinder

This problem has often been used in establishing the viability of shell elements in linear analysis [2,25]. But to our knowledge, there is currently no "benchmark" solution available for its nonlinear response. Only one-eighth of the shell was modeled (because of symmetry) using a 16x16 element mesh for both HMSH5 and HMSH3 (Fig. 3). Ten equal load increments were applied to arrive at the full load of 750 (about four times the load utilized in linear solution). A total of 39 iterations were needed for HMSH5 and 44 for HMSH3.

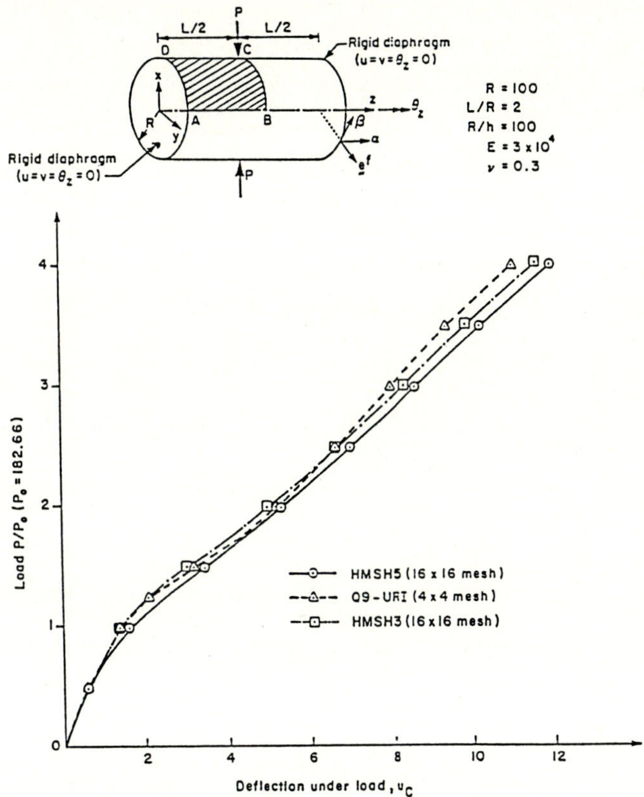

Fig. 3. A Pinched Cylinder With Rigid Diaphragms

6.3 A Pinched Hemisphere

This is another "obstacle test" in linear analysis. It is also an excellent test of the ability of an element to handle "truly" three-dimensional finite rotations. Using symmetry, one quadrant of the shell is modeled with 10x10 mesh for HMSH5 and 16x16 mesh for HMSH3. The total load F=100 is considered for small-strain/large-rotation analysis (recall that F=1 is typically used for the linear case). Only two loading steps are required for HMSH5 from F=0→10 and F=10→100 with a total of 13 iterations.

For HMSH3, three loading steps were necessary for convergence (F=0→10, F=10→50, and F=50→100) with a total of 16 iterations. Results are shown in Fig. 4, together with the solution reported in [23] using the resultant-based shell element with a 16x16 mesh.

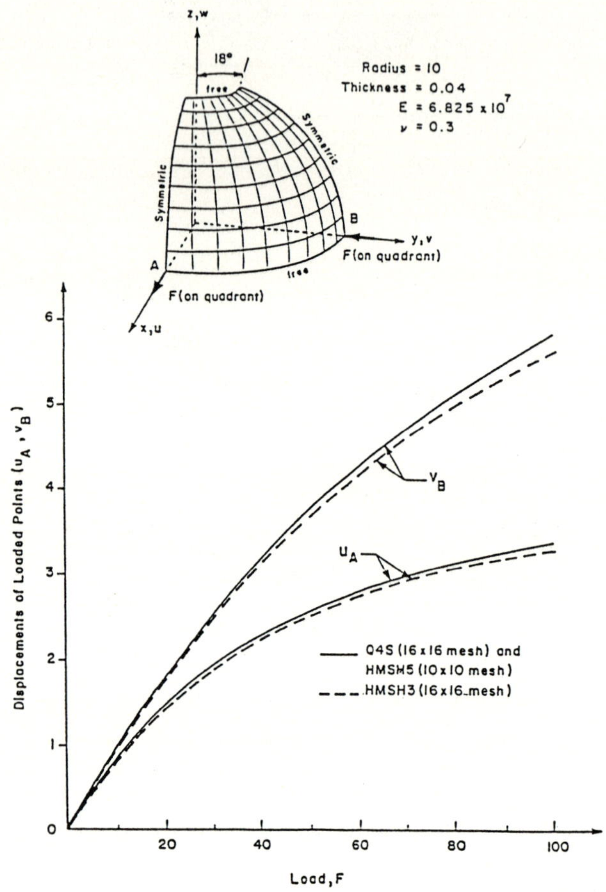

Fig. 4. A Pinched Hemispherical Shell

In addition, a large-strain analysis using the algorithm of Eqs. (5.4) and (5.6) and the same 10x10 mesh for HMSH5 was also performed for the above shell subjected to a very high load level F=900. We utilized two and 8 equal load steps for F=0→100 and F=100→900, respectively. Significant deformations occurred in this case; e.g. at the final load level (F=900), u_A=4.922 and v_B=12.709, with a total rotation of nearly 123 degrees at B.

6.4 Finite-Strain Analysis of Homogeneous Deformations

Finally, we consider here two of the commonly-used tests for the analysis of large homogeneous deformations [34,38,39]; i.e., (i) finite extension (Fig. 5); and (ii) finite "simple" shear (Fig. 6). In both cases, *plane-strain* conditions were imposed. The numerical results were obtained by a 2x2 finite element mesh of HMSH5 for the initial unit cube. The *load* control was employed, with 12 load increments corresponding to stretch ratio $\lambda_1 = 1 \rightarrow 2.2$ in (i), and 10 increments for shear "strains" $\tan \gamma = 0 \rightarrow 1$ in case (ii), requiring a total of 20 and 12 iterations, respectively. When used with the same mesh and load-incrementing scheme, HMSH3 gave almost identical results (not shown in figures).

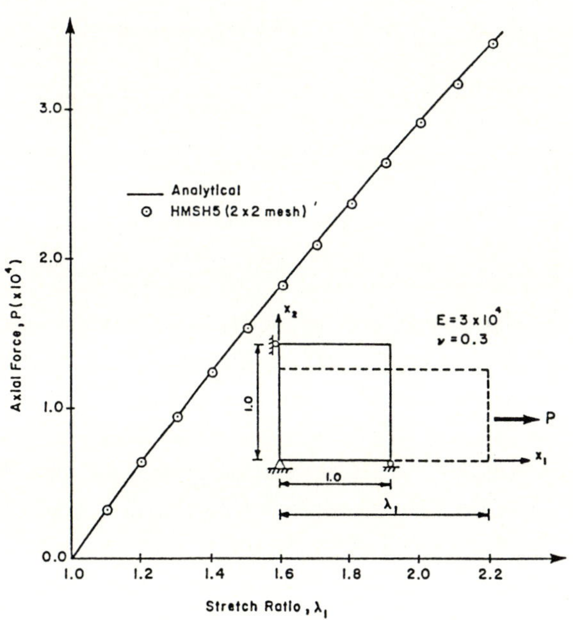

Fig. 5. Homogeneous Finite Extension

7. Conclusions

We presented here the fully-nonlinear formulation of curved shells using the simple mixed models HMSH3 and HMSH5. Several noteworthy aspects are included. A careful selection is made for the polynomial functions in the strain assumption, thus leading to robust elements (i.e., kinematically stable, free from locking, etc.). A geometrically-exact procedure is

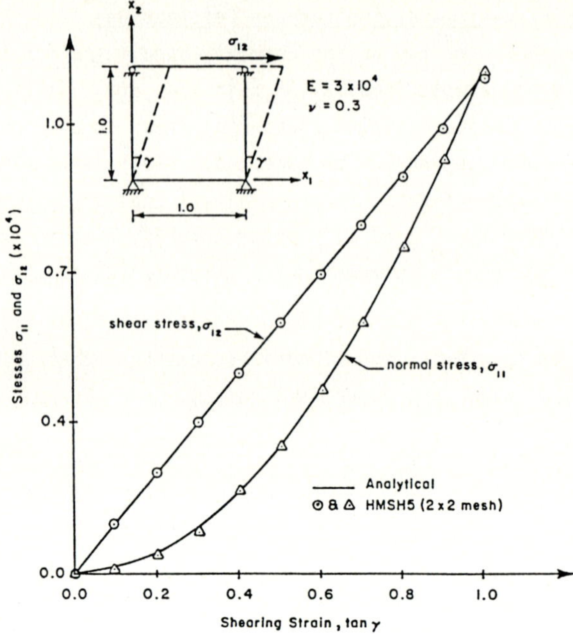

Fig. 6. Homogeneous Finite Simple Shear

utilized for element configuration update with finite nodal rotations. Even with a "single" geometric stiffness matrix, this was shown to be capable of attaining quadratic convergence rate in practical shell applications [27]. For updating of the spatial stress field in the presence of large strains, "objective" generalized midpoint schemes were developed by making use of the polar-decomposition of deformation gradient, which are in keeping with the underlying mixed method.

Several numerical simulations have been presented to demonstrate the effectiveness and practical usefulness of the proposed formulation.

<u>Acknowledgement</u>

This work is supported by National Science Foundation, under Research Grant Number EET-8714628.

<u>References</u>

1. K.J. Bathe, <u>Finite Element Procedures in Engineering Analysis</u>, Prentice Hall, Englewood Cliffs, NJ, 1982.

2. T.J.R. Hughes, <u>The Finite Element Method</u>, Prentice Hall, Englewood Englewood Cliffs, NJ, 1987.

3. J.H. Argyris, P.C. Dunne, G.A. Malejannakis, and E. Shelke, "A Simple Triangular Facet Shell Element with Applications to Linear and Nonlinear Equilibrium and Elastic Stability Problems", <u>Comp. Meth. Appl. Mech. Engr.</u>, Vol. 10, No. 3, 371-403, and Vol. 11, No. 1, 97-131, 1977.

4. J.H. Argyris, P.C. Dunne, and D.W. Scharpf, "On Large Displacement-Small Strain Analysis of Structures with Rotational Degrees of Freedom", <u>Comp. Meth. Appl. Mech. Engr.</u>, Vol. 14, 401-451, and Vol. 15, 99-135, 1978.

5. P.L. Boland and T.H.H. Pian, "Large Deflection Analysis of Thin Elastic Structures by the Assumed Stress Hybrid Finite Element Method", <u>Comp. and Struct.</u>, Vol. 1, 1-12, 1977.

6. G. Horrigmoe, "Hybrid Stress Finite Element Model for Nonlinear Shell Problems", <u>Int. J. Num. Meth. Engr.</u>, Vol. 12, 1819-1839, 1978.

7. G. Horrigmoe and P.G. Bergan, "Instability Analysis of Free Form Shells by Flat Finite Elements", <u>Comp. Meth. Appl. Mech. Engr.</u>, Vol. 16, 11-35, 1978.

8. W. Kanok-Nukulchai, "A Simple and Efficient Finite Element for General Shell Analysis", <u>Int. J. Num. Meth. Engr.</u>, Vol. 14, 179-200, 1979.

9. W. Kanok-Nukulchai and W.K. Wong, "Element-Based Lagrangian Formulation for Large-Deformation Analysis", <u>Comp. Struct.</u>, 30, 967-974, 1988.

10. B. Krakeland, "Nonlinear Analysis of Shells Using Degenerate Isoparametric Elements", in <u>Finite Elements in Nonlinear Mechanics</u>, Vol. 1, P.G. Bergan et al (Eds.), Tapir Publishers, Trondheim, Norway, 265-284, 1978.

11. A.K. Noor and S.J. Hartley, "Nonlinear Shell Analysis via Mixed Isoparametric Elements", <u>Comp. and Struct.</u>, Vol. 7, 615-626, 1977.

12. E. Ramm, "A Plate/Shell Element for Large Deflections and Rotations", in <u>Formulations and Computational Algorithms in Finite Element Analysis</u>, K.J. Bathe, J.T. Oden and W. Wunderlich (Eds.), M.I.T. Press, Cambridge, MA, 1977.

13. J.H. Argyris and Sp. Symeonides, "Nonlinear Finite Element Analysis of Elastic Systems Under Nonconservative Loading-Natural Formulation. Part I. Quasistatic Problems", <u>Comp. Meth. Appl. Mech. Engr.</u>, Vol. 26, 75-123 and 373-383, 1981.

14. K.J. Bathe and S. Bolourchi, "A Geometric and Material Nonlinear Plate and Shell Element", <u>Comp. and Struct.</u>, Vol. 11, 23-48, 1980.

15. P.G. Bergan and M.K. Nygard, "Nonlinear Shell Analysis Using Free-Formulation Finite Elements", in _Finite Element Methods for Nonlinear Problems_, P.G. Bergan, K.J. Bathe and W. Wunderlich (Eds.), Springer-Verlag, Berlin, 1985.

16. T.J.R. Hughes and W.K. Liu, "Nonlinear Finite Element Analysis of Shells Part I - Three Dimensional Shells", _Comp. Meth. Appl. Mech. Engr._, Vol. 26, 331-362, 1981.

17. T.J.R. Hughes and E. Carnoy, "Nonlinear Finite Element Shell Formulation Accounting for Large Membrane Strains", _Comp. Meth. Appl. Mech. Engr._, Vol. 39, 69-82, 1983.

18. J. Oliver and E. Onate, "A Total Lagrangian Formulation for the Geometrically Nonlinear Analysis of Structures Using Finite Elements, Part I. Two-Dimensional Problems: Shell and Plate Structures", _Int. J. Num. Meth. Engr._, Vol. 20, 2253-2281, 1984.

19. H. Parisch, "Large Displacements of Shells Including Material Nonlinearities", _Comp. Meth. Appl. Mech. Engr._, Vol. 27, 183-214, 1981.

20. K.S. Surana, "Geometrically Nonlinear Formulation for Curved Shell Elements", _Int. J. Num. Meth. Engr._, Vol. 19, 581-615, 1983.

21. A.K. Noor and C.M. Anderson, "Mixed Models and Reduced/Selective Integration Displacement Models for Nonlinear Shell Analysis", _Int. J. Num. Meth. Engr._, Vol. 18, 1429-1454, 1982.

22. J.J. Rhiu and S.W. Lee, "A Nine Node Finite Element for Analysis of Geometrically Nonlinear Shells", to appear, _Int. J. Num. Meth. Engr._, 1988.

23. J.C. Simo, D.D. Fox and M.S. Rifai, "Formulation and Computational Aspects of a Stress Resultant Geometrically Exact Shell Model", in _Computational Mechanics '88: Theory and Applications, Vol. 1_, S.N. Atluri and G. Yagawa (Eds.), Springer-Verlag, Berlin, 26.ii.1-26.ii.9, 1988.

24. A.F. Saleeb and T.Y. Chang, "An Efficient Quadrilateral Element for Plate Bending Analysis", _Int. J. Num. Meth. Engr._, Vol. 24, 1123-1155, 1987.

25. A.F. Saleeb, T.Y. Chang and W. Graf, "A Quadrilateral Shell Element Using a Mixed Formulation", _Comp. and Struct._, Vol. 26, No. 5, 787-803, 1987.

26. A.F. Saleeb, T.Y. Chang and S. Yingyeunyong, "A Mixed Formulation of C^o Linear Triangular Plate/Shell Element - The Role of Edge Shear Constraints", _Int. J. Num. Meth. Engr._, Vol. 26, 1101-1128, 1988.

27. A.F. Saleeb, T.Y. Chang, W. Graf and S. Yingyeunyong, "Hybrid/Mixed Model for Nonlinear Shell Analysis and Its Applications to Large Rotation Problems", to appear, _Int. J. Num. Meth. Engr._, 1989.

28. T. Belytschko, W.K. Liu and B.E. Englemann, "The Gamma Elements and Related Developments", in _Finite Element Methods for Plates and Shell Structures-Vol. 1: Element Technology_, T.J.R. Hughes and E. Hinton (Eds.), Pineridge Press, Swansea, 316-347, 1986.

29. J.E. Marsden and T.J.R. Hughes, _Mathematical Foundations of_ Elasticity, Prentice Hall, Englewood Cliffs, NJ, 1983.

30. J.H. Argyris, "An Excursion into Large Rotations", _Comp. Meth. Appl. Mech. Engr._, Vol. 32, 85-155, 1982.

31. E. Dvorkin, E. Onate, and J. Oliver, "On a Non-Linear Formulation for Curved Timoshenko Beam Elements Considering Large Displacement/Rotation Increments", _Int. J. Num. Meth. Engr._, 26, 1597-1613, 1988.

32. A. Cardona and M. Geradin, "A Beam Finite Element Non-Linear Theory with Finite Rotations", _Int. J. Num. Meth. Engr._, 26, 2403-2438, 1988.

33. J.C. Simo and L. Vu-Quoc, "A Three-Dimensional Finite Strain Rod Model, Part II: Computational Aspects", _Comp. Meth. Appl. Mech. Engr._, Vol. 58, 79-116, 1986.

34. S.N. Atluri, "On Constitutive Relations at Finite Strain: Hypoelasticity and Elastoplasticity with Isotropic or Kinematic Hardening", _Comp. Meth. Appl. Mech. Engr._, Vol. 43, 137-171, 1984.

35. J.C. Simo and K.S. Pister, "Remarks on Rate Constitutive Equations for Finite Deformation Problems: Computational Implications", _Comp. Meth. Appl. Mech. Engr._, Vol. 44, 201-215, 1984.

36. J.C. Simo, "A Framework for Finite Strain Elastoplasticity Based on Maximum Plastic Dissipation and the Multiplicative Decomposition: Part I Continuum Formulation", _Comp. Meth. Appl. Mech. Engr._, Vol. 66, 199-219, 1988.

37. T.J.R. Hughes and J. Winget, "Finite Rotation Effects in Numerical Integration of Rate Constitutive Equations Arising in Large-Deformation Analysis", _Int. J. Num. Meth. Engr._, Vol. 15, 1862-1867, 1980.

38. J.C. Nagtegaal and J.E. De Jong, "Some Computational Aspects of Elastic-Plastic Large Strain Analysis", _Int. J. Num. Meth. Engr._, Vol. 17, 15-41, 1981.

39. P.M. Pinsky, M. Ortiz, and K.S. Pister, "Numerical Integration of Rate Constitutive Equations in Finite Deformation Analysis", _Comp. Meth. Appl. Mech. Engr._, Vol. 40, 137-158, 1983.

40. J.H. Argyris, J.S. Doltsinis, P.M. Pimenta and H. Wustenberg, "Thermomechanical Response of Solids at High Strains-Natural Approach", _Comp. Meth. Appl. Mech. Engr._, Vol. 32, 3-57, 1982.

41. K.W. Reed and S.N. Atluri, "Analysis of Large Quasistatic Deformations of Inelastic Bodies by a New Hybrid-Stress Finite Element Algorithm", _Comp. Meth. Appl. Mech. Engr._, Vol. 39, 245-295, and Vol. 40, 171-198, 1983.

42. K.W. Reed and S.N. Atluri, "Constitutive Modeling and Computational Implementations for Finite Strain Plasticity", <u>Int. J. Plasticity</u>, Vol. 3, 163-191, 1985.

43. R. Rubinstein and S.N. Atluri, "Objectivity of Incremental Constitutive Relations over Finite Time Steps in Computational Finite Deformation Analysis", <u>Comp. Meth. Appl. Mech. Engr.</u>, Vol. 36, 277-290, 1983.

44. S.N. Atluri and H. Murakawa, "New General and Complementary Energy Theorems, Finite Strain, Rate Sensitive Inelasticity and Finite Elements: Some Computational Studies", in <u>Nonlinear Finite Element Analysis in Structural Mechanics</u>, W. Wunderlich, E. Stein, and K.J. Bathe (Eds.), Springer, Berlin, 28-48, 1981.

45. T.H.H. Pian, "Variational Principles for Incremental Finite Element Methods", <u>Journal of the Franklin Institute</u>, Vol. 302, 473-488, 1976.

A New Mixed Finite Element for Analysis of Axisymmetric Inelastic Shells

F. G. KOLLMANN

Fachgebiet Maschinenelemente und Getriebe
Technische Hochschule Darmstadt, Darmstadt, West Germany

V. BERGMANN

Department of Theoretical and Applied Mechanics
Cornell University, Ithaca, N.Y., USA

Summary

Using a general geometrically linear theory of inelastic shells by Kollmann and
Mukherjee a mixed finite element model is formulated. This element contains strains
and displacements as basic unknowns. It is specialized for an axisymmetrically loaded
conical geometry as originally proposed by Zienkiewicz and coworkers for elastic shells.
Numerical results for elastically and inelastically deformed shells are presented, where
inelastic deformations are described by Hart's constitutive model.

Introduction

In this paper a new finite element is presented which is designed for inelastic analysis of

axisymmetric plates and shells. The element is based on a general theory of inelastic

shells by Kollmann and Mukherjee [1]. It is of the mixed type and contains displace-

ments and strains as the primary unknowns. The inelastic material behavior is gov-

erned by a rate–type constitutive model as proposed e.g. by Hart [2], Miller [3,4],

Robinson [5] and Walker [6]. Due to the mathematical structure of such models it is

important to capture the stresses with high accuracy since small changes of the stresses

can result in large changes of the inelastic strain rates.

Almost all prior research on numerical analysis of inelastic shells is based on the finite

element method (FEM). Cormeau [7] uses a displacement formulation and describes

the inelastic behavior by an associated von Mises type flow rule. Hughes and Liu [8,9]

start from three–dimensional nonlinear continuum mechanics and construct a specific

shell theory which is tailored toward FE procedures. They assume a very general anisotropic viscoplastic model and solve an impressive number of examples. Parisch [10] also starts from three–dimensional continuum mechanics and develops a layered model with piecewise linearly distributed stresses through the thickness.

A shell theory in the sense of this paper is a strictly two–dimensional theory which contains only vectors and tensor fields which are functions of curvilinear coordinates on the midsurface S of the undeformed shell. Therefore, the covariant or contravariant basis vectors of the shell midsurface and the unit normal vector on S are employed. In every shell theory, some basic assumptions leading to a consistent formulation have to be specified. For linear elastic shells these assumptions concern the displacement vector (leading to an approximation of the strain tensor) and the stress tensor. Specifically, a linear distribution of stresses over the shell thickness is assumed a priori. However, such an assumption is not admissible for inelastic shells. The distribution of stresses over the shell thickness in an inelastically deformed shell can deviate considerably from the elastic one. Due to inelastic effects the spatial stress field is redistributed with time and changes from a linear to a nonlinear function of the thickness coordinate.

FE models in solid mechanics are frequently based on a variational principle. The rate form of the potential energy principle leads in the FE discretization to a standard displacement formulation where the total strain rates are computed from the velocities by the discrete strain–displacement operator. It is a well known fact that in any displacement formulation the strains and stresses are less accurate than the displacements. Therefore, in this paper we apply a functional which contains only kinematical quantities, velocities and strain rates, as primary variables to be varied independently. Such a principle has been suggested for elastic bodies by Oden and Reddy [11]. A corresponding formulation for elastic shells has been suggested by Lee and Pian [12]. They mention the advantages of this principle for inelastic analysis of structures but do not include the inelastic terms. Mukherjee and Kollmann [13] have independently rediscovered this variational principle and extended it to inelastic deformation. Kollmann and

Mukherjee [1] have developed a shell theory starting from this principle which is the basis for the present paper. They give further reasoning for the advantages of this variational principle in inelastic shell analysis.

Basic Shell Theory

In the following, as far as possible, vectors and tensors are written in direct notation. Where necessary, vector and tensor components referred to the covariant and contravariant basis vectors of the undeformed shell midsurface and the unit normal vector are introduced. Latin indexes range from 1 to 3, Greek ones from 1 to 2. Superposed dots indicate time derivatives. A single vertical slash denotes covariant differentiation with respect to the fundamental metric tensor of the undeformed shell midsurface.

A fundamental constitutive assumption is that the total strain rate tensor $\dot{\gamma}$ can be decomposed additively into an elastic and inelastic part,

$$\dot{\gamma} = \dot{\gamma}^E + \dot{\gamma}^N . \tag{1}$$

The tensor $\dot{\gamma}^E$ of the elastic strain rates is related to the stress rate tensor $\dot{\sigma}$ by Hooke's law. In the isothermal case the inelastic strain rates obey an evolution law of the following form

$$\dot{\gamma}^N = \dot{\gamma}^N (\sigma, q) . \tag{2}$$

σ is the current value of the stress tensor and q denotes a set of suitably selected, but otherwise unspecified, internal state variables for which evolution laws also exist.

$$\dot{q} = \dot{q} (\sigma, q) \tag{3}$$

We consider an arbitrary shell and denote its undeformed midsurface by S and its lateral faces by S^+ and S^-. For simplicity we assume a constant shell thickness h. We introduce curvilinear coordinates θ^α on S. Then covariant and contravariant basis vectors a_α and a^β can be defined on S. The first and second fundamental tensor on S are denoted as a and b, respectively, and b is the determinant of b. The covariant basis vectors g_α of the shell space B can be obtained from the basis vector a_α by the linear transformation

$$g_\alpha = \mu\, a_\alpha \tag{4}$$

where the shifter tensor μ is given by

$$\mu := I - \xi\, b \ . \tag{5}$$

I is the identity tensor and ξ a normal coordinate measured in the direction of the unit normal vector a_3 on S. The determinant μ of the shifter tensor μ is given by

$$\mu = 1 - \xi\, \mathrm{tr}\, b + \xi^2\, b \ , \tag{6}$$

where tr denotes the trace operator.

Kollmann and Mukherjee [1] assume the following representation of the displacement vector u^* of the shell space

$$u^* := u + \xi\, w \ . \tag{7}$$

Here u is the displacement vector of the shell midsurface and w the difference vector. This is the simplest representation of the displacement vector u^* which can account for transverse shear strains of the shell.

Denoting the strain tensor of the shell space by $\overset{*}{\gamma}$, Kollmann and Mukherjee [1] have derived the following representations for its components

$$\mu \, \overset{*}{\gamma}_{\alpha\beta} = e_{\alpha\beta} + \xi \, \kappa_{\alpha\beta} \, ,$$

$$\mu \, \overset{*}{\gamma}_{\alpha 3} = \gamma_\alpha + \xi \, \rho_\alpha \, , \tag{8}$$

$$\overset{*}{\gamma}_{33} = \gamma_3 \, .$$

They also give the strain–displacement relations.

As in classical linear elastic shell theory it is assumed that the normal stress σ_{33} and normal strain γ_{33} vanish, where the latter implies $w_3 = 0$.

To obtain a compact representation [1] lines of principal curvature on the shell midsurface as coordinate lines are used and physical components (denoted by $< \cdot >$) are introduced. Since matrix representation is convenient, the following vectors are defined in matrix form.

$$\dot{u}^T : = \left[\dot{u}_{<i>} \right] \qquad \dot{w}^T : = \left[\dot{w}_{<i>} \right]$$

$$\dot{e}^T : = \left[\dot{e}_{<\alpha\beta>} \right] \qquad \dot{\kappa}^T : = \left[\dot{\kappa}_{<\alpha\beta>} \right] \qquad \dot{\phi}^T : = \left[\dot{\gamma}_{<\alpha>}, \dot{\rho}_{<\alpha>} \right] \tag{9}$$

The strain–displacement relations for the shell can be written as

$$\dot{e} = L_{eu} \dot{u} \, , \tag{10}$$

$$\dot{\kappa} = L_{\kappa u} \dot{u} + L_{\kappa w} \dot{w} \, , \tag{11}$$

$$\dot{\phi} = L_{\phi u}\dot{u} + L_{\phi w}\dot{w} .$$

(12)

Explicit expressions of the operator matrices L_{eu}, $L_{\kappa u}$, $L_{\kappa w}$, $L_{\phi u}$ and $L_{\phi w}$ have been derived by Kollmann and Mukherjee [11].

Next, the boundary conditions must be specified. The boundary ∂B of the shell B can be decomposed as

$$\partial B = S^+ \cup S^- \cup S_e ,$$

(13)

where S_e is the edge of the shell. Its intersection with the shell midsurface S gives the edge curve s. In shell theory, only tractions can be prescribed on S^+ and S^-. But on disjoint subsets s_{eu} and $s_{e\sigma}$ of the edge curve s_e displacements and tractions can be prescribed, respectively.

On the edge curve s_{eu} we prescribe

$$\dot{u} = \dot{\tilde{u}} , \qquad \dot{w} = \dot{\tilde{w}} ,$$

(14)

where $\dot{\tilde{u}}$ and $\dot{\tilde{w}}$ are prescribed functions of space and time.

The traction boundary conditions are of a more involved nature. Here we only consider the lateral surface S^+ and S^-. For convenience the following resultant loads and couple loads are introduced

$$\dot{l}_{<i>} := \left[\dot{\tilde{\sigma}}_{<3i>}\mu \right]_{\xi = -h/2}^{\xi = h/2}$$

(15)

$$\dot{m}_{<i>} := \left[\dot{\tilde{\sigma}}_{<3i>} \xi \mu \right]_{\xi = -h/2}^{\xi = h/2} \tag{16}$$

These quantities can be represented in the following vector form

$$i^T := \left[i_{<i>} \right], \qquad \dot{m}^T := \left[\dot{m}_{<i>} \right]. \tag{17}$$

The inelastic strain rates, $\dot{\gamma}^{*N}_{<ij>}$, contribute to the following rates of inelastic pseudo-forces and pseudo–moments

$$\dot{N}^N_{<ij>} := 2G \int_{-h/2}^{h/2} \dot{\gamma}^{*N}_{<ij>} \, d\xi \,, \tag{18}$$

$$\dot{M}^N_{<ij>} := 2G \int_{-h/2}^{h/2} \dot{\gamma}^{*N}_{<ij>} \, \xi d\xi \,. \tag{19}$$

The rates of the inelastic pseudo-forces and pseudo–moments contribute to

$$\dot{N}^{NT} := \left[\dot{N}^N_{<\alpha\beta>} \right], \qquad \dot{M}^{NT} := \left[\dot{M}^N_{<\alpha\beta>} \right],$$

$$Q^{NT} := \left[\dot{N}^N_{<\alpha3>}, \dot{M}^N_{<\alpha3>} \right]. \tag{20}$$

To obtain a more concise representation, generalized displacement rates $\dot{v}$ and strain rates $\dot{\gamma}$ are introduced

$$\dot{v}^T := [\dot{u}, \dot{w}] \,, \tag{21}$$

$$\dot{\gamma}^T := [\dot{e}, \dot{\kappa}, \dot{\phi}] \,. \tag{22}$$

The concise version of the variational principle reads

$$\delta\left\{\int_S \left[\tfrac{1}{2}\,\dot{\gamma}^T D_{\gamma\gamma}\dot{\gamma} - \dot{\gamma}^T D_{\gamma\gamma} L_{\gamma v}\dot{v} + \dot{F}_L^T\dot{v} + \dot{F}^{N^T} L^N_{\gamma v}\dot{v}\right] dS\right\} = 0 \,. \tag{23}$$

Here $D_{\gamma\gamma}$ is a generalized elasticity matrix, $L_{\gamma v}$ a generalized strain–displacement operator and $L^N_{\gamma v}$ a linear operator. Details are given by Kollmann and Bergmann [14,15].

Mixed Finite Element Model

First, the discrete version of eqn.(23) is derived. We admit different sets of shape functions for the strain rates and displacement rates. However, at this stage no specific choice of shape functions for displacements and strains is introduced.

Denoting the nodal displacement rates and strain rates by $\dot{\hat{v}}$ and $\dot{\hat{\gamma}}$, respectively, leads to representations analogous to eqns.(21) and (22). We introduce the matrix N of shape functions for the displacement rates and the matrix of shape functions $\bar{N}$ for the strain rates. The discrete representations of the displacement rate and strain rate field, respectively, are given by

$$\dot{v} = N\,\dot{\hat{v}}\,, \tag{24}$$

$$\dot{\gamma} = \bar{N}\,\dot{\hat{\gamma}}\,. \tag{25}$$

The discrete version of eqn.(23) on the element level follows by standard methods as

$$K_{\gamma\gamma}\dot{\hat{\gamma}} - K_{\gamma v}\dot{\hat{v}} = 0_{\text{sx1}}\,, \tag{26}$$

$$-K_{\gamma v}^T\dot{\hat{\gamma}} = -\dot{P}_L - \dot{P}_N\,. \tag{27}$$

In eqns.(26) and (27) the following quantities have been introduced

$$K_{\gamma\dot\gamma} = \int_S \bar{N}^T D_{\gamma\gamma} \bar{N}\, dS \tag{28}$$

$$K_{\gamma\dot v} = \int_S \bar{N}^T D_{\gamma\gamma} L_{\gamma v} N\, dS \tag{29}$$

$$\dot{P}_L = \int_S N^T \dot{F}_L\, dS \tag{30}$$

$$\dot{P}_N = \int_S N^T L_{\gamma v}^{N^T} \dot{F}^N\, dS \tag{31}$$

The left–hand side of the system (26) and (27) it is not necessarily positive–definite, since the functional in eqn.(23) becomes stationary but does not take a minimum. On the other hand, the matrix $K_{\gamma\gamma}$ is positive definite, since the term $\gamma^T K_{\gamma\gamma}\, \gamma$ represents the discrete approximation of the strain energy density of the shell.

<u>Axisymmetric Conical Shell Element</u>

In this section we give details of an axisymmetric conical element. A similar element for a displacement formulation has been used by Zienkiewicz, Bauer, Morgan and Onate [16]. Figure 1 shows a cross-section of the conical element. L is the length of the element, h the constant thickness of the shell, R the mean radius, β the angle of aperture of the cone, and s the arclength on the generator of the cone measured from the apex. Finally ϕ is an angle measured on a circle formed by the intersection of the conical surface and any plane orthogonal to the axis. We introduce s and ϕ as curvilinear coordinates on the conical surface S. The geometric objects and strain displacement operators for the axisymmetric case are presented in [14].

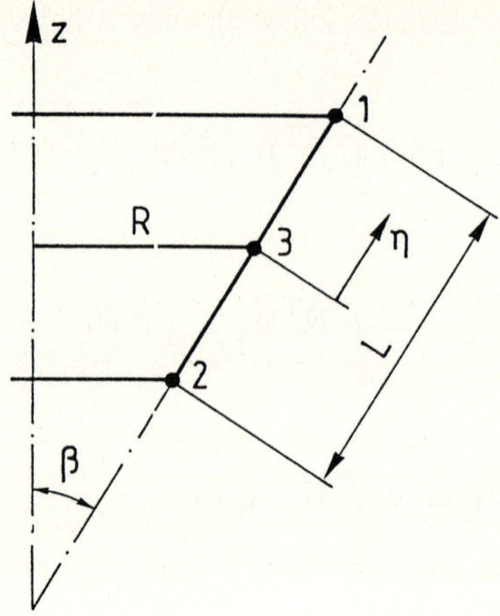

Fig.1. Conical finite shell element

It is convenient to introduce a dimensionless coordinate η by the transformation

$$s = L \left[\frac{\eta}{2} + \frac{R}{L \sin\beta} \right] . \tag{32}$$

We use quadratic shape functions $N^i(\eta)$ both for the displacements and the strains

$$N^1(\eta) = \frac{1}{2} (\eta^2 + \eta) ,$$

$$N^2(\eta) = \frac{1}{2} (\eta^2 - \eta) , \tag{33}$$

$$N^3(\eta) = 1 - \eta^2 .$$

This leads to the following interpolation schemes

$$\dot{u} = N\hat{\dot{u}} , \qquad \dot{w} = N\hat{\dot{w}} , \tag{34}$$

$$\dot{e} = N\dot{\hat{e}}, \qquad \dot{\kappa} = N\dot{\hat{\kappa}}, \qquad \dot{\phi} = N\dot{\hat{\phi}}. \tag{35}$$

The matrices N and $\bar{N}$ have standard representations.

The computation of the matrices $K_{\gamma\gamma}$ and $K_{\gamma v}$ and the load term $\dot{P}_L$ is standard. Details of the computation of the inelastic pseudo–force rate $\dot{P}_N$ are given by Kollmann and Bergmann [14]. The implementation of mixed inelastic shell element is completely analogous to the numerical procedure for a hybrid strain element as described by Kollmann and Bergmann [15]. The only difference is that for time integration of Hart's inelastic constitutive model we did not use the implicit algorithm by Cordts and Kollmann [17] but an explicit routine by Banthia and Mukherjee [18].

<u>Hart's Inelastic Constitutive Model</u>

Hart [2] has presented a constitutive model which possesses a sound experimental basis. In a later paper Hart [19] also gives a micromechanical foundation for important features of his model.

In addition to eqn.(1) the inelastic strain rate tensor is decomposed into a recoverable anelastic strain $\dot{\gamma}^A_{ij}$ and a completely irrecoverable strain γ^P_{ij}.

$$\dot{\gamma}^N_{ij} = \dot{\gamma}^A_{ij} + \dot{\gamma}^P_{ij} \tag{36}$$

where all inelastic strain rates are purely deviatoric. It is further assumed that the deviator of the applied stress s_{ij} can be decomposed into two auxiliary deviatoric tensors.

$$s_{ij} = s^A_{ij} + s^F_{ij} \tag{37}$$

Next, invariants of the stress and strain rate quantities are defined as e.g.

$$s^A = \sqrt{\tfrac{3}{2}\, s^A_{ij}\, s^A_{ij}}$$

$$\dot{\gamma}^A = \sqrt{\tfrac{2}{3}\, \dot{\gamma}^A_{ij}\, \dot{\gamma}^A_{ij}} \tag{38}$$

Similar invariants are introduced for the tensors s^f_{ij}, $\dot{\gamma}^N_{ij}$ and $\dot{\gamma}^P_{ij}$. The three–dimensional deformation behavior is governed by the following isotropic flow rules

$$\dot{\gamma}^A_{ij} = \frac{3}{2}\,\frac{\dot{\gamma}^A}{s^A}\, s^A_{ij} \tag{39}$$

$$\dot{\gamma}^N_{ij} = \frac{3}{2}\,\frac{\dot{\gamma}^N}{s^F}\, s^F_{ij} \tag{40}$$

$$\dot{\gamma}^P_{ij} = \frac{3}{2}\,\frac{\dot{\gamma}^P}{s^A}\, s^A_{ij} \tag{41}$$

The scalar invariants occurring in eqns.(39) to (41) are related by uniaxial evolution equations.

$$s^A = \mathcal{M}\,\gamma^A \tag{42}$$

$$\dot{\gamma}^N = \dot{\gamma}_0 \left[\frac{s^F}{s_0}\right]^M \tag{43}$$

$$\dot{\gamma}^P = \dot{\gamma}^* \left[\ln\frac{\sigma^*}{s^A}\right]^{-1/\lambda} \tag{44}$$

$$\dot{\gamma}^* = \dot{\gamma}^*_{ST} \left[\frac{\sigma^*}{\sigma^*_s}\right]^m \exp\left[\frac{Q}{R}\left[\frac{1}{T_B} - \frac{1}{T}\right]\right] \tag{45}$$

$$\dot{\sigma}^* = \dot{\gamma}^P \sigma^* \ \Gamma \left(\sigma^*, s^A \right) \tag{46}$$

Here $\mathcal{M}$ is an anelastic modulus and therefore eqn.(42) represents a linear anelastic spring. Eqn.(43) describes a nonlinear dashpot characterized by the flow parameters $\dot{\gamma}_0$ and M. The quantity s_0 is a reference stress. The quantity σ^* is called hardness by Hart and must not be confused with the metallurgical hardness. Instead it describes isotropic hardening. Eqn.(44) describes the evolution of the completely irrecoverable inelastic strain rate $\dot{\gamma}^P$. The evolution of the flow parameter $\dot{\gamma}^*$ is governed by eqn.(45) where σ_S^* and T_B are reference values of the hardness and absolute temperature, respectively. The quantity $\dot{\gamma}_{ST}^*$ is a flow parameter. The activation energy for atomic self diffusion is denoted as Q and R is the universal gas constant. Finally, eqn.(46) describes the evolution of the hardness σ^*. Here $\Gamma \left(\sigma^*, s^A \right)$ is a hardening function which has to be determined empirically. We use a formulation proposed by Kumar, Mukherjee, Huang and Li [20].

$$\Gamma \left(\sigma^*, s^A \right) = \left[\frac{\beta}{\sigma^*} \right]^{\delta} \left[\frac{s^A}{\sigma^*} \right]^{\beta / \sigma^*} \tag{47}$$

The evolution eqns.(43) to (46) form a system of autonomous ordinary differential equations of first order. Unfortunately this system is not only highly nonlinear, but it becomes mathematically stiff in some regions which are essential for applications. Therefore, stable time integration algorithms [17] or the viscoplastic limit [21] have to be applied. In the latter eqn.(44) is replaced by

$$\dot{\gamma}^P = \frac{s_{ij}^A \ \dot{\gamma}_{ij}^N}{\gamma^A (\mathcal{M} + \sigma^* \ \Gamma)} \tag{48}$$

whenever $s^A \geq \nu \sigma^*$. Here ν is number close to unity. In the present numerical analysis $\nu = 0.99$ was used.

<u>Numerical results</u>

As a test problem we consider a purely elastic cylindrical shell under internal pressure with simply supported ends. The shell is 1.000 mm in total length and 500 mm in diameter, with a wall thickness of 10 mm. The material is stainless steel at 400° C and the material parameters (E,ν) are given in table 1. One—half of the cylindrical shell is modeled with an uniform mesh of 10 elements along the midsurface. The internal pressure is 6.0 MPa.

In figures 2 and 3 the results obtained with the mixed finite element are compared with an analytical solution [22]. A very good correspondence of the numerical with the analytical solution can be observed. Further elastic test examples are given by Bergmann [23].

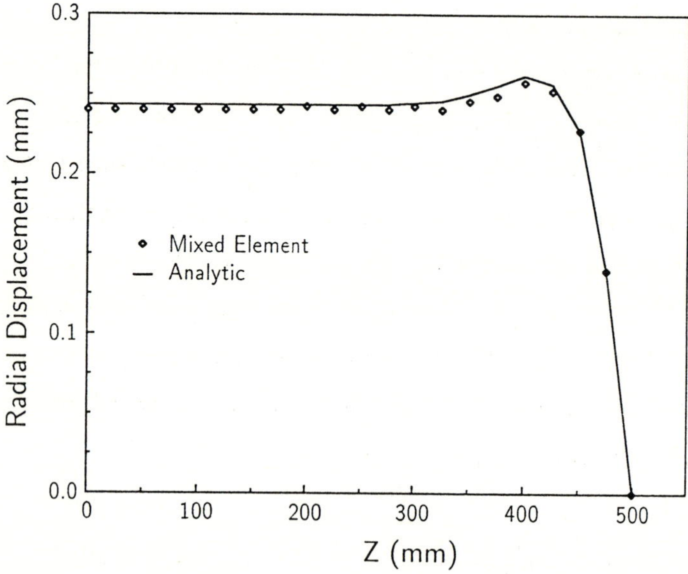

Fig.2. Radial deflections for simply—supported cylinder unter internal pressure

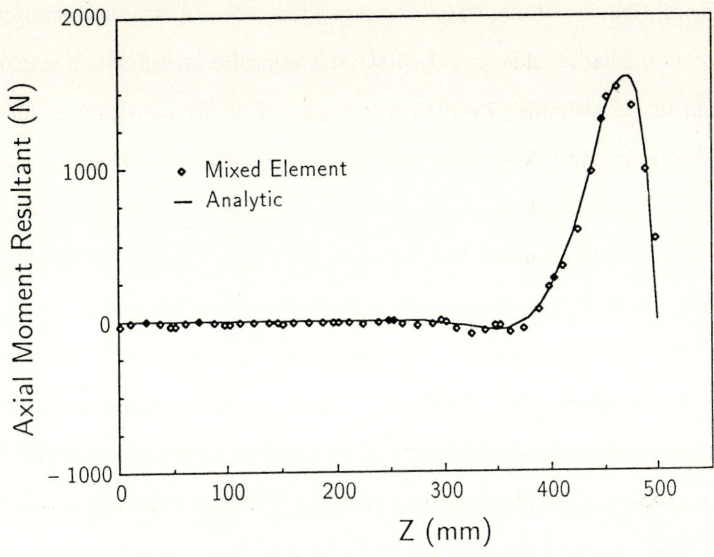

Fig.3. Axial bending moments for simply–supported cylinder under internal pressure

Finally, the ability of the axisymmetric mixed element to perform under conditions of inelastic behavior is tested with the same cylindrical shell as for the elastic example. However, the inelastic effects are included now. The material parameters for elastic and inelastic deformation are given in table 1. For this analysis, one–half of the cylinder is discretized into 25 elements.

λ	=	0.15	M	=	7.8
m	=	5	$\mathcal{M}$	=	$9.1 \cdot 10^{4}$ MPa
E	=	$1.68 \cdot 10^{5}$ MPa	ν	=	0.298
$\dot{\epsilon}_o$	=	$3.15 \mathrm{s}^{-1}$	σ_o	=	68.95 MPa
$\dot{\epsilon}_{ST}$	=	$1.269 \cdot 10^{-24} \mathrm{s}^{-1}$	σ_S^*	=	68.95 MPa
T_B	=	673 K			
β	=	$1.2346 \cdot 10^{3}$ MPa	δ	=	0.133
R	=	8,314.3 J/(kmol K)	Q	=	$0.272 \cdot 10^{9}$ J/kmol

Table 1. Material parameters for Hart's model for stainless steel SS304 at 400⁰

The cylindrical shell is initially stress and strain free, with an assumed homogeneous distribution of the hardness $\sigma^* = 117.26$ MPa. The shell is loaded with a rate of pressurization of 12 MPa/s until a maximum pressure of 6.60 MPa is attained. This pressure is held constant until T = 5.0 s, and then the cylinder is unloaded at the rate of 12 MPa/s.

In order to evaluate the performance of the finite element during inelastic deformation, the numerical results are compared with a semi–analytical solution by Kollmann and Mukherjee [24]. In figure 4, the results for displacements are compared at two different times during the deformation history. It can be seen that, although the results compare very well at the end of the active loading phase (T = 0.55 s), some divergence occurs by the end of the constant pressure state (T = 5.00 s). This discrepancy is probably due to the fact that for the semi–analytical solution the classical Love–Kirchhoff theory has been used, in which transverse shear strains are neglected while the mixed element is based on a shell theory which includes transverse shear strains. In [15] further details are given.

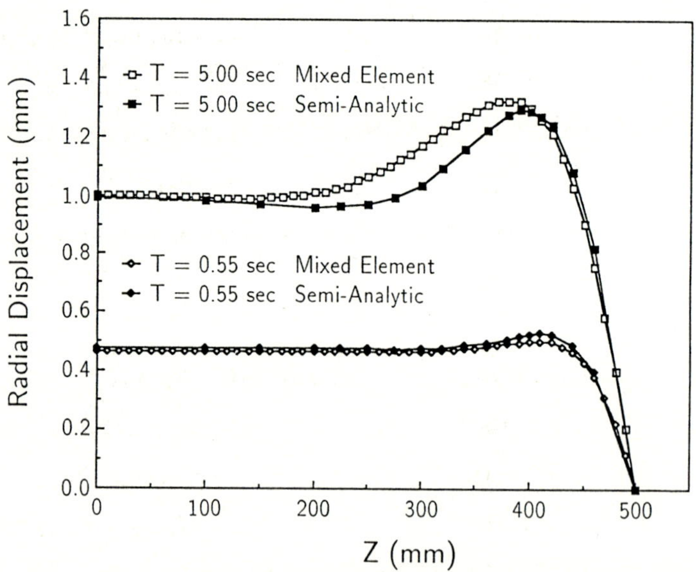

Fig.4. Comparison of radial displacements for inelastic cylinder

Finite element results for the radial displacements throughout the deformation history are shown in figure 5. This figure readily demonstrates the recovery of elastic deformation during the unloading phase (from T = 5.00 s to T = 5.55 s). Additional radial deformation is recovered after the cylinder has been completely unloaded. As demonstrated by Kollmann and Mukherjee [24], this effect is due to the recovery of anelastic strains in Hart's model. Finally, figure 6 presents a comparison in the bending region of the distributions of circumferential and axial stresses over the shell thickness in the bending region at the end of the active loading phase. For both stresses, the finite element results are within 2.5% of the semi–analytical solution thoughout the shell thickness.

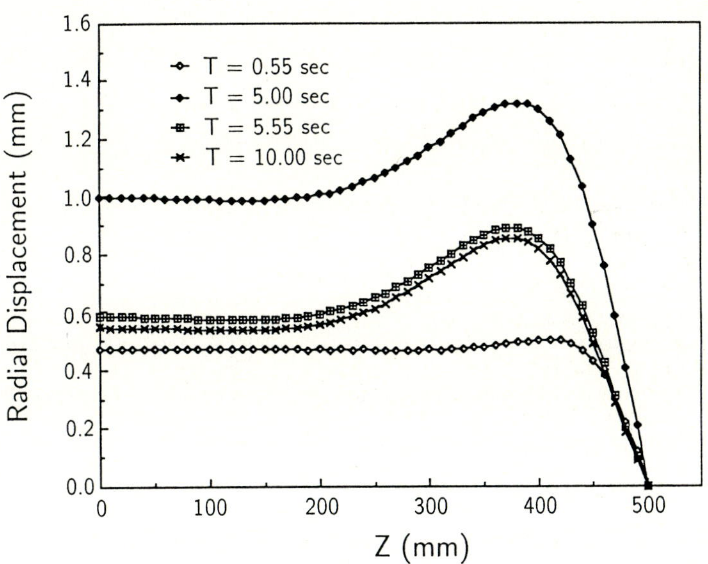

Fig.5. Finite element results for radial displacements for inelastic cylinder

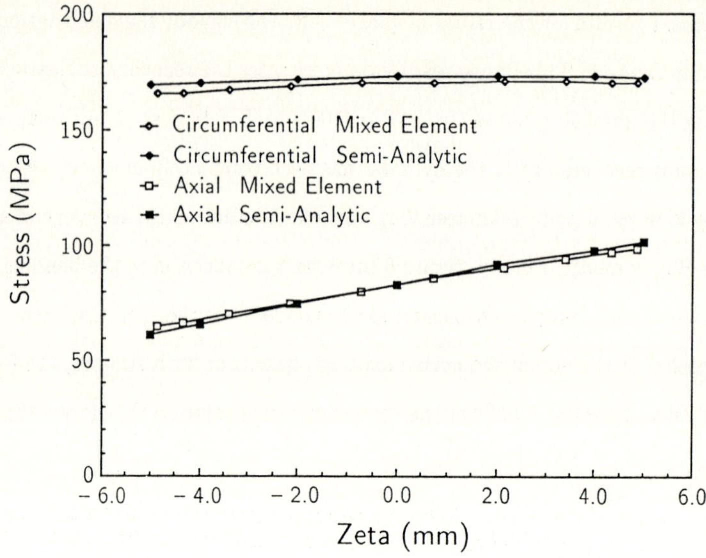

Fig.6. Comparison of stress distributions at the end of the active loading phase
(T = 0.55 s)

Conclusions

A new mixed finite element for analysis of elastic and inelastic axisymmetric shells is presented. This element contains displacements and strains as primary unknowns. In this paper an element with quadratic interpolation for displacements and strains is presented. For elastic test examples this element shows very good correspondence with known analytical solutions. No locking phenomena have been observed. For an inelastically deformed cylindrical shell also a good correspondence between the numerical and a semi–analytical solution could be obtained. Discrepancies of the deflections in the bending region can be explained from differences of the underlying shell theories. The advantages of the new element should be explored further.

References

[1]   Kollmann, F.G.; Mukherjee, S.: A general, geometrically linear theory of inelastic thin shells. Acta Mechanica. 57 (1985) 41–67.

[2]   Hart, E.W.: Constitutive relations for the non–elastic deformation of metals. Trans. ASME, J. Engng. Mat. Technol. 98 (1976) 193–202.

[3]   Miller, K.A.: An inelastic constitutive model for monotonic cyclic and creep deformation. Part 1: Equations development and analytical procedures. Trans. ASME, J. Engng. Mat. Technol. 98 (1976) 97–105.

[4]   Miller, K.A.: An inelastic constitutive model for monotonic, cyclic and creep deformation. Part 2: Applications to type 304 stainless steel. Trans. ASME, J. Engng. Mat. Technol. 98 (1976) 106–113.

[5]   Robinson, D.H.: A candidate creep recovery model for 2–1/4 Cr–1 Mo steel and its experimental implementation. Oak Ridge: Oak Ridge National Laboratory, report ORNL–TM–5110, 1975.

[6]   Walker, K.P.: Research and development program for nonlinear structural modeling with advanced time–dependent constitutive relation–ships. Lewisville: NASA, NASA–report no. CR–165533, 1981.

[7]   Cormeau, I.: Elastoplastic thick shell analysis by viscoplastic solid finite elements. Int. J. Num. Meth. 12 (1978) 203–227.

[8]   Hughes, T.J.R.; Liu, W.K.: Nonlinear finite element analysis of shells. Part 1: Three–dimensional shells. Comp. Meth. Appl. Mech. Engng. 26 (1981) 333–362.

[9]   Hughes, T.J.R.; Liu, W.K.: Nonlinear finite element analysis of shells. Part 2: Two–dimensional shells. Comp. Meth. Appl. Mech. Engng. 27 (1981) 167–181.

[10]   Parisch, H.: Large displacements of shells including material nonlinearities. Comp. Meth. Appl. Mech. Engng. 27 (1981) 183–204.

[11]   Oden, J.T.; Reddy, J.N.: Variational methods in theoretical mechanics. Berlin–Heidelberg–New York: Springer 1976.

[12]   Lee, S.W.; Pian, T.H.H.: Improvement of plate and shell finite elements by mixed formulations. AIAA J. 16 (1978) 29–34.

[13]   Mukherjee, S.; Kollmann, F.G.: A new rate principle suitable for analysis of inelastic deformation of plates and shells. Trans. ASME, J. Appl. Mech. 52 (1985) 533–535.

[14]   Kollmann, F.G.; Bergmann, V.: A new finite element for geometrically linear, inelastic analysis of axisymmetric plates and shells. Ithaca: Cornell University, Department of Theoretical and Applied Mechanics. Unpublished report 1986.

[15]   Kollmann, F.G.; Bergmann, V.: A new hybrid strain finite element for geometrically linear, inelastic analysis of axisymmetric shells, to appear in Computational Mechanics.

[16]   Zienkiewicz, O.C.; Bauer, J.; Morgan, K.; Onate, E.: A simple and efficient
element for axisymmetric shells. Int. J. Num. Meth. Engng. 11 (1977) 1545–1558.

[17]   Cordts, D.; Kollmann, F.G.: An implicit time integration scheme for inelastic
constitutive equations with internal state variables. Int. J. Num. Meth. Engng.
23 (1986) 533–554.

[18]   Banthia, V.; Mukherjee, S.: On an improved time integration for stiff constitutive models of inelastic deformation. Trans. ASME, J. Engng. Mat. Technol. 107
(1985) 282–285.

[19]   Hart, E.: A micromechanical basis for constitutive equations with internal state
variables. Trans. ASME, J. Engng. Mat. Technol. 106 (1984) 322–325.

[20]   Kumar, V.; Mukherjee, S.; Huang, F.H.; Li, C.–Y.: Deformation in type 304
austenitic stainless steel. Final report EPRI NP–1276, Electric Power Research
Institute, Palo Alto, Calif. 1979.

[21]   Kumar, V.; Mukherjee, S.: A boundary–integral equation formulation for time-
dependent inelastic deformation in metals. Int. J. mech. Sci. 19 (1977) 713–724.

[22]   Timoshenko, S.P.; Woinowsky–Krieger, P.: Theory of plates and shells. 2nd
edition. Tokyo: Mc Graw–Hill International Book Company 1970.

[23]   Bergmann, V.: A finite element formulation for inelastic plates and shells based
on a mixed variational principle. PhD dissertation, Cornell University, Ithaca,
N.Y. 1989.

[24]   Kollmann, F.G.; Mukherjee, S.: Inelastic deformation of thin cylindrical shells
under axisymmetric loading. Inch.–Arch. 54 (1984) 355–367.

An Assumed Strain Mixed Formulation
for Nonlinear Shells

J. J. RHIU and S. W. LEE

Department of Aerospace Engineering
The University of Maryland
College Park, MD 20742, U. S. A.

Summary

A mixed finite element formulation with stabilization matrix is presented for geometrically nonlinear thin shells. This formulation is based on the degenerate solid shell element concept and the Hellinger–Reissner principle with independent strain field. The independent strain field is divided into a lower order part and a higher order part. Using the equivalence between displacement formulation and mixed formulation, the lower order part is replaced by the displacement–dependent strain at reduced integration points. The higher order independent strain terms are selected to stabilize the spurious kinematic modes which exist in the element stiffness matrix obtained by using reduced integration. One set of the higher order assumed independent strain components is presented for a four–node degenerate solid shell element. For a nine–node degenerate solid shell element, two versions of the higher order assumed independent strain field are proposed. Numerical results demonstrate that the nine–node shell element based on the present formulation produces very accurate and reliable solutions even for very thin shells undergoing large rotations.

Introduction

Since it was introduced in 1970 [1], the degenerate solid shell element approach has been very popular in the finite element analysis of linear thin shell structures. This concept adopts the basic assumptions of a shell theory which incorporates transverse shear deformation. Thus the degenerate solid shell element requires only C^o continuity and can utilize the isoparametric representation for a curved shell geometry. Therefore, the requirement of interelement compatibility is satisfied without any difficulty. In addition, this shell element can well describe the rotational behavior of shells. Consequently, the degenerate solid shell element concept has also been applied to model geometrically and/or materially nonlinear thin shell structures, especially based on the conventional assumed displacement approach [2–8].

However, the degenerate solid shell element suffers from the combined effect of membrane locking and transverse shear locking for thin curved shells when a full Gaussian quadrature is used to generate element stiffness matrices [9,10]. The locking effect can be relieved in some degree by using a reduced and/or selective integration [11,12]. As an example, a nine–node degenerate shell element can use the 2×2 reduced quadrature rule on all strain energy terms or the 2×2 rule on the membrane and the transverse shear terms while the 3×3 full quadrature on the bending term. However, the reduced and/or selective integration scheme will result in spurious kinematic modes which produce no strain [13,14]. Among these kinematic modes, the compatible modes, which prevail even after assemblying elements, may lead to an unstable shell element model. Therefore, it is essential to develop a degenerate solid shell element which does not exhibit locking and the undesirable effect of compatible spurious kinematic modes.

Considerable research efforts have been devoted to construct non–locking and reliable shell elements. One scheme is the introduction of assumed strain which is interpolated and/or extrapolated from the displacement–dependent strain values at certain integration points [15–20]. Recently, a mode decomposition technique [21] was used by introducing inextensional bending. One method to suppress the kinematic modes which result from a reduced and/or selective integration is to utilize a stabilization matrix as reported in References [22] to [24]. The stabilization matrix in these references is usually controlled by a small stabilization parameter which must be carefully chosen. To remedy this difficulty, a combination of assumed strain stabilization procedure and strain interpolation was introduced in Reference [25].

The hybrid/mixed models [9,26,27] with either assumed stress or assumed strain have also been used. In particular, a mixed shell element formulation introduced by Lee and Pian in 1978 [9] is is based on the Hellinger–Reissner principle with assumed independent strain. The assumed independent strain field in this formulation is chosen to suppress the compatible spurious kinematic modes. With a proper assumed strain field, this formulation can produce shell finite element models which are free of locking and kinematically stable. A nine–node mixed element was formulated in conjunction with the degenerate solid shell element concept and tested for isotropic or composite thin shells undergoing small or large deflections [28–30]. In addition, a three–dimensional solid shell element [31] was also developed by using the mixed formulation. In this conventional mixed formulation, however, it is required to invert a matrix for the generation of an element stiffness matrix. The size of the matrix to be inverted depends directly on the number of polynomial terms used in the assumed strain field. For example, the nine–node degenerate solid shell element based on the formulation requires inversion of a matrix of size 38×38. Therefore, this conventional mixed finite element model requires more computing time than the corresponding assumed displacement model to generate element stiffness matrix. This appears to be one of the shortcomings of the conventional mixed formulation.

In Reference [32], a new efficient mixed formulation was developed. As in the case of the conventional mixed formulation, this new mixed model is also based on the Hellinger–Reissner principle with independent strain. The key aspect of the new formulation is that the assumed independent strain is split into two parts; the lower order polynomial part and the higher order polynomial part. Based on the equivalence theorem given in Reference [33] or [34], the lower order independent strain is replaced by the displacement–dependent strain evaluated at lower order integration points. Thus this generates the element stiffness matrix which is equivalent to that of the assumed displacement formulation element obtained by the reduced integration. On the other hand, The higher order part of the assumed independent strain field is chosen to suppress the compatible spurious kinematic modes. This leads to a stabilization stiffness matrix. These procedures make the new mixed formulation element much more efficient than the conventional mixed formulation element. In addition, this new formulation provides a rational basis of introducing stabilization matrix into element stiffness matrix. Following this approach, a nine–node and a sixteen–node degenerate solid shell elements [35,36] and a eighteen–node and a thirty two–node three–dimensional solid elements were proposed [37,38]. The numerical results of various shell example problems show that the new mixed formulation elements are computationally much more efficient than the conventional mixed formulation elements. Moreover, the new mixed formulation leads to reliable and very accurate solutions even for extremely thin shells.

In the present paper, the new mixed formulation is extended to the analysis of geometrically nonlinear shells. Considering the efficiency in computing element stiffness matrices, the new mixed formulation is expected to be more effective for nonlinear shell problems than the conventional mixed formulation. In addition, it will be useful to investigate the effectiveness of the aforementioned stabilization procedure for thin shells undergoing large deflection. The next section describes the new mixed finite element formulation for geometrically nonlinear shells. Then the discussion on the choice of the higher order part of the assumed strain field is given for four–node and nine–node degenerate solid shell elements. Finally, several geometrically nonlinear shell example problems are solved to test performance of the new mixed formulation.

Finite Element Formulation

Kinematics of Deformation

Figure 1 illustrates the motion of a material point P in a curved shell. For the description of geometry and kinematics of deformation, we employ a local Cartesian coordinate system with components x, y, z and a global Cartesian coordinate system with components X, Y and Z. The local coordinate axes x, y and z are parallel to the orthogonal unit vectors a_1, a_2 and a_3, respectively. The a_1 and a_2 vectors are tangent to the shell midsurface while a_3 vector is normal to the shell midsurface. In addition, the nondimensional coordinates ξ, η and ζ are also used.

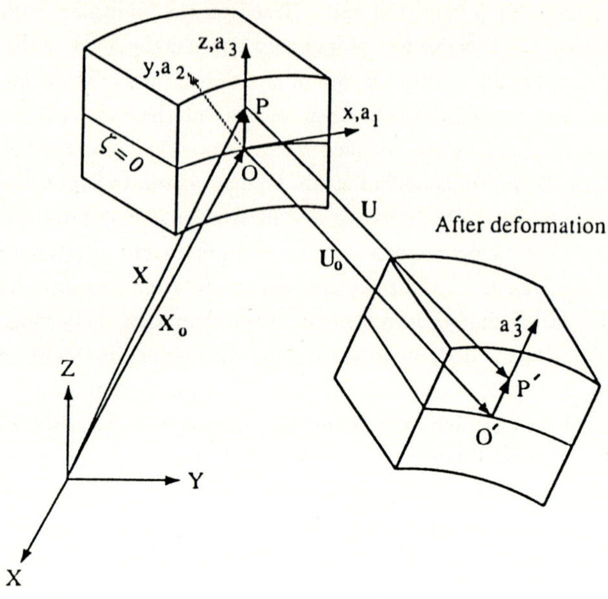

Figure 1: Kinematics of shell deformation

Under the basic thin shell assumptions, displacement vector $\mathbf{U}$ of the generic point P defined on the global coordinate system can be written as

$$\mathbf{U} = \mathbf{U_o} + \zeta\frac{t}{2}(\mathbf{a_3}' - \mathbf{a_3}) \tag{1}$$

where $\mathbf{U_o}$ is the global displacement vector of the point O on the shell midsurface, t is the shell thickness, and the nondimensional coordinate ζ which varies from -1 to $+1$ is defined such that

$$z = \frac{t}{2}\zeta \tag{2}$$

In addition, $\mathbf{a_3}'$ in equation (1) denotes $\mathbf{a_3}$ in the deformed configuration. The description of rotation between $\mathbf{a_3}$ and $\mathbf{a_3}'$ has been a classical problem, which has attracted the interest of many researchers [39]. Among the various methods, Euler's finite rotation formula or its modified versions have been frequently used in the finite element analysis of shells. In the present paper, the rotation of $\mathbf{a_3}$ to $\mathbf{a_3}'$ are accomplished by two successive body–fixed rotations as shown in Figure 2; the θ_1 rotation around $\mathbf{a_1}$ and then the θ_2 rotation around $\mathbf{a_2}'$. Then, $\mathbf{a_3}'$ is expressed as

$$\mathbf{a_3}' = sin\theta_2\,\mathbf{a_1} + sin\theta_1 cos\theta_2\,\mathbf{a_2} + cos\theta_1 cos\theta_2\,\mathbf{a_3} \tag{3}$$

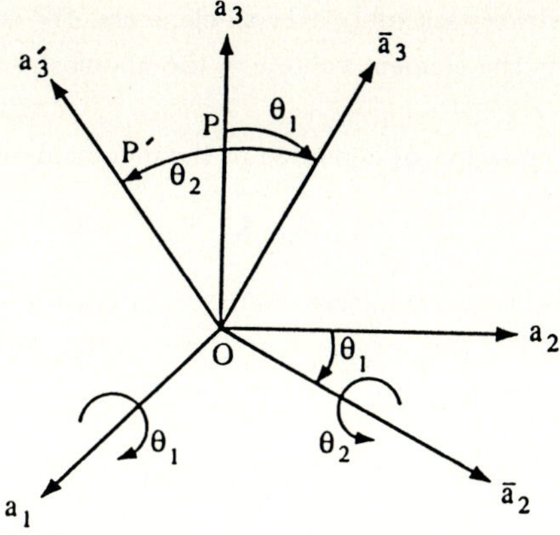

Figure 2: Rotation from a_3 to a_3'

With equation (1), the Largrangian strain vector defined on the global coordinate system can be derived from the strain–displacement relationship. Then, the local displacement–dependent strain vector $\overline{\mathbf{E}}$ is obtained through the strain transformation. On the other hand, the equilibrium equation resulting from the finite element approximation for geometrically nonlinear thin shell structures is solved by an iteration scheme such as the Newton–Raphson method. Accordingly, incremental quantities are introduced for $\overline{\mathbf{E}}$ as follows :

$$\overline{\mathbf{E}} = \lfloor \overline{E}_{xx}, \overline{E}_{yy}, \overline{E}_{xy}, \overline{E}_{yz}, \overline{E}_{zx} \rfloor^{T} = \overline{\mathbf{E}}^{o} + \Delta \overline{\varepsilon} + \Delta \overline{\eta} \tag{4}$$

where $\overline{\mathbf{E}}^{o}$ is the strain vector obtained from the previous iteration step and the superscript T represents transpose. The incremental strain vector $\Delta \overline{\varepsilon}$ is linear while the incremental vector $\Delta \overline{\eta}$ is quadratic in incremental displacements.

Incremental Equilibrium Equation

For a solid in equilibrium, the stationary condition of the Hellinger–Reissner principle can be expressed as

$$\delta \pi_{R} = \sum_{e} \int [\delta \overline{\mathbf{E}}^{T} \mathbf{S} + \delta \mathbf{S}^{T}(\overline{\mathbf{E}} - \mathbf{E})]\, dV_{e} \; - \; \delta W = 0 \tag{5}$$

where $\mathbf{S}$ is the second Piola–Kirchhoff stress vector, $\delta \overline{\mathbf{E}}$ is the virtual strain vector, $\delta \mathbf{S}$ is the virtual stress vector, and $\mathbf{E}$ is the independent strain vector. In addition, the

242

summation sign $\sum$ indicates assembly over all elements, δW is the virtual work term for applied loads, and V_e is the element volume in the undeformed configuration.

The stress vector $\mathbf{S}$ in equation(5) is related to the independent strain vector $\mathbf{E}$ such that

$$\mathbf{S} = \lfloor S_{xx}, S_{yy}, S_{xy}, S_{yz}, S_{zx} \rfloor^T = \mathbf{CE} \tag{6}$$

where $\mathbf{C}$ denotes an elastic coefficient matrix. In addition, in accordance with the Newton–Raphson method, the independent strain vector $\mathbf{E}$ is also expressed in incremental form as follows :

$$\mathbf{E} = \lfloor E_{xx}, E_{yy}, E_{xy}, E_{yz}, E_{zx} \rfloor^T = \mathbf{E}^\circ + \Delta\varepsilon \tag{7}$$

where $\mathbf{E}^\circ$ is the value of $\mathbf{E}$ at the previous iteration step and $\Delta\varepsilon$ is the incremental independent strain vector. Furthermore, the virtual strain and stress vectors are written as

$$\delta\overline{\mathbf{E}} \;=\; \delta\overline{\varepsilon} + \delta\overline{\eta} \tag{8}$$

$$\delta\mathbf{S} \;=\; \mathbf{C}\delta\varepsilon \tag{9}$$

Substituting equations (4) and (6) to (9) into equation (5), $\delta\pi_R$ becomes

$$\delta\pi_R \;=\; \sum_e \{ \int [\delta\overline{\varepsilon}^T \mathbf{C}\Delta\varepsilon + \delta\varepsilon^T \mathbf{C}(\Delta\overline{\varepsilon} - \Delta\varepsilon) + \delta\overline{\eta}^T \mathbf{S}^\circ] \, dV_e$$
$$+ \; \int \delta\overline{\varepsilon}^T \mathbf{S}^\circ dV_e + \int \delta\varepsilon^T \mathbf{C}(\overline{\mathbf{E}}^\circ - \mathbf{E}^\circ) \, dV_e \} - \delta W + A_n = 0 \tag{10}$$

where $\mathbf{S}^\circ$ is the value of $\mathbf{S}$ at the previous iteration step and thus from equation (6)

$$\mathbf{S}^\circ = \mathbf{CE}^\circ \tag{11}$$

On the other hand, the term A_n in equation (10) represents terms nonlinear in $\Delta\mathbf{q_e}$, which will be discarded hereafter. In addition, the third integral associated with $(\overline{\mathbf{E}}^\circ - \mathbf{E}^\circ)$ term can be neglected.

Following the approach used in References [35] and [36], the independent local incremental strain vector $\Delta\varepsilon$ in the present formulation is divided into as follows :

$$\Delta\varepsilon = \Delta\varepsilon_L + \Delta\varepsilon_H \tag{12}$$

where $\Delta\varepsilon_L$ is the independent local incremental strain vector with lower order polynomial terms in ξ, η and ζ. On the other hand, $\Delta\varepsilon_H$ is the higher order component of the independent local incremental strain vector. Similarly, $\mathbf{E}^\circ$ can be split into two parts such as

$$\mathbf{E}^\circ = \mathbf{E}^\circ_L + \mathbf{E}^\circ_H \tag{13}$$

and

$$S^o = S^o_L + S^o_H \tag{14}$$

where, using equation (11),

$$S^o_L = CE^o_L \tag{15}$$

$$S^o_H = CE^o_H \tag{16}$$

Introducing equations (11) to (16) into equation (10), the stationary condition can be rewritten as

$$\begin{aligned}
\delta\pi_R =\ & \sum_e \{ \int [\delta\bar{\varepsilon}^T C \Delta\varepsilon_L + \delta\varepsilon_L{}^T C(\Delta\bar{\varepsilon} - \Delta\varepsilon_L) + \delta\bar{\eta}^T S^o_L] dV_e + \int \delta\bar{\eta}^T S^o_H dV_e \\
& + \int (\delta\bar{\varepsilon}^T C \Delta\varepsilon_H + \delta\varepsilon_H{}^T C \Delta\bar{\varepsilon}) dV_e - \int (\delta\varepsilon_L{}^T C \Delta\varepsilon_H - \delta\varepsilon_H{}^T C \Delta\varepsilon_L) dV_e \\
& - \int \delta\varepsilon_H C \Delta\varepsilon_H dV_e + \int \delta\bar{\varepsilon}^T S^o_L dV_e + \int \delta\bar{\varepsilon}^T S^o_H dV_e \} - \delta W = 0
\end{aligned} \tag{17}$$

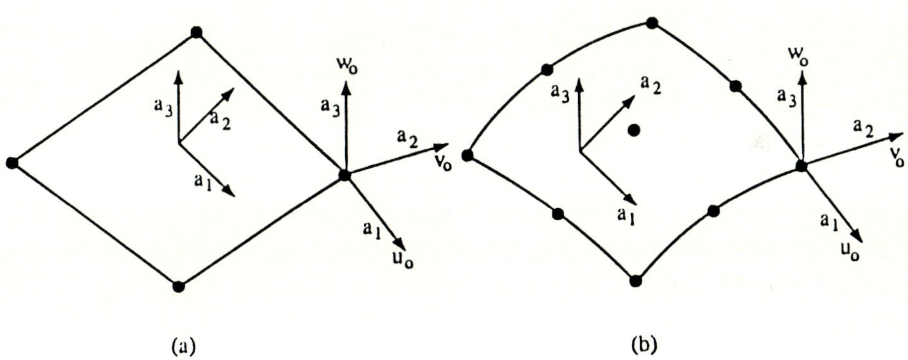

Figure 3: Shell mid-surface : (a) four-node element (b) nine-node element

Now we consider the four–node and the nine–node degenerate solid shell elements whose midsurfaces are shown in Figures 3(a) and 3(b), respectively. For a four–node shell element of flat rectangular geometry, the displacement–dependent strain vector $\Delta\bar{\varepsilon}$ is at most bilinear in ξ, η and linear in ζ. Assuming the lower order independent strain vector $\Delta\varepsilon_L$ to be constant in ξ, η and linear in ζ, the first, fourth and sixth integrals in

points used in the ζ direction. The remaining terms are integrated by the $2 \times 2 \times 2$ point rule. Following a similar observation, for a nine–node shell element of flat rectangular geometry, the $2 \times 2 \times 2$ point rule is adopted for the first, fourth and sixth integrals in equation (17) and the $3 \times 3 \times 2$ point rule is used for other integral terms. Even though these integration rules hold only for flat rectangular element, in the present formulation they will be used even for elements with arbitrary geometry.

On the other hand, according to the equivalence between the assumed strain and the displacement–dependent strain as discussed in References [33] and [34], it is possible to choose $\Delta \varepsilon_L$ such that

$$\Delta \varepsilon_L = \Delta \bar{\varepsilon} \tag{18}$$

at the lower order integration points. Then applying the integration rules and the equivalence given in equation (18), equation (17) becomes

$$
\begin{aligned}
\delta \pi_R = {} & \sum_e [\int_L \delta \bar{\varepsilon}^T \mathbf{C} \delta \bar{\varepsilon} dV_e + \int_L \delta \bar{\eta}^T \mathbf{S}_L^\circ dV_e + \int_H \delta \bar{\eta}^T \mathbf{S}_H^\circ dV_e \\
& + \int_H \delta \bar{\varepsilon}^T \mathbf{C} \Delta \varepsilon_H dV_e - \int_L \delta \bar{\varepsilon}^T \mathbf{C} \Delta \varepsilon_H dV_e \\
& + \int_H \delta \varepsilon_H{}^T \mathbf{C} \Delta \bar{\varepsilon} dV_e - \int_L \delta \varepsilon_H{}^T \mathbf{C} \Delta \bar{\varepsilon} dV_e - \int_H \delta \varepsilon_H{}^T \mathbf{C} \Delta \varepsilon_H dV_e] \\
& - [\delta W - \sum_e (\int_L \delta \bar{\varepsilon}^T \mathbf{S}_L^\circ dV_e + \int_H \delta \bar{\varepsilon}^T \mathbf{S}_H^\circ dV_e)] = 0
\end{aligned}
\tag{19}
$$

where the letters L and H under the integral signs indicate the lower order integration and the higher order integration, respectively.

For the degenerate solid shell elements, $\Delta \bar{\varepsilon}$ can be expressed in terms of the incremental element nodal degrees of freedom vector $\Delta \mathbf{q_e}$ which consists of incremental values of three translational displacements and two rotational angles on the shell midsurface. In matrix form, it can be written as

$$\Delta \bar{\varepsilon} = \mathbf{B}(\xi, \eta, \zeta) \Delta \mathbf{q_e} \tag{20}$$

where $\mathbf{B}$ is a matrix relating $\Delta \bar{\varepsilon}$ to $\Delta \mathbf{q_e}$.

Symbolically, the higher order strain component is written as

$$\Delta \varepsilon_H = \mathbf{P}(\xi, \eta, \zeta) \Delta \boldsymbol{\alpha_e} \tag{21}$$

where $\mathbf{P}$ is the shape function matrix of the higher order assumed strain and $\Delta \boldsymbol{\alpha_e}$ is the vector of the higher order assumed strain parameters. For the four–node shell element, $\mathbf{P}$ is at most bilinear in ξ and η and linear in ζ. For the nine–node shell element, $\mathbf{P}$ is assumed to be biquadratic in ξ and η and linear in ζ.

On the other hand, from equations (20) and (21), the virtual strain vectors $\delta\bar{\varepsilon}$ and $\delta\varepsilon_H$ can be expressed as

$$\delta\bar{\varepsilon} \;=\; \mathbf{B}\delta\mathbf{q}_e \tag{22}$$

$$\delta\varepsilon_H \;=\; \mathbf{P}\delta\alpha_e \tag{23}$$

In addition, each component of $\delta\bar{\eta}$ is linear in $\delta\mathbf{q}_e$ and $\Delta\mathbf{q}_e$. For example, symbolically

$$\delta\bar{\eta}_{xx} = \delta\mathbf{q}_e^T \mathbf{R}_{xx} \Delta\mathbf{q}_e \tag{24}$$

and other terms in $\delta\bar{\eta}$ are similarly written.

Introducing equations (20) to (24) into equation (19), the following expression is obtained

$$\sum_e \{\delta\mathbf{q}_e^T[(\mathbf{K}_L + \mathbf{K}_N)\Delta\mathbf{q}_e + \delta\mathbf{q}_e^T\mathbf{G}^T\Delta\alpha_e - \Delta\mathbf{Q}_e] + \delta\alpha_e^T(\mathbf{G}^T\Delta\mathbf{q}_e - \mathbf{H}\Delta\alpha_e)\} = 0 \tag{25}$$

where

$$\mathbf{K}_L \;=\; \int_L \mathbf{B}^T\mathbf{C}\mathbf{B}\,dV_e \tag{26}$$

$$\mathbf{K}_N \;=\; \int_L [(S_{xx}^o)_L\mathbf{R}_{xx} + \cdots + (S_{zx}^o)_L\mathbf{R}_{zx}]dV_e$$

$$\quad + \int_H [(S_{xx}^o)_H\mathbf{R}_{xx} + \cdots + (S_{zx}^o)_H\mathbf{R}_{zx}]dV_e \tag{27}$$

$$\mathbf{G} \;=\; \int_H \mathbf{P}^T\mathbf{C}\mathbf{B}\,dV_e - \int_L \mathbf{P}^T\mathbf{C}\mathbf{B}\,dV_e \tag{28}$$

$$\mathbf{H} \;=\; \int_H \mathbf{P}^T\mathbf{C}\mathbf{P}\,dV_e \tag{29}$$

In addition, the incremental element load vector $\Delta\mathbf{Q}_e$ is given by

$$\Delta\mathbf{Q}_e = \mathbf{F}_e - \mathbf{F}_i \tag{30}$$

where

$$\mathbf{F}_i = \int_L \mathbf{B}^T\mathbf{S}_L^o\,dV_e + \int_H \mathbf{B}^T\mathbf{S}_H^o\,dV_e \tag{31}$$

and $\mathbf{F}_e$ is the element load vector due to the external load.

Noting that $\delta\alpha_e$ is arbitrary, equation (25) leads to the following incremental compatibility equation between $\Delta\alpha_e$ and $\Delta\mathbf{q}_e$ in each element :

$$\mathbf{H}\Delta\alpha_e = \mathbf{G}\Delta\mathbf{q}_e \tag{32}$$

or solving for $\Delta\alpha_e$

$$\Delta\alpha_e = \mathbf{H}^{-1}\mathbf{G}\Delta\mathbf{q}_e \tag{33}$$

Substituting equation (33) into equation (25), we obtain the incremental equilibrium equation such that

$$\sum_e \delta q_e^T (K_e \Delta q_e - \Delta Q_e) = 0 \tag{34}$$

The tangential element stiffness matrix K_e in this equation is expressed as

$$K_e = K_L + K_N + K_S \tag{35}$$

where K_S is given by

$$K_S = G^T H^{-1} G \tag{36}$$

In equation (35), the first term K_L, which is expressed in equation (26), is equivalent to the element stiffness matrix of the assumed displacement model with the indicated lower order integration rule; $1 \times 1 \times 2$ point rule for the four–node element and $2 \times 2 \times 2$ point rule for the nine–node element. Thus the K_L matrix has the spurious kinematic modes which will be shown in the next section. The undesirable kinematic modes in K_L are suppressed by the K_S matrix. Thus the matrix K_S in equation (36) plays the role of a stabilization matrix in the present formulation. Note that the size of G and H matrices in the present formulation is very small because they involve only higher order assumed strain field.

Kinematic Modes and Higher Order Assumed Strain

The spurious kinematic modes in the K_L matrix produce zero strain at the lower order integration points. Under the assumption of small displacement, it is relatively easy to analytically determine the kinematic modes for a degenerate solid shell element with flat rectangular geometry. A detailed procedure can be found in Reference [29] or [36].

Four–node Shell Element

For the four–node flat square element with the sides along $x = \pm 1$ and $y = \pm 1$, the spurious kinematic modes can be identified as follows :

$$\theta_1 = c_1 x; \quad \theta_2 = -c_1 y \tag{37}$$

$$u_o = c_2 xy \tag{38}$$

$$v_o = c_3 xy \tag{39}$$

$$w_o = c_4 xy \tag{40}$$

$$\theta_1 = c_5 xy \tag{41}$$

$$\theta_2 = c_6 xy \tag{42}$$

where $c_1, \cdots c_6$ are arbitrary constants, u_o, v_o, w_o are three translational displacements, and θ_1, θ_2 are two rotational angles. Among the six spurious kinematic modes, the mode in equation (37) is incompatible and disappears for an assembly of only two elements. All other modes are compatible and may result in an unstable finite element even after assembly of elements. In the present formulation, these compatible modes are to be suppressed by properly chosen higher order strain terms.

The linear strain terms corresponding to the compatible kinematic modes in equations (38) to (42) are given by

$$\begin{aligned}
\overline{\varepsilon}_{xx} &= c_2 y + c_6 z y \\
\overline{\varepsilon}_{yy} &= c_3 x + c_5 z x \\
\overline{\varepsilon}_{xy} &= c_2 x + c_3 y + z(c_6 x + c_5 y) \\
\overline{\varepsilon}_{yz} &= c_4 x + c_5 x y \\
\overline{\varepsilon}_{zx} &= c_4 y + c_6 x y
\end{aligned} \tag{43}$$

Noting that the kinematic modes in $\mathbf{K}_L$ are suppressed by $\Delta \boldsymbol{\varepsilon}_H$ in the present formulation, the above equation can provides a basis for selecting the higher order assumed incremental strain terms of the four–node shell element. For example, the kinematic mode associated with c_2 can be suppressed by including either y in $\Delta \varepsilon_{xx}{}^H$ or x in $\Delta \varepsilon_{xy}{}^H$, where the superscript H represents the higher order linear strain component. The kinematic mode corresponding to c_4 is suppressed by including either the x term in $\Delta \varepsilon_{yz}{}^H$ or the y term in $\Delta \varepsilon_{zx}{}^H$. However, both terms are included in the higher order assumed strain components to avoid a directional imbalance to the element stiffness matrix. Therefore, more than one choice of higher order assumed strain field can be employed to suppress the compatible kinematic modes. On the other hand, it is essential to select the z–independent terms as simple as possible to avoid locking [9,34].

With these considerations, we may choose the following higher order strain field for the four–node degenerate solid shell element :

$$\begin{aligned}
\Delta \varepsilon_{xx}{}^H &= \alpha_1 y + \alpha_2 z y \\
\Delta \varepsilon_{yy}{}^H &= \alpha_3 x + \alpha_4 z x \\
\Delta \varepsilon_{xy}{}^H &= 0 \\
\Delta \varepsilon_{yz}{}^H &= \alpha_5 x \\
\Delta \varepsilon_{zx}{}^H &= \alpha_6 y
\end{aligned} \tag{44}$$

where $\alpha_1, \alpha_2, \cdots \alpha_6$ are unknown assumed strain parameters. Then the size of $\mathbf{P}$ matrix in equation (21) is 6×6. In this assumed strain field, the kinematic modes represented by c_2 and c_6 are suppressed by the polynomial terms in $\Delta \varepsilon_{xx}{}^H$. On the other hand, the terms in $\Delta\varepsilon_{yy}{}^{H}$ suppress the modes associated with c_3 and c_5. The remaining mode is suppressed by the terms in either $\Delta\varepsilon_{yz}{}^{H}$ or $\Delta\varepsilon_{zx}{}^{H}$.

Nine–node Shell Element

As shown in Reference [29] or [35], the kinematic modes for the nine–node degenerate solid shell element with flat square geometry with the sides along $x = \pm 1$ and $y = \pm 1$ are given as follows :

$$u_o = d_1(x - 3xy^2) \; ; \quad v_o = -d_1(y - 3x^2y) \tag{45}$$

$$\theta_1 = d_2(y - 3x^2y) \; ; \quad \theta_2 = -d_2(x - 3xy^2) \tag{46}$$

$$u_o = d_3(x^2 + y^2 - 3x^2y^2) \tag{47}$$

$$v_o = d_4(x^2 + y^2 - 3x^2y^2) \tag{48}$$

$$w_o = d_5(x^2 + y^2 - 3x^2y^2) \tag{49}$$

$$\theta_1 = d_6(x^2 + y^2 - 3x^2y^2) \; ; \quad w_o = -\frac{1}{3}c_6y \tag{50}$$

$$\theta_2 = d_7(x^2 + y^2 - 3x^2y^2) \; ; \quad w_o = -\frac{1}{3}c_7x \tag{51}$$

Among the above kinematic modes, the modes in equations (45) and (46) need not be suppressed at an element level since they are incompatible. The remaining modes are compatible and are to be suppressed by the higher order assumed strain terms.

Following similar observations given in the four–node solid element case, the following set of the higher order assumed strain field, called A version, can be used to suppress the compatible kinematic modes associated with equations (47) to (51) :

$$\begin{aligned}
\Delta\varepsilon_{xx}{}^{H} &= 0 \\
\Delta\varepsilon_{yy}{}^{H} &= 0 \\
\Delta\varepsilon_{xy}{}^{H} &= \alpha_1 xy^2 + \alpha_2 x^2 y + z(\alpha_3 xy^2 + \alpha_4 x^2 y) \\
\Delta\varepsilon_{yz}{}^{H} &= \alpha_5 x^2 y \\
\Delta\varepsilon_{zx}{}^{H} &= \alpha_6 xy^2
\end{aligned} \tag{52}$$

In this set, the higher order terms in $\Delta\varepsilon_{xy}{}^{H}$ suppress the kinematic modes related to d_3, d_4, d_6, and d_7. The term in either $\Delta\varepsilon_{yz}$ or $\Delta\varepsilon_{zx}$ suppresses the mode represented by d_5.

An alternative choice, referred to as B version, is given as

$$\begin{aligned}
\Delta\varepsilon_{xx}{}^{H} &= \alpha_1 xy^2 + \alpha_2 zxy^2 \\
\Delta\varepsilon_{yy}{}^{H} &= \alpha_3 x^2 y + \alpha_4 zx^2 y \\
\Delta\varepsilon_{xy}{}^{H} &= 0 \\
\Delta\varepsilon_{yz}{}^{H} &= \alpha_5 x^2 y \\
\Delta\varepsilon_{zx}{}^{H} &= \alpha_6 xy^2
\end{aligned} \tag{53}$$

It should be mentioned that this B version of the higher order assumed strain was investigated in Reference [35]. In this version, the compatible kinematic modes associated with d_3, d_4, d_6, and d_7 are suppressed by the terms in $\Delta\varepsilon_{xx}{}^H$ and $\Delta\varepsilon_{yy}{}^H$. Note that, even for the nine–node shell element, the size of $\mathbf{P}$ matrix in equation (21) will also be 6×6.

Invariance of Element Stiffness Matrix

It should be noted that, for an element with an arbitrary curved, irregular shape geometry, the higher order assumed strain components are chosen by replacing x, y and z in equations (44), (52) and (53) with the parent coordinates ξ, η and ζ, respectively. For example, the higher order assumed strain $\Delta\varepsilon_{xx}{}^H$ in equation (44) is expressed as

$$\Delta\varepsilon_{xy}{}^H = \alpha_1\eta + \alpha_2\zeta\eta \tag{54}$$

Other strain components are similarly expressed. Then the higher order assumed strain fields selected in this manner consist of incomplete polynomial sets in ξ and η. For an element with an arbitrary shape geometry, this leads to an element stiffness matrix which is in general not invariant. The invariance of an element stiffness matrix is enforced by assigning a specific orthogonal local coordinate system at each integration point for a given element geometry. In the present formulation, we adopt the local coordinate system used in Reference [30] or [38].

Numerical Tests

Simple geometrically nonlinear shell example problems were solved to test the present nine–node degenerate shell element. The numerical results of the present four–node shell element is not reported in this paper. The original Newton–Raphson method or a modified Newton–Raphson method combined with the arc–length constraint in Reference [40] was used to solve the nonlinear iterative equilibrium equations. All computations were carried out in double precision on the Sun Microsystems machine at the University of Maryland.

A Shallow Cylindrical Shell under A Concentrated Load

In Figure 4, a cylindrical shell subjected to a point load P at the central point C is shown. The longitudinal straight boundaries are hinged and immovable, whereas the the circular edges are free. Elastic material data are Young's modulus $E = 3.10275 \ kN/mm^2$ and Poisson's ratio $\nu = 0.3$. Geometrical data are radius $R = 2540 \ mm$, axial length $L = 508 \ mm$, thickness $t = 12.7 \ mm$, and angle $\theta = 0.2 \ rad$. Since the geometry is symmetric, only one quarter of the shell was modeled by evenly divided 2×2 and 3×3 meshes.

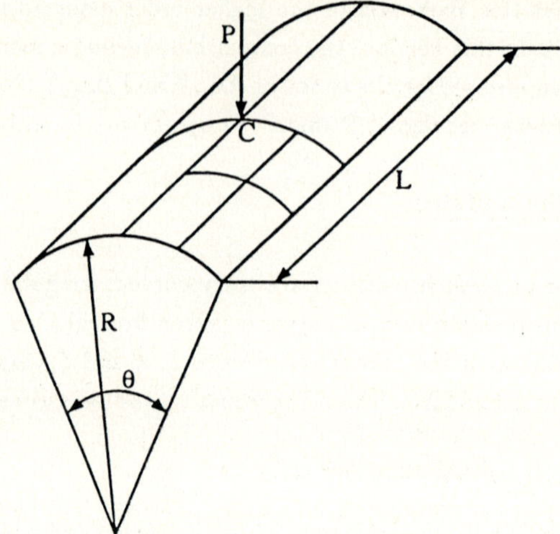

Figure 4: A shallow cylindrical shell under a concentrated load

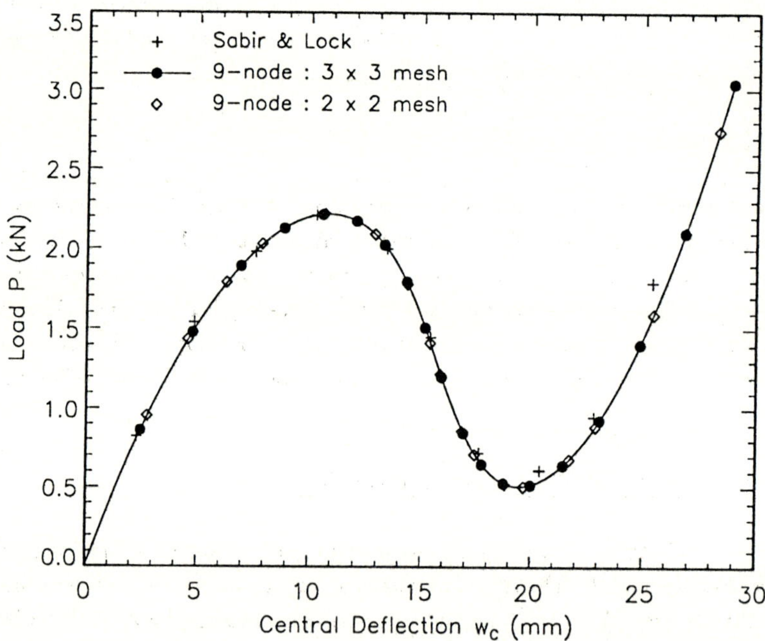

Figure 5: Load-deflection curve at point C of the shallow cylindrical shell

The load–deflection curve at the load point C for the present nine–node shell element formulation with the A or B version higher order assumed strain is given in Figure 5 and is compared with Reference [41]. The figure demonstrates that the present nine–node shell element describes the snap–through behavior of this example problem very well.

A Cylindrical Shell with Free Edges

The cylindrical shell with free edges is subjected to concentrated outward normal point forces at two opposite sides of the cylinder as shown in Figure 6. Elastic and geometric data are $E = 10.5 \times 10^6 \ psi$, $\nu = 0.3125$, $R = 4.953 \ in.$, $L = 10.35 \ in.$, and $t = 0.094 \ in.$ Due to the symmetry of geometric and loading conditions, the octant portion $ABCD$ of the cylinder was modeled by the uniformly divided 6 × 2 mesh and the 8 × 4 refined mesh. The 8 × 4 mesh which is shown in the figure is obtained from the uniformly divided 7 × 3 mesh by equally dividing elements along the side CD and the arc CB. The reason for using this mesh is to represent the steep gradient of deflection near the load point. The analytical solution in Reference [42] shows the deflection at the loading point up to 17 times of the shell thickness. Figure 7 shows the load–deflection curve at

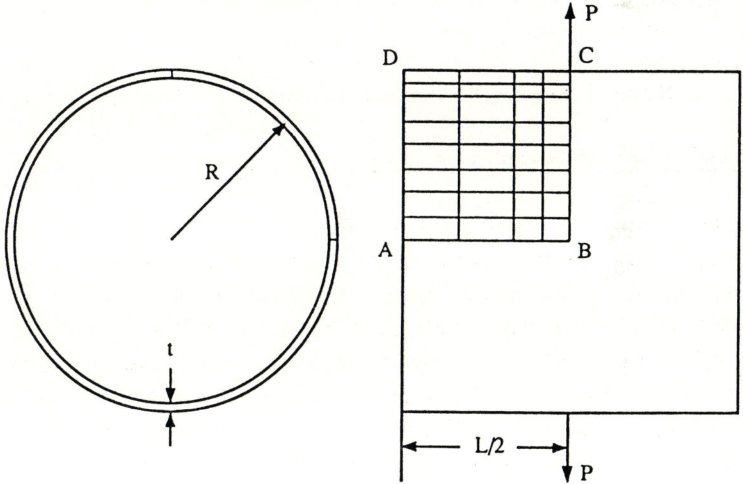

Figure 6: A cylindrical shell with free edges

the loading point obtained by using the present nine–node shell element. Both A and B versions gave almost identical numerical results. The numerical solutions are in very good agreement with the analytical solutions. In the present test, the computation was carried out until the deflection at the loading point reached more than 25 times the shell thickness.

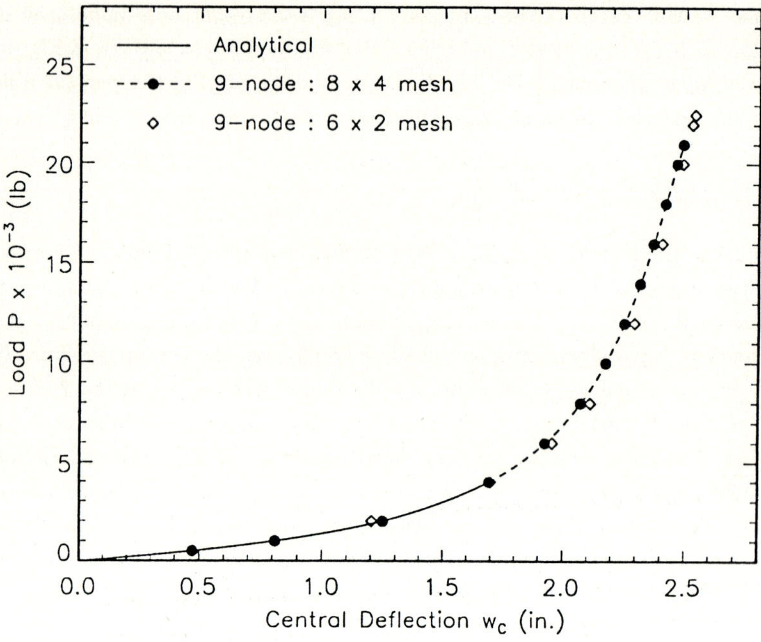

Figure 7: Load–deflection curve at point C of the cylinder

A Spherical Shell under Concentrated Loads

A spherical shell subjected to two opposite concentrated loads as shown in Figure 8 was also considered to test performance of the present nine–node shell element formulation. Elastic and geometrical data are $E = 10^7$ psi, $\nu = 0.3$, $R = 100$ $in.$, and $t = 1$ $in.$ Owing to the symmetry of geometry and loading conditions, only a 2° portion of the the upper half of the sphere was modeled by 15–element and 18–element meshes. These meshes were also used for the linear analysis in Reference [29]. For the convenience of modeling, a small region at loading the point was excluded. Two different cases were tested. In one case, a region within 0.05^o was excluded and replaced with proper boundary conditions while in another case a region within 0.05^o was excluded. Numerical values were identical to each other. Figure 9 shows the deflection behavior at the loading point. Again the version A and the version B of the present nine–node shell element gave identical numerical results.

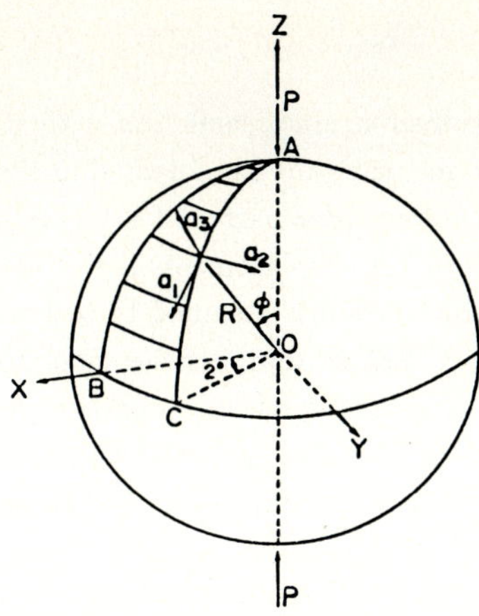

Figure 8: A spherical shell under concentrated loads

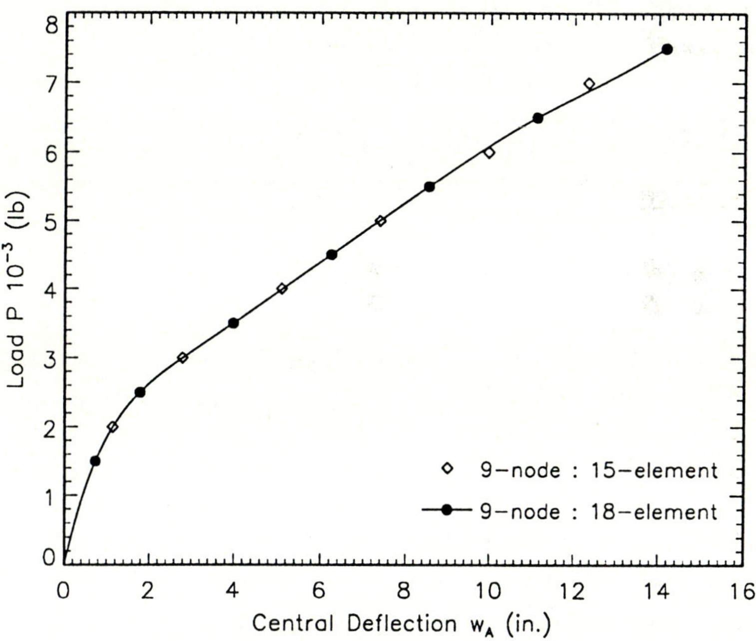

Figure 9: Load-deflection curve at point A of the sphere

Conclusions

In this paper, a new efficient assumed strain formulation with stabilization matrix has been extended to the analysis of geometrically nonlinear thin shell structures. The assumed strain field is split into a lower order part and a higher order part. The lower order assumed strain is replaced by the displacement–dependent strain at lower order integration points. The higher order strain is assumed to include only a small number of higher order polynomial terms enough to suppress the compatible spurious kinematic modes. In conjunction with the use of proper lower order and higher order Gauss quadrature rules, the present formulation results in an element stiffness matrix composed of the portion based on the assumed displacement formulation with reduced integration and the stabilization matrix which controls the kinematic modes.

In the present paper, four–node and nine–node elements are proposed in conjunction with the degenerate solid shell element concept. One set of the higher order assumed independent strain terms is considered for the four–node shell element, while for the nine–node element, two versions of higher order assumed independent strain field are proposed. The results of numerical tests for the nine–node shell element indicate that both versions are almost identical. The numerical results also indicate the present formulation provides very reliable solutions which are in good agreement with other known solutions. It is also observed that the present nine–node element can describe thin shells experiencing considerably large rotation. The new efficient formulation presented in this paper can be extended to geometrically and/or materially nonlinear composite as well as isotropic shells.

Acknowledgement

The support of the present work by the Office of Naval Research under contract N00014–84–K–0385 is gratefully acknowledged.

References

1. Ahmad, S.; Irons, B. M.; Zienkiewicz, O. C.: Analysis of thick and thin shell structures by curved elements. Int. J. Numer. Methods Eng. 2 (1970) 419–451.

2. Ramm, E.: A plate/shell element for large deflections and rotations. In Formulations and Computational Algorithms in Finite Element Analysis. U.S.–Germany Symp. Bathe, K. J.; Oden, J. T.; Wunderlich, W. (eds.) Cambridge MA M.I.T. Press 1977 264–293.

3. Parisch, H.: Geometrical nonlinear analysis of shells. Comp. Methods Appl. Mech. Eng. 14 (1978) 159–178.

4. Parisch, H.: Large displacements of shells including material nonlinearities. Comp. Methods Appl. Mech. Eng. 27 (1081) 183–214.

5. Hughes, T. J. R.; Liu, W. K.: Nonlinear finite element analysis of shells: Part I. Three–dimensional shells. Comp. Methods Appl. Mech. Eng. 26 (1981) 331–362.

6. Hughes, T. J. R.; Liu, W. K.: Nonlinear finite element analysis of shells: Part II. Two–dimensional shells. Comp. Methods Appl. Mech. Eng. 27 (1981) 167–181.

7. Surana, K. S.: Geometrically nonlinear formulation for the curved shell elements. Int. J. Numer. Methods Eng. 19 (1983) 581–615.

8. Oliver, J.; Onate, E.: A total Largrangian formulation for the geometrically nonlinear analysis of structures using finite elements. Part I. Two dimensional problems: shell and plate structures. Int. J. Numer. Methods Eng. 20 (1984) 2253–2281.

9. Lee, S. W.; Pian T. H. H.: Improvement of plate and shell finite elements by mixed formulations. AIAA J. 16 (1978) 29–34.

10. Stolarski H.; Belytschko T.: Membrane locking and reduced integration for curved elements. J. of Appl. Mech. ASME 49 (1982) 172–176.

11. Zienkiewicz, O. C.; Taylor, R. L.; Too J. M.: Reduced integration technique in general analysis of plates and shells. Int. J. Numer. Methods Eng. 3 (1971) 275–290.

12. Hughes, T. J. R.; Cohen, M.; Haroun, M.: Reduced and selective integration techniques in the finite element analysis of plates. Nucl. Eng. Des. 46 (1978) 203–222.

13. Pugh, E. D. L.; Hinton, E.; Zienkiewicz, O. C.: A study of quadrilateral plate bending elements with reduced integration. Int. J. Numer. Methods Eng. 12 (1978) 1059–1079.

14. Parish H.: A critical survey of the 9–node degenerate shell element with special emphasis on thin shell application and reduced integration. Comp. Methods Appl. Mech. Eng. 20 (1979) 323–350.

15. Hughes, T. J. R.; Tezduyar T. E.: Finite elements based upon Mindlin plate theory with particular reference to the four–node isoparametric element. J. of Appl. Mech. ASME 48 (1981) 587–596.

16. MacNeal, R. H.: Derivation of element stiffness matrices by assumed strain distributions. Nucl. Eng. Des. 70 (1982) 3–12.

17. Bathe, K. J.; Dvorkin, E. N.: A four–node plate bending element based on Mindlin /Reissner theory and a mixed interpolation. Int. J. Numer. Methods Eng. 21 (1985) 367–383.

18. Huang, H. C.; Hinton, E.: A new nine node degenerated shell element with enhanced membrane and shear interpolation. Int. J. Numer. Methods Eng. 19 (1986) 73–92.

19. Park, K. C.; Stanley, G. M.: A curved C^o shell element based on assumed natural-coordinate strains. J. Appl. Mech. 53 (1986) 278–290.

20. Jang, J.; Pinsky, P. M.: An assumed covariant strain based 9–node shell element. Int. J. Numer. Methods Eng. 24 (1987) 2389–2411.

21. Stolarski, H. K.; Chiang, M. M.: The mode–decomposition, C^o formulation of curved, two–dimensional structural elements. Int. J. Numer. Methods Eng. 28 (1989) 145–154.

22. Belytschko, T.; Ong, J. S. J.; Liu, W. K.: A consistent control of spurious singular modes in the 9–node Largrange element for the Laplace and Mindlin plate equations. Comp. Methods Appl. Mech. Eng. 44 (1984) 269–295.

23. Belytschko, T.; Liu, W. K.; Ong, J. S. J.; Lam, J. S. L.: Implementation and application of a 9–node Largrange shell element with spurious mode control. Comp. Struct. 20 (1985) 121–128.

24. Belytschko, T.; Liu, W. K.; Ong, J. S. J.: Mixed variational principles and stabilization of spurious mode in the 9–node element. Comp. Methods Appl. Mech. Eng. 62, (1987), 275–292.

25. Belytschko, T.; Wong, B. L.: Assumed strain stabilization procedure for the 9–node Largrange shell element. Int. J. Numer. Methods Eng. 28 (1989) 385–414.

26. Spilker, R. L.: Invariant 8–node hybrid stress elements for thin plates and shells. Int. J. Numer. Methods Eng. 18 (1982) 1153–1178.

27. Saleeb, A. F.; Chang, T. Y.; Graf, W.: A quadrilateral shell element using a mixed formulation. Comp. Struct. 19 (1987) 787–803.

28. Lee, S. W.; Wong, S. C.; Rhiu, J. J.: Study of a nine–node mixed formulation finite element for thin plates and shells. Comp. Struct. 21 (1985) 1325–1334.

29. Rhiu, J. J.; Lee, S. W.: A nine node finite element for analysis of geometrically non–linear shells. Int. J. Numer. Methods Eng. 26 (1988) 1945–1962.

30. Yeom, C. H.; Lee, S. W.: An assumed strain finite element model for large deflection composite shells. accepted for publication in Int. J. Numer. Methods Eng. (1989).

31. Kim, Y. H.; Lee, S. W.: A solid element formulation for large deflection analysis of composite shells. Comp. Struct. 30 (1988) 269–274.

32. Lee, S. W.; Rhiu, J. J.: A new efficient approach to the formulation of mixed finite element models for structural analysis. Int. J. Numer. Methods Eng. 23 (1986) 1629–1641.

33. Malkus, D. S.; Hughes, T. J. R.: Mixed finite element methods–reduced and selective integration techniques : a unification of concepts. Comp. Methods Appl. Mech. Eng. 15 (1978) 31–81.

34. Lee, S. W.: Finite element methods for reduction of constraints and creep analysis. Ph. D. dissertation. Dept. Aero. and Astro. M.I.T. U.S.A. February 1978.

35. Rhiu, J. J.; Lee, S. W.: A new efficient mixed formulation for thin shell finite element models. Int. J. Numer. Methods Eng. 24 (1987) 581–604.

36. Rhiu, J. J.; Lee, S. W.: A sixteen node shell element with a matrix stabilization scheme. Comput. Mech. 3 (1988) 99–113.

37. Ausserer, M. F.; Lee, S. W.: A eighteen–node solid element for thin shell analysis. Int. J. Numer. Methods Eng. 26 (1988) 1345–1365.

38. Rhiu, J. J.; Lee, S. W.: Two higher order shell finite elements with stabilization matrix. Accepted for publication in AIAA J. (1989).

39. Cheng, H.; Gupta, K. C.: An historical note on finite rotations. ASME Journal of Applied Mechanics March 56 (1989) 139–145.

40. Crisfield, M. A.: A fast incremental/iterative solution procedure that handles snap–through. Comp. Struct. 13 (1981) 55–62.

41. Sabir, A. B.; Lock A. C.: The application of finite elements to the large deflection geometrically non–linear behavior of cylindrical shells. *Variational Methods in Engineering.* Brebbia, C. A.; Tottenham, H. (eds.) Southampton University Press 1973.

42. Harte, R.; Doppelt gekrümmte finite Dreieckelemente für die lineare und geometrsch nichtlineare Berechnung allgemeiner Flächentragwerke. Institute für Konstruktiven Ingenieurbau. Ruhr-Universitä in Bochum, Mitteilung Nr. 82–10, Nov. 1982.

Refined Finite Element Laminated Models for the Static and Dynamic Elasto-Plastic Analysis of Anisotropic Shells

D. R. J. Owen and Z. H. Li

University College of Swansea, U.K.

Summary

A refined finite element shell model for the numerical analysis of thick or thin anisotropic laminated shells under static or dynamic loading is presented. A layered approach is adopted for solution with displacement variables assumed at each laminate interface. Elastic–plastic numerical analysis is performed based on flow theory and a Huber–Mises yielding surface extended by Hill for fully three dimensional anisotropic materials. Dynamic analysis is based on Newmark's algorithm used in conjunction with the Hughes and Liu predictor corrector scheme. Numerical results obtained for laminated shells are presented and the effects of boundary constraints on the load/boundary deformation characteristics and on the spread of plastic zones are discussed. Comparisons are made to show the effects of anisotropy and bending/stretching coupling on the elastic–plastic response. The effects of lamina angle sequence on the structural characteristics are also illustrated.

Introduction

Increased interest in laminated construction has led to the development of many numerical models for analysis of the stress and boundary deformation behaviour of composite shells. These include (a) Global models, based on assumed displacement fields, which can adequately predict displacements, natural frequencies and buckling loads but are not sufficiently accurate for stress field computations [1–3] and (b) Local models, encompassing elasticity solutions [4,5], hybrid–stress elements [6,7] and analytical solutions [8], where each layer is represented as an anisotropic continuum and hence become economically unacceptable as the number of layers become even moderately large. The need for a numerical model which is both accurate and economical is therefore apparent. It has been shown that laminated plates composed of advanced fibrous composite materials are susceptible to thickness effects for high rates of elastic modulus to transverse shear modulus and theories based on either the Kirchhoff or Reissner–Mindlin hypotheses are no longer sufficiently accurate for predicting the behaviour of such laminates. It has also been demonstrated in [4,5] and [8,9], through the exact three-dimensional analysis of simply supported laminates, that the in-plane deformations are piecewise linear and the lateral displacement is nearly constant across the thickness. The present work adopts the above displacement characteristics for the development of a refined finite element model for the static and dynamic elasto–plastic analysis of thick or thin anisotropic laminated shells. A layered approach is adopted and displacement variables are assumed at each laminate interface. The "tangent" displacements are assumed to be linear across each layer thickness and the "normal" displacement is taken to be constant through the depth of the shell cross–section. The deformation components, strains and stresses are defined in terms of local coordinate systems which are naturally suited to the modelling of practical laminated anisotropic structures. A substructuring technique [10,11] is adopted to reduce the computational effort. For dynamic analysis the Newmark algorithm with the Hughes and Liu predictor–corrector scheme is adopted resulting in an "effective static problem" which is solved using a Newton–Raphson type solution scheme.

The element model

The element formulation is based on the use of four coordinate systems:

Nodal coordinate system. A nodal coordinate system $V_{il}(i = \alpha, \beta, \gamma)$ is defined at 'normal' l with $V_{\gamma lk-1} = V_{\gamma lk+1}$ and is used to define the components of nodal displacements $(u_\alpha, v_\beta, w_\gamma)$, where subscripts lk+1 and lk−1 denote the top and bottom nodal points on the 'normal' l, respectively (see Fig.1). The unit vector $V_{\gamma l}$ is constructed from the nodal coordinates of the top nodal point k+1 and the bottom nodal point k−1 as

$$V_{\gamma l} = x_l^{k+1} - x_l^{k-1} / |x_l^{k+1} - x_l^{k-1}| = \Delta x_l / |\Delta x_l| \qquad (1)$$

where $\Delta x_l = [\Delta x_l, \Delta y_l, \Delta z_l]^T$. The unit vector $V_{\gamma l}$ defines the direction of the 'normal' across the nodal line k+1 and k−1. The vector $V_{\alpha l}$ is perpendicular to $V_{\gamma l}$ and parallel to the global x–z plane in the Cartesian coordinate system.

$$V_{\alpha l} = j \times V_{\gamma l} / |j \times V_{\gamma l}| \qquad (2)$$

or if $V_{\gamma l}$ is in the y direction

$$V_{\alpha l} = V_{\gamma l} \times i / |V_{\gamma l} \times i| \qquad (3)$$

where i and j are the unit vectors along the x and y directions in the Cartesian coordinate system, respectively. The vector $V_{\beta l}$ is normal to the plane defined by $V_{\alpha l}$ and $V_{\gamma l}$ so that

$$V_{\beta l} = V_{\gamma l} \times V_{\alpha l} / |V_{\gamma l} \times V_{\alpha l}| \qquad (4)$$

Global Cartesian coordinate set (x,y,z). The global Cartesian coordinate system is used to define the nodal coordinates.

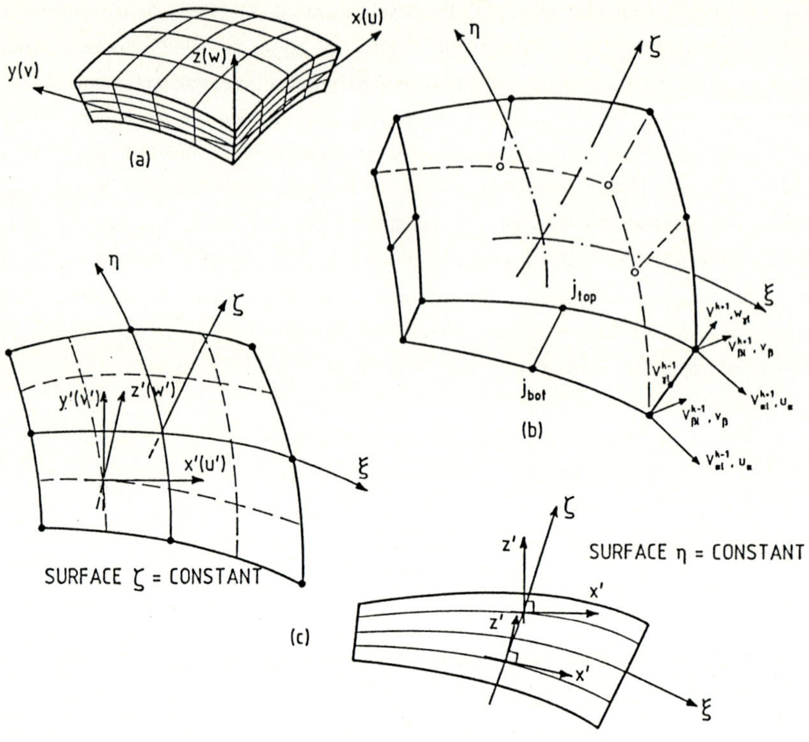

Fig. 1   Coordinate system: a) Global coordinate system, b) Nodal and curvilinear systems, c) Local system of axes

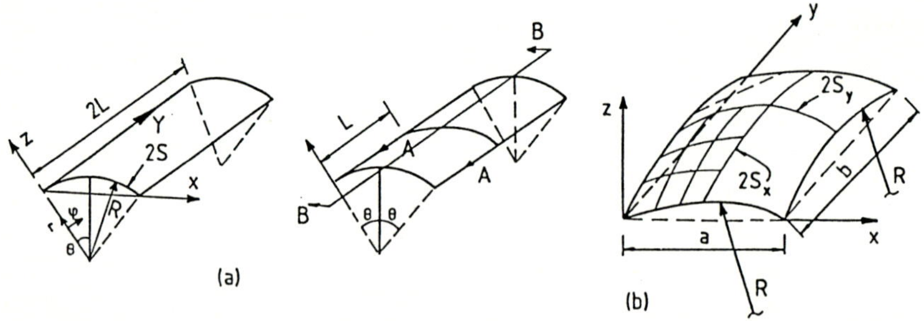

Fig. 2   Test problem geometries

Natural coordinate set (ξ,η,ζ). In this system ξ and η are two curvilinear coordinates based at the centre of the element and ζ is a linear coordinate in the thickness direction. The shape functions N are expressed in terms of this curvilinear coordinate system.

Local coordinate set—x'_i. To calculate the stresses and strains defined at the element sampling points, a local Cartesian coordinate system is introduced. The direction $x'_3(z')$ is taken perpendicular to the surface $\zeta=$constant (where the sampling points are located), and is obtained from the cross product of vectors in the ξ and η directions

$$x'_3 = \begin{Bmatrix} \dfrac{\partial x}{\partial \xi} \\[2mm] \dfrac{\partial y}{\partial \xi} \\[2mm] \dfrac{\partial z}{\partial \xi} \end{Bmatrix} \times \begin{Bmatrix} \dfrac{\partial x}{\partial \eta} \\[2mm] \dfrac{\partial y}{\partial \eta} \\[2mm] \dfrac{\partial z}{\partial \eta} \end{Bmatrix} \tag{5}$$

To define the material directions in relation to the local system of axes in a consistent manner, which is a practical requirement for the analysis of anisotropic structures, fibre composites and reinforced concrete shells etc., a definition of the x'_1 and x'_2 directions are adopted in a similar manner to $\mathbf{V}_{\alpha l}$ and $\mathbf{V}_{\beta l}$. The direction cosine matrix $[\theta]$ which relates the transformations between the local and the global coordinate system is defined by

$$[\theta] = \begin{bmatrix} \bar{x}' & \bar{y}' & \bar{z}' \end{bmatrix} \tag{6}$$

where $\bar{x}'$, $\bar{y}'$ and $\bar{z}'$ are unit vectors in the direction of the x', y' and z' axes respectively.

Displacement and geometry fields. The element displacements within the kth element layer can be expressed as

$$u^e_\alpha = N_j(\xi,\eta)\left[(1-\zeta)u^{k-1}_{\alpha j} + (1+\zeta)u^{k+1}_{\alpha j}\right]/2 = \bar{N}_m(\xi,\eta,\zeta)u_{\alpha m}$$

$$v^e_\beta = N_j(\xi,\eta)\left[(1-\zeta)v^{k-1}_{\beta j} + (1+\zeta)v^{k+1}_{\beta j}\right]/2 = \bar{N}_m(\xi,\eta,\zeta)v_{\beta m} \tag{7}$$

$$w^e_\gamma = N_j(\xi,\eta)w_{\gamma j}$$

in matrix form, we have

$$\{u^e_{\alpha,\beta,\gamma}\} = [\bar{N}]\{\delta^e\} \tag{8}$$

262

in which

$$\{u^e_{\alpha,\beta,\gamma}\} = \{u^e_\alpha, v^e_\beta, w^e_\gamma\} \tag{9}$$

$$\{\delta^e\} = \{u^{k-1}_{\alpha 1}\, u^{k-1}_{\alpha 2}\, \cdots\, u^{k+1}_{\alpha 1}\, u^{k+1}_{\alpha 2}\, \cdots\, v^{k-1}_{\beta 1}\, v^{k-1}_{\beta 2}\, \cdots\, v^{k+1}_{\beta 1}\, v^{k+1}_{\beta 2}\, \cdots\, w_{\gamma 1} w_{\gamma 2}\cdots\}$$

$$= \{\delta^e_\alpha, \delta^e_\beta, \delta^e_\gamma\} \tag{10}$$

and where subscripts α, β and γ denote directions expressed in the nodal coordinate system. The term N_j represents a quadratic Serendipity interpolation for $j=1,2,\ldots,8$, a quadratic Lagrangian interpolation for $j=1,2,\ldots,9$, or alternatively Heterosis type shape functions. The repeated indices j and m imply summation over the element nodes and superscripts $k-1$ and $k+1$ denote the bottom and the top surfaces of the kth element layers, respectively.

The element nodal displacements $\{u_{\alpha j}, v_{\beta j}, w_{\gamma j}\}$ can be transformed to the global coordinate system $x_i (i=1,2,3)$ as $\{u_j, v_j, w_j\}$, so that

$$\left\{\begin{array}{c} u^e_j \\[2mm] v^e_j \\[2mm] w^e_j \end{array}\right\} = \left\{\begin{array}{c} u^e_{\alpha j} V^x_{\alpha j} + v^e_{\beta j} V^x_{\beta j} + w^e_{\gamma j} V^x_{\gamma j} \\[2mm] u^e_{\alpha j} V^y_{\alpha j} + v^e_{\beta j} V^y_{\beta j} + w^e_{\gamma j} V^y_{\gamma j} \\[2mm] u^e_{\alpha j} V^z_{\alpha j} + v^e_{\beta j} V^z_{\beta j} + w^e_{\gamma j} V^z_{\gamma j} \end{array}\right\} \tag{11}$$

where V^x_{ij}, V^y_{ij} and $V^z_{ij} (i=\alpha,\beta,\gamma)$ denote the components of the unit base vectors of the nodal coordinate system expressed in the global system. Substituting (8) into (11), the relation between displacements in the nodal and global coordinate systems can be written as

$$\left\{\begin{array}{c} u^e \\[2mm] v^e \\[2mm] w^e \end{array}\right\} = \left[V_j\right]\left[\bar{N}\right]\{\delta^e_j\} \tag{12}$$

Element Geometry. In the isoparametric formulation, the coordinates of a point within an element are expressed in terms of the nodal coordinates as

$$x^e = N_j(\xi,\eta)\left[(1-\zeta)x^{k-1}_j + (1+\zeta)x^{k+1}_j\right]/2 = \bar{N}_m(\xi,\eta,\zeta)x_m$$

$$y^e = N_j(\xi,\eta)\left[(1-\zeta)y^{k-1}_j + (1+\zeta)y^{k+1}_j\right]/2 = \bar{N}_m(\xi,\eta,\zeta)y_m \tag{13}$$

$$z^e = N_j(\xi,\eta)\left[(1-\zeta)z_j^{k-1} + (1+\zeta)z_j^{k+1}\right]/2 = \bar{N}_m(\xi,\eta,\zeta)z_m$$

where again the subscripts j and m indicate summation and $k-1$ and $k+1$ represent the bottom and top surfaces of the element layer.

Definition of strains. The strain components are defined in terms of the local system of axes $x'_i(i=1,2,3)$ employing the usual shell assumption $\epsilon'_z = 0$. Then

$$\{\epsilon'\} = \begin{Bmatrix} \epsilon_{x'} \\ \epsilon_{y'} \\ \gamma_{x'y'} \\ \gamma_{x'z'} \\ \gamma_{y'z'} \end{Bmatrix} = \begin{Bmatrix} \dfrac{\partial u'}{\partial x'} \\[2mm] \dfrac{\partial v'}{\partial y'} \\[2mm] \dfrac{\partial u'}{\partial y'} + \dfrac{\partial v'}{\partial x'} \\[2mm] \dfrac{\partial u'}{\partial z'} + \dfrac{\partial w'}{\partial x'} \\[2mm] \dfrac{\partial v'}{\partial z'} + \dfrac{\partial w'}{\partial y'} \end{Bmatrix} \tag{14}$$

where u', v' and w' are the displacement components in the local system x'_i. These local derivatives are related to the global derivatives by the following operation

$$\begin{bmatrix} \dfrac{\partial u'}{\partial x'} & \dfrac{\partial v'}{\partial x'} & \dfrac{\partial w'}{\partial x'} \\[2mm] \dfrac{\partial u'}{\partial y'} & \dfrac{\partial v'}{\partial y'} & \dfrac{\partial w'}{\partial y'} \\[2mm] \dfrac{\partial u'}{\partial z'} & \dfrac{\partial v'}{\partial z'} & \dfrac{\partial w'}{\partial z'} \end{bmatrix} = [\theta]^T \begin{bmatrix} \dfrac{\partial u}{\partial x} & \dfrac{\partial v}{\partial x} & \dfrac{\partial w}{\partial x} \\[2mm] \dfrac{\partial u}{\partial y} & \dfrac{\partial v}{\partial y} & \dfrac{\partial w}{\partial y} \\[2mm] \dfrac{\partial u}{\partial z} & \dfrac{\partial v}{\partial z} & \dfrac{\partial w}{\partial z} \end{bmatrix} [\theta] \tag{15}$$

where $[\theta]$ is a transformation matrix defined by (6). The derivatives of the displacements with respect to the global coordinates are given by

$$\begin{bmatrix} \dfrac{\partial u}{\partial x} & \dfrac{\partial v}{\partial x} & \dfrac{\partial w}{\partial x} \\[2mm] \dfrac{\partial u}{\partial y} & \dfrac{\partial v}{\partial y} & \dfrac{\partial w}{\partial y} \\[2mm] \dfrac{\partial u}{\partial z} & \dfrac{\partial v}{\partial z} & \dfrac{\partial w}{\partial z} \end{bmatrix} = J^{-1} \begin{bmatrix} \dfrac{\partial u}{\partial \xi} & \dfrac{\partial v}{\partial \xi} & \dfrac{\partial w}{\partial \xi} \\[2mm] \dfrac{\partial u}{\partial \eta} & \dfrac{\partial v}{\partial \eta} & \dfrac{\partial w}{\partial \eta} \\[2mm] \dfrac{\partial u}{\partial \zeta} & \dfrac{\partial v}{\partial \zeta} & \dfrac{\partial w}{\partial \zeta} \end{bmatrix} \tag{16}$$

where J is the Jacobian matrix of the transformation in (13).

The strain matrix B, relating the strain components in the local system to the element nodal variables, can be constructed by coordinate transformation as

$$\{\epsilon'\} = B\{\delta\} \tag{17}$$

where $\{\bar{\epsilon}\}$ and $\{\delta\}$ are defined in (14) and (10) respectively.

Elastic Constitutive Relations

Although for the most general cases of anisotropy the number of independent elastic constants is 21, this number is considerably reduced if the material internal composition possesses symmetry of certain types [8]. As in Refs. [10-12], the following state of anisotropy will be assumed. Considering a state of anisotropy which possesses three mutually orthogonal planes of symmetry, if the reference system of the orthogonal local axes (x'_j) is parallel to the principal material axes $(1,2,3)$ and noting the assumptions w_γ = constant across the shell normals and that the normal stress $\sigma'_{zk} = \sigma_{3k} = 0$, then the following stress-strain relationship is obtained

$$\{\sigma_{1,2,3}\}_k = \bar{D}\{\epsilon_{1,2,3}\}_k \tag{18}$$

where

$$\{\sigma_{1,2,3}\}_k = \{\sigma_1 \sigma_2 \tau_{12} \tau_{13} \tau_{23}\}_k \tag{19}$$

$$\{\epsilon_{1,2,3}\}_k = \{\epsilon_1 \epsilon_2 \gamma_{12} \gamma_{13} \gamma_{23}\}_k \tag{20}$$

$$\bar{D}_k = \begin{bmatrix} \bar{D}_1 & \bar{D}_{12} & 0 & 0 & 0 \\ \bar{D}_{12} & \bar{D}_2 & 0 & 0 & 0 \\ 0 & 0 & \bar{D}_3 & 0 & 0 \\ 0 & 0 & 0 & \bar{D}_4 & 0 \\ 0 & 0 & 0 & 0 & \bar{D}_5 \end{bmatrix}_k \tag{21}$$

$$\bar{D}_{1k} = E_{1k}/\left[1 - \mu_{12k}\mu_{21k}\right]; \qquad \bar{D}_{3k} = G_{12k}$$

$$\bar{D}_{2k} = E_{2k}/\left[1 - \mu_{12k}\mu_{21k}\right]; \qquad \bar{D}_{4k} = G_{13k} \tag{22}$$

$$\bar{D}_{12k} = E_{2k}\mu_{12k}/\left[1 - \mu_{12k}\mu_{21k}\right]; \qquad \bar{D}_{5k} = G_{23k}$$

Subscript k denotes the kth particular lamina to which expressions (18-20) relate. If the principal axes of anisotropy 1,2 do not coincide with the reference axes x', y', but are rotated by a certain angle β, the new elasticity matrix D_k is determined from the following transformation:

$$D_k = T_k'^T \bar{D}_k T_k' \tag{23}$$

where transformation matrix T' is defined in Ref. [13], and $\bar{D}_k$ is the general elasticity matrix of lamina k given as

$$D_k = \begin{bmatrix} D_1 & D_{12} & D_{13} & 0 & 0 \\ D_{12} & D_2 & D_{23} & 0 & 0 \\ D_{13} & D_{23} & D_3 & 0 & 0 \\ 0 & 0 & 0 & D_4 & D_{45} \\ 0 & 0 & 0 & D_{45} & D_5 \end{bmatrix}_k \tag{24}$$

Expression (24) defines the elasticity matrix D_k for any particular lamina k in terms of the local coordinate set (x',y',z'). It is obvious that this matrix is dependent on fibre orientation and in general is different for each lamina. The stress components for a particular lamina k can be written in terms of local strains $\{\epsilon'\}$ and elasticity constant D_k as

$$\{\sigma'\}_k = D_k \{\epsilon'\}_k \tag{25}$$

Flow Theory of Plasticity

The generalized Huber–mises Law. The Huber–Mises yield criterion, generalised for anisotropic materials, is adopted which can be expressed in the form

$$f_k = \bar{\sigma}_k^2 = \alpha_{12k}(\sigma_{11k} - \sigma_{22k})^2 + \alpha_{23k}(\sigma_{22k} - \sigma_{33k})^2 +$$
$$\alpha_{13k}(\sigma_{33k} - \sigma_{11k})^2 + 3\alpha_{44k}\tau_{12k}^2$$
$$+ 3\alpha_{55k}\tau_{13k}^2 + 3\alpha_{66k}\tau_{23k}^2 \tag{26}$$

where the σ's (τ's) are stress components, the α's are parameters of anisotropy, the subscripts 1, 2 and 3 refer to directions of the three principal material axes of anisotropy and the subscript k refers to the kth lamina.

Noting the assumptions that w_γ = constant across the thickness and that the normal stress $\sigma_{z'k} = \sigma_{3k} = 0$, yield function (26) can be rewritten

266

$$\bar{\sigma}_k^2 = a_{1k}\sigma_{1k}^2 + a_{2k}\sigma_{2k}^2 + 2a_{12k}\sigma_{1k}\sigma_{2k} + a_{3k}\tau_{12k}^2 +$$

$$a_{4k}\tau_{13k}^2 + a_{5k}\tau_{23k}^2 \qquad (27)$$

where $\sigma_{1k} = \sigma_{11k}, \sigma_{2k} = \sigma_{22k}, \tau_{12k}, \tau_{13k}$ and τ_{23k} are the non-zero stress components existing in the kth lamina and $a_{1k}, a_{2k}, a_{12k}, a_{3k}, a_{4k}$ and a_{5k} are the anisotropic parameters for the kth lamina, which can be determined experimentally. In matrix form, (27) can be written

$$\bar{\sigma}_k^2 = \{\sigma_{1,2,3}^T\}_k \, A_k \{\sigma_{1,2,3}\}_k \qquad (28)$$

where parameter matrix A_k is given by

$$A_k = \begin{bmatrix} a_{1k} & a_{12k} & 0 & 0 & 0 \\ a_{12k} & a_{2k} & 0 & 0 & 0 \\ 0 & 0 & a_{3k} & 0 & 0 \\ 0 & 0 & 0 & a_{4k} & 0 \\ 0 & 0 & 0 & 0 & a_{5k} \end{bmatrix} \qquad (29)$$

Once again if the principal axes of anisotropy 1 and 2 do not coincide with the reference axes x' and y' (the local coordinate set) but are rotated by a certain angle θ, the equation must be transformed. The equation of stress transformation is

$$\{\sigma_{1,2,3}\}_k = T_k \{\sigma_{x',y',z'}\}_k \qquad (30)$$

where T is the transformation matrix defined in Ref.[13]. The effective stress expressed in the reference system x', y' and z' is then

$$\bar{\sigma}_k^2 = \{\sigma_{x',y',z'}^T\}_k \bar{A}_k \{\sigma_{x',y',z'}\}_k \qquad (31)$$

in which $A_k = T^T_k A_k T_k$ is the matrix of the new anisotropic parameters given by

$$\bar{A}_k = \begin{bmatrix} \bar{a}_{11k} & \bar{a}_{12k} & \bar{a}_{13k} & 0 & 0 \\ \bar{a}_{12k} & \bar{a}_{22k} & \bar{a}_{23k} & 0 & 0 \\ \bar{a}_{13k} & \bar{a}_{23k} & \bar{a}_{33k} & 0 & 0 \\ 0 & 0 & 0 & \bar{a}_{44k} & \bar{a}_{45k} \\ 0 & 0 & 0 & \bar{a}_{45k} & \bar{a}_{55k} \end{bmatrix} \qquad (32)$$

The parameters in equation (29) can be determined by six independent yield tests. By successively allowing all stress components to be zero in (27) except the one under consideration, the initial parameters (no hardening effects) are obtained. Interested readers are referred to Refs. [12,13] for details.

Numerical Solution for Static Analysis

The solution of the quasistatic nonlinear elastic–plastic problem is reached by the usual incremental and iterative procedure [13,14]. The incremental elastic–plastic constitutive relation for the general layer is given by

$$\{d\sigma_k\} = D_{epk} \{d\epsilon_k\} \tag{33}$$

where

$$D_{epk} = D_k - \frac{D_k b_k b_k^T D_k}{H_k + b_k^T D_k b_k} \tag{34}$$

and in which

$$b_k^T = \frac{\partial f_k}{\partial \{\sigma_k\}} \tag{35}$$

The term H_k is the tangent to the effective plastic stress–strain curve and is a function of the accumulated effective plastic strain $\bar{\epsilon}^p_k$.

The current element stiffness matrix is then evaluated by assembling all layer contributions and takes the form

$$[K] = \int_{-1}^{+1} \int_{-1}^{+1} \int_{-1}^{+1} [B(\xi,\eta,\zeta)]^T [D_{ep}] [B(\xi,\eta,\zeta)] \det J \, d\xi \, d\eta \, d\zeta \tag{36}$$

Elimination of variables after assembly at the layer level – Substructuring

Let the top and bottom surface of the laminate be loaded by normal stresses, q(1) and q(2), respectively. With the assumption that the normal displacement is constant across the thickness, the normal load q (where q = q(1) + q(2)) can be assumed to act at the top face. In contrast to other approaches [15], where elimination is undertaken at internal nodes and nodeless variables, in this work elimination of variables is performed at the 'layer' level [10,11]. The element is assembled for a typical layer p ($p = 1,2,\ldots,n$, where n is the number of layers discretised across the thickness of the laminate). The number of layers may be equal to or greater than the number of layers of the laminate. In the usual way

$$\partial \pi^P / \partial \delta^P = K^P \delta^P + f^P \tag{37}$$

Since δP can be subdivided into parts which are common with the upper layer, $\delta^{P,i+1}$, and others which occur in the particular layer p only, $\delta^{P,i}$, we can immediately write

$$\frac{\partial \pi}{\partial \delta^{P,i}} = \frac{\partial \pi^P}{\partial \delta^{P,i}} = 0 \tag{38}$$

and eliminate $\delta^{P,i}$ from further consideration. Writing (37) in a partitioned form, we have

$$
\partial \pi^P / \partial \delta^P =
\begin{bmatrix} \partial \pi^P / \partial \delta^{P,i} \\ \partial \pi^P / \partial \delta^{P,i+1} \end{bmatrix}
$$

$$
= \begin{bmatrix} K^P_{11} & K^P_{12} \\ K^P_{21} & K^P_{22} \end{bmatrix}
\begin{bmatrix} \delta^{P,i} \\ \delta^{P,i+1} \end{bmatrix}
+ \begin{bmatrix} f^{P,i} \\ f^{P,i+1} \end{bmatrix}
$$

$$
= \begin{bmatrix} 0 \\ \partial \pi^P / \partial \delta^{P,i+1} \end{bmatrix}, \tag{39}
$$

where $f^{P,i} = \{0\}$, ($p = 1,2,\ldots,n$), $f^{P,i+1} = \{0\}$ ($p = 1,2,\ldots,n-1$).

From the first set of equations given above, we have

$$\delta^{P,i} = - \left[K^P_{11}\right]^{-1} \left[K^P_{12}\right] \delta^{P,i+1} \tag{40}$$

which, on substituting in the second set, yields

$$\partial \pi^P / \partial \delta^{P,i+1} = [\bar{K}^P] \delta^{P,i+1} + \bar{f}^{P,i+1} \tag{41}$$

in which

$$\left[\bar{K}^P\right] = \left[K^P_{22}\right] - \left[K^P_{21}\right]\left[K^P_{11}\right]^{-1}\left[K^P_{12}\right] \tag{42}$$

$$\bar{f}^{P,i+1} = \begin{cases} 0 & p \neq n \\ f^{P,i+1} & p = n \end{cases}$$

It is obvious that, for $p = n$, the term $[\overline{KP}]$ is the global stiffness, and the total number of variables to be evaluated is the same as for one layer. This means that no matter how many layers the laminate contains, just one layer's variables need to be solved for. This leads to a considerable saving in the equation–solving effort and makes this local displacement–based model both practical and economical.

Dynamic Equilibrium Equations

For dynamic analysis the equilibrium of a body in motion can be described by the principle of virtual work and expressed by the following equation at time station t_n, irrespective of material behaviour

$$\int_{\Omega} \left[\delta\epsilon_n\right]^T\{\sigma_n\}d\Omega - \int_{\Omega}\left[\delta u_n\right]^T\left[b_n - \rho_n u_{n,tt} - c_n u_{n,t}\right]d\Omega$$

$$- \int_{\Gamma_t}\left[\delta u_n\right]^T t_n d\Gamma = 0 \qquad (43)$$

where δu_n is the vector of virtual displacements, $\delta\{\epsilon_n\}$ the vector of associated virtual strains, δb_n the vector of applied body forces, δt_n the vector of surface tractions, $\{\sigma_n\}$ the vector of stresses, ρ_n the mass density, c_n the damping parameter and the subscripts $(,t)$ denote differentiation with respect to time. The domain of interest Ω has two boundaries: Γ_t on which boundary tractions t_n are specified and Γ_u on which displacements u_n are specified.

The dynamic equilibrium equation (43) at the nodes j of any discrete set of structural elements can be expressed as follows:

$$\left\{P_j\right\}_n - \left\{F^b_j\right\}_n - \left\{F^I_j\right\}_n - \left\{F^d_j\right\}_n - \left\{F^t_j\right\}_n = 0 \qquad (44)$$

where $\{P_j\}_n$ is the vector of nodal internal resisting or restoring forces, $\{F^b_j\}_n$ is the vector of consistent nodal forces for the applied body forces, $\{F^I_j\}_n$ the vector of nodal inertia forces, $\{F^d_j\}_n$ the vector of nodal damping forces and $\{F^t_j\}_n$ the vector of consistent nodal forces associated with the boundary tractions.

We can rewrite equation (44) in matrix form, neglecting body forces, so that at time station t_n, for element e, we have

$$\left[M^e\right]\{a_n\} + \left[C^e\right]\{v_n\} + \left\{P^e\left[d_n\right]\right\} = \left\{f^e_n\right\} \qquad (45)$$

where $[M^e]$ is the mass matrix, $[C^e]$ the damping matrix, $\{P^e\}$ the internal resisting forces, and $\{d_n\}$, $\{v_n\}$, $\{a_n\}$ are, respectively, the nodal displacement, velocity and acceleration vectors.

Mass matrix. Employing the same displacement field assumptions as used for element stiffness evaluation, leads to a consistent mass matrix formulation and an implicit time integration scheme and this approach is employed in the present work.

Damping matrix. Rayleigh damping, which is a combination of mass proportional and stiffness proportional damping, is assumed and can also be considered as a two term approximation to the Caughey series [16]. The damping matrix is given as

$$[C] = \alpha_o [M] + \alpha_1 [K] \qquad\qquad (46)$$

where α_o and α_1 are the mass proportional and stiffness proportional damping factors respectively.

Internal resisting forces. These forces can be easily found by use of the principle of virtual work.

$$\left\{P\left[d_n\right]\right\} = \Sigma \int_{\Omega_e} [\delta\epsilon]^T \left\{\sigma^e\right\} d\Omega$$

The summation indicates that the internal resisting forces {P} for the entire structure are obtained by a direct assembly of the element force vectors.

Time integration algorithm

To solve Eq. (45), the Newmark algorithm with the Hughes and Liu predictor–corrector scheme [14,17] is used as follows:

1. To predict the solution at time station n+1, set iteration counter i=0
2. Begin the predictor phase in which we set

$$d_{n+1}^{(i)} = \bar{d}_{n+1} = d_n + \Delta t v_n + \Delta t^2 (1-2\beta) a_n/2$$

$$v_{n+1}^{(i)} = \bar{v}_{n+1} = v_n + \Delta t (1-\gamma) a_n \qquad\qquad (47)$$

$$a_{n+1}^{(i)} = d_{n+1}^{(i)} - \bar{d}_{n+1}/(\Delta t^2 \beta) = 0$$

3. Evaluate the residual forces

$$\Psi^{(i)} = f_{n+1} - Ma_{n+1}^{(i)} - Cv_{n+1}^{(i)} - P(d_{n+1}^{(i)})$$

4. If required, form the effective stiffness matrix using the expression

$$K^* = M/(\Delta t^2 \beta) + \gamma C_T/(\Delta t \beta) + K_T(d_{n+1}^{(i)})$$

otherwise use a previously calculated K^*.

5. Evaluate the correction to the displacements

$$\Delta d^{(i)} = [K^*]^{-1} \psi^{(i)}$$

6. Enter the corrector phase in which we set

$$d_{n+1}^{(i+1)} = d_{n+1}^{(i)} + \Delta d^{(i)}$$
$$a_{n+1}^{(i+1)} = \left[d_{n+1}^{(i+1)} - \bar{d}_{n+1} \right]/(\Delta t^2 \beta) \qquad (48)$$
$$v_{n+1}^{(i+1)} = \bar{v}_{n+1} + \Delta t \gamma a_{n+1}^{(i+1)}$$

7. Check for solution convergence; if the condition $\|\Delta d^{(i)}{}_{n+1}\| < \|d^{(i)}{}_{n+1}\|$ is not satisfied then update iteration counter i = i+1 and go to step (3), otherwise continue.

8. Set

$$d_{n+1} = d_{n+1}^{(i+1)}$$
$$v_{n+1} = v_{n+1}^{(i+1)} \qquad (49)$$
$$a_{n+1} = a_{n+1}^{(i+1)}$$

for use in the next time step. Also set n = n+1, form P and begin the next time step.

Numerical Examples

The following material properties which correspond to a high modulus graphite/epoxy composite [18-20] are adopted for all examples:

$$E_1 = 25 \times 10^6 \, psi, \quad E_t = 10^6 \, psi, \quad G_{1t} = 0.5 \times 10^6 \, psi$$
$$G_{tt} = 0.2 \times 10^6, \quad \mu_{1t} = \mu_{tt} = 0.25, \quad \rho = 0.14316 \times 10^{-3} \, Lb - f \, s^2/in^4$$

where 1 signifies the direction parallel to the fibres, t the transverse direction and μ_{1t} is the Poisson's ratio measuring strain in the t direction under uniaxial normal stress in the 1 direction. The anisotropic plastic data assumed are detailed in Ref. [12].

For the convenience of numerical analysis and comparison purposes, the names cross–ply and angle–ply, are introduced in a similar way to the case of laminated plates to describe the lamina stack sequences and lamina angle sequences. The cases considered are: a) cross–ply with fibre orientations changing between 90° and 0° with respect to the y axis in the global coordinate system and the 90° lamina at the outer surfaces; namely in lamina stack sequence $90^{\circ}/0^{\circ}/90^{\circ}...90^{\circ}$; b) angle–ply with fibre orientations alternating between $+\varphi$ and $-\varphi$ with respect to the y axis and the $+\varphi$ lamina at the bottom surface, and in which lamina of the same orientations have equal thickness. In the case of cylindrical laminated shells the y axis is the generator of the shell. Normalized quantities, $u = uE_T/s^3h$, $(s=2L/h$, h the shell thickness), $w = w/h$ and $z = z/h$ are defined to present the numerical results.

With reference to Fig. 2(a), the following boundary conditions are considered for cylindrical shells.

Case A:

$$v_\beta = w_\gamma = 0 \qquad \text{at } \varphi = 0 \text{ and } \varphi = 2\theta$$

$$u_\alpha = 0 \qquad \text{at } \varphi = 0, \ \varphi = 2\theta \text{ and } r = R$$

$$u_\alpha = w_\gamma = 0 \qquad \text{at } y = 0 \text{ and } y = 2L$$

$$v_\beta = 0 \qquad \text{at } y = 0, \ y = 2L \text{ and } r = R$$

This is to simulate a simply supported condition in which the tangent direction to the mid–surface on the boundary is constrained.

Case B:

$$u_\alpha = w_\gamma = 0 \qquad \text{at } \varphi = 0 \text{ and } \varphi = 2\theta$$

$$v_\beta = w_\gamma = 0 \qquad \text{at } y = 0 \text{ and } y = 2L$$

The laminated cylindrical shells are subjected to a normally applied distributed Sine load

$$q(x,y) = q_o \sin(\pi x/2a)\sin(\pi y/2L)$$

where q_o is the load parameter and L and a are illustrated in Fig. 2(a).

Elastic–plastic cylindrical laminated anisotropic shells. Figs. 3 and 4 show the effect of boundary support on the elastic–plastic behaviour of a 3 lamina cross–ply cylindrical shell subjected to a sine surface load. The boundary conditions in Fig. 3 and Fig. 4 correspond to case A and case B respectively and the geometry characteristics are 2S = 2L=10, h=2.5 (s=4) and $\theta = 15^{\circ}$ and a 3 x 3 x 6 mesh configuration is employed.

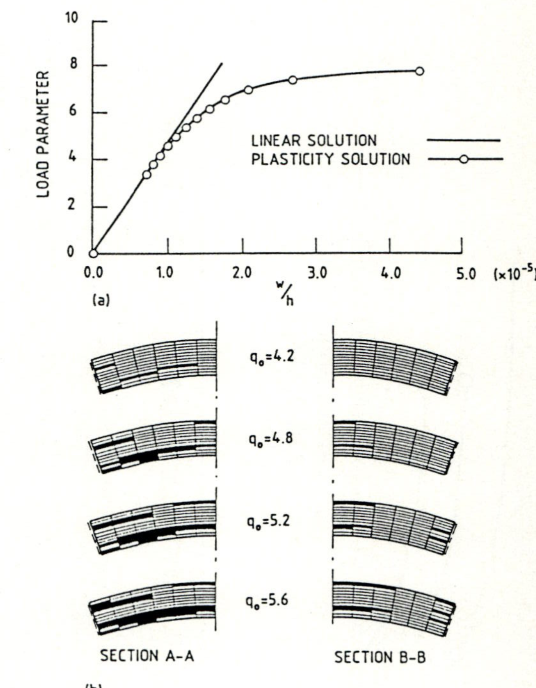

Fig. 3 Elastic-plastic numerical response of a Sine-loaded simply supported (case A, mid-surface support) three layer (h/3, h/3, h/3) cross-ply laminated cylindrical shell (with s=4, 3x3x6 mesh) showing:

(a) load-central deflection characteristics
(b) spread of plastic zones

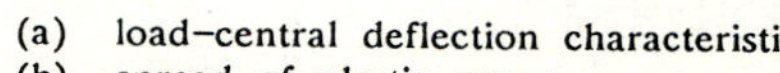

Fig. 4 Elastic-plastic numerical response of a Sine-loaded simply supported (case B, surface support) three layer (h/3, h/3, h/3) cross-ply laminated cylindrical shell (with s=4, 3x3x6 mesh) showing:

(a) load-central deflection characteristics
(b) spread of plastic zones

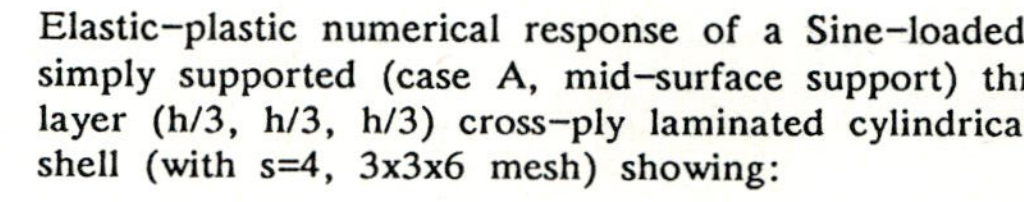

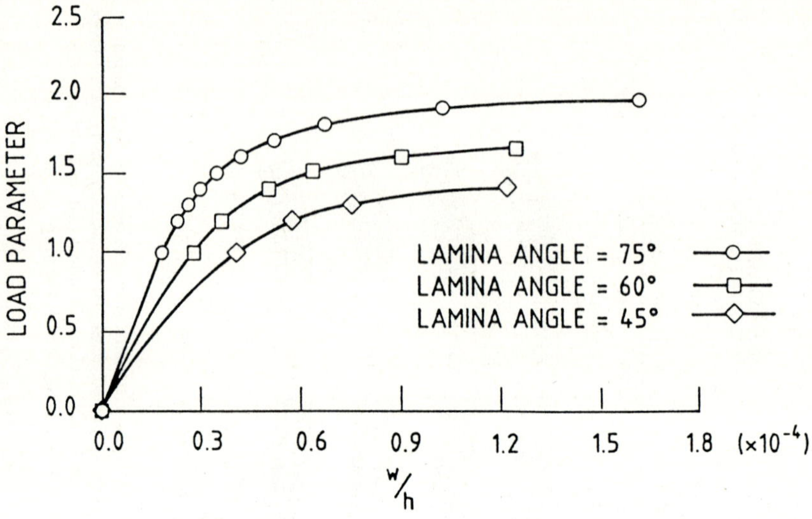

Fig. 5   Elastic–plastic numerical response of a uniformly loaded simply supported (case B, surface support) four layer (h/4, h/4, h/4, h/4) angle-ply ($\theta/-\theta/\theta/-\theta$) laminated cylindrical shell (with s=10, 3x3x4 mesh) showing load–central deflection characteristics.

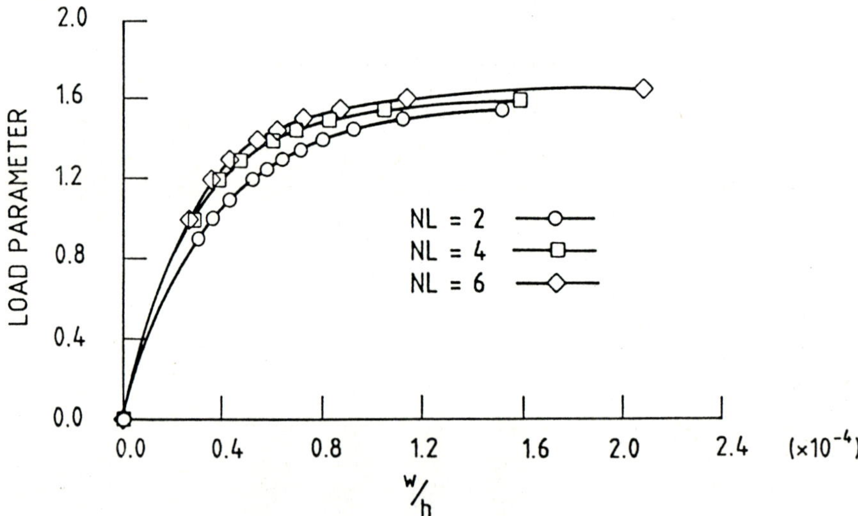

Fig. 6   Elastic–plastic numerical response of a uniformly loaded simply supported (case B, surface support) cross-ply laminated cylindrical shell (with s=10) showing load–central deflection characteristics: NL = Number of Lamina

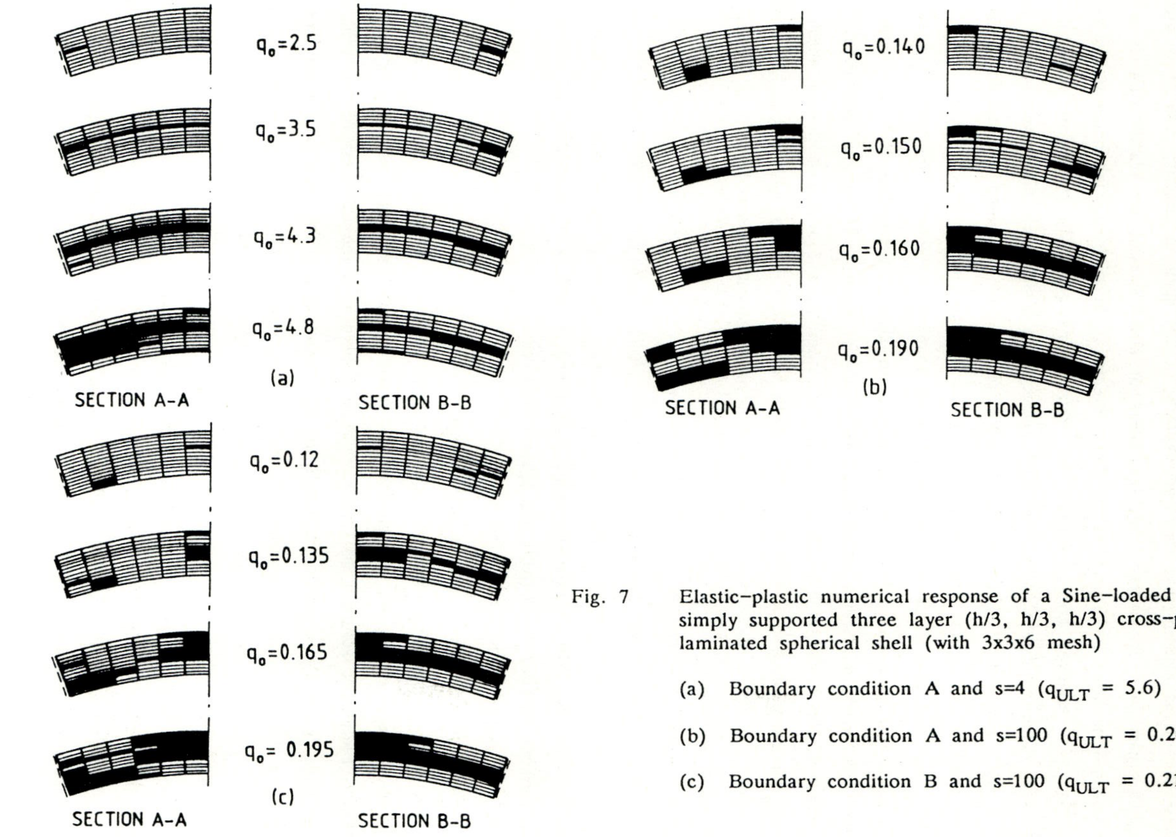

Fig. 7 Elastic–plastic numerical response of a Sine–loaded simply supported three layer (h/3, h/3, h/3) cross–ply laminated spherical shell (with 3x3x6 mesh)

(a) Boundary condition A and s=4 (q_{ULT} = 5.6)

(b) Boundary condition A and s=100 (q_{ULT} = 0.24)

(c) Boundary condition B and s=100 (q_{ULT} = 0.215)

Because of the midplane support on the boundary face in Fig. 3, the occurence of plasticity is first detected at the boundary near the middle point. However, yielding is first detected at the bottom lamina at the boundary region in Fig. 4 for a surface support (Case B). It is evident that the load capacity of the shell in support case B is higher than for case A. It is seen that the effect of shear stresses on plastic flow is more pronounced in case A. The effect of fibre orientation in angle-ply laminates is illustrated in Fig. 5. Analyses are performed for four lamina angle-ply cylindrical shells under boundary condition case B with s=2L/h=10 subjected to a uniformly distributed normal load q_o. The results show that the elastic-plastic response of laminated anisotropic shells is more sensitive to lamina angle variation than is the case for composite plates Ref. [12]. Fig. 6 indicates the effect of bending/stretching coupling on the elastic-plastic behaviour. All cases in Fig. 6 have the same geometry with 2L=10, h=1.0 and $\theta = 15^\cdot$ with the same boundary conditions (case B) and only the number of lamina is varied. A quadrant of the cross-ply laminated shells is discretised by a 3 x 3 x 4 element mesh for a two lamina shell, a 3 x 3 x 4 element mesh for a four lamina shell and a 3 x 3 x 6 element mesh for a six lamina shell. A uniformly distributed normal force is applied to the upper surface.

Elastic-plastic spherical laminated anisotropic shells. A set of 3 lamina cross-ply spherical shells with square base are considered whose geometry characteristics are illustrated in Fig. 2(b). Boundary support conditions A and B corresponding to mid-plane and surface restraints respectively are again considered.

Fig. 7 shows the effects of boundary support and shell thickness on the elastic-plastic behaviour of a 3 lamina cross-ply spherical shell subjected to a sine surface load. To give a consistent basis for comparison the ratio of span to thickness, s, used in the previous analyses is adopted, so that $s = 2S_x/h$, where h is the thickness of the spherical shell and $2S_x$ is the circular arc length in the x-direction. The boundary conditions in Fig. 7(a) and Fig. 7(b) correspond to case A and those of Fig. 7(c) correspond to case B. The geometry characteristics are $2S_x = 2S_y = 10$, $\theta = 15^\cdot$ with h=2.5 (s=4) in Fig. 7(a) and h=0.1 (s=100) in both Fig. 7(b) and Fig. 7(c) and a 3 x 3 x 6 mesh configuration is employed. Because of the midplane support to the boundary face and thick shell action in Fig. 7(a), the occurence of plasticity is first detected at the boundary near the midpoint. In Fig. 7(c) yielding is first detected at the bottom lamina near the boundary (where a maximum shear force exists) and also at the top surface of the shell in the central region.

Elastic-plastic dynamic response of laminated spherical shells. The dynamic response of a series of laminated anisotropic spherical shells is considered. A set of three lamina (h/3,h/3,h/3) symmetric cross-ply $(0^\cdot/90^\cdot/0^\cdot)$ laminated spherical shells with laminate thickness 5 inch, i.e. s=10, is analysed by adopting a 3 x 3 x 6 element mesh for one quadrant of the shell and the results are presented in Fig. 8. Simply supported boundary condition, case A, is adopted and the time stepping increment employed is 0.5×10^{-4} sec. when the shell angle, θ, is $10^{\cdot}$, 0.4×10^{-4} when the shell angle, θ, is $20^{\cdot}$ and 0.3×10^{-4} sec. when the shell angle θ is $30^{\cdot}$. The laminates are subjected to a suddenly applied Sine load at the upper surface of the structure with load parameter $q_o = 1.0$. The results in Fig. 8 show the effects of the shell angle (different structure design) on the bending stretching coupling behaviour which decreases with increasing shell angle and the variation of the shell angle is seen to significantly affect the dynamic response.

A set of laminated angle-ply spherical shells with laminate thickness 5 inch is analysed and presented in Figs. 9 and 10 by adopting a 4 x 4 x nl element mesh for whole shell analysis, where nl denotes the number of element layers which is the number of laminae if the shell is composed of more than two laminae and is four if the shell is constructed from two laminae. Clamped boundary conditions are adopted and the time stepping increment employed is 0.3×10^{-4} sec. when the shell angle, θ, is $10^{\cdot}$ or $20^{\cdot}$ and is 0.15×10^{-4} sec. when the shell angle, θ, is greater.

The dynamic response of a set of four ply (h/4,h/4,h/4,h/4) angle-ply laminated spherical shells with lamina angles ($\pm\varphi$) of 10,20,30 and 45 degrees, subjected to a suddenly applied Sine load at the upper surface of the structures with load parameter $q_o = 2.0$ are presented in Fig. 9 where the effects of the lamina stacking angle $\pm\varphi$ on the coupling of bending and stretching is illustrated. The dynamic response of a set of angle-ply laminated spherical shells with lamina angle ($\varphi=45^{\cdot}$, $\varphi/-\varphi/...$) and lamina number = 2, 4, and 6, subjected to a suddenly applied Sine load at the upper surface of the structure with load parameter $q_o = 2.0$, are presented in Fig. 10. The lamina sequences and thicknesses in the shells are (h/2,h/2),(h/4,h/4,h/4,h/4) and (h/6,h/6,h/6,h/6,h/6,h/6), respectively and the effects of the lamina stacking number on the bending and stretching coupling are shown.

<u>Discussion and Conclusions</u>

A local element model based on a refined degenerated shell theory for thick and thin anisotropic laminated shells has been developed in which a series of local and global coordinate systems are introduced to make the model particularly suitable for the solution of practical composite shell structures. The significance of boundary support conditions in anisotropic laminated shell structures is illustrated and the results presented support the conclusion arrived at in Ref. [12] that plastic zones have a tendency to spread along weaker layers in both thick and thin laminates with this effect being more noticeable in composite shells than plate structures. It is evident from Fig. 5 that composite shells are

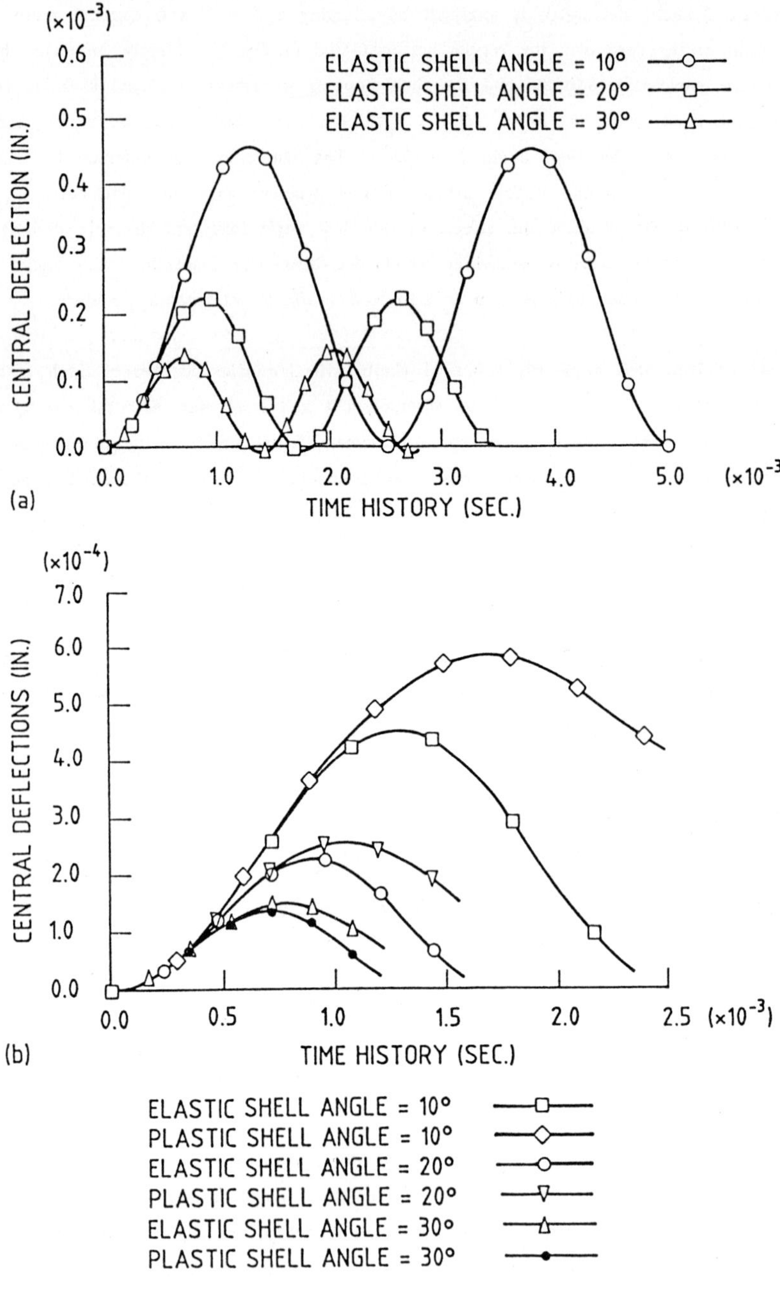

Fig. 8   Dynamic response of a set of cross–ply laminated spherical shells subjected to a suddenly applied Sine surface load with load parameter 1.0 – showing the effect of variation of shell angle on the dynamic response
(a)   elastic response
(b)   elastic and elastic–plastic response

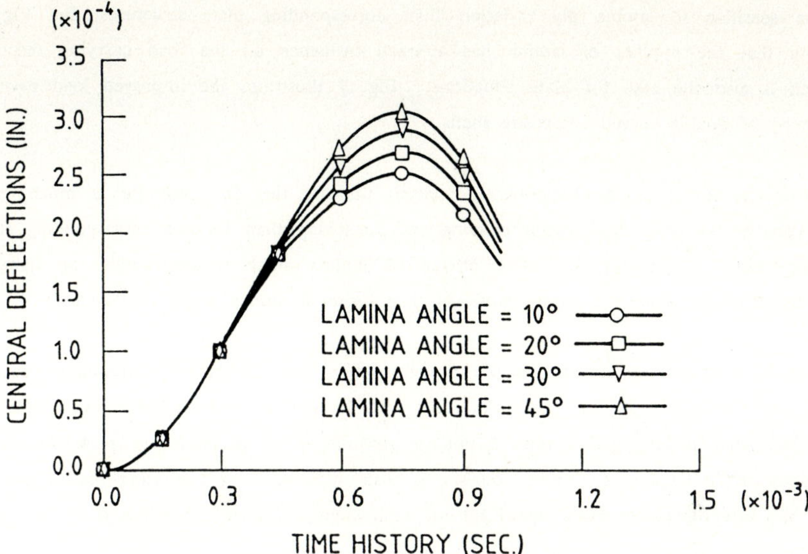

Fig. 9   Dynamic response of a set of angle–ply laminated spherical shells subjected to a suddenly applied Sine surface load with load parameter 2.0 – showing the effect of lamina stack angle on the dynamic response

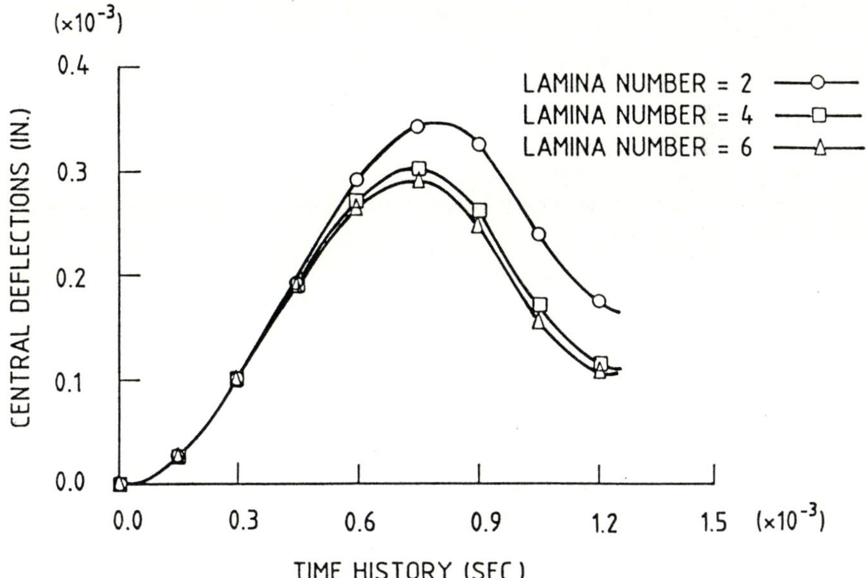

Fig. 10   Dynamic response of a set of angle–ply laminated spherical shells subjected to a suddenly applied Sine surface load with load parameter 2.0 – showing the effect of lamina stack number on the bending stretching coupling behaviour

more sensitive to lamina ply variation than corresponding plate structures, but Fig. 6 shows that the number of lamina has a small influence on the load carrying response which is also the case for plate situations. Fig. 7 illustrates the improved load carrying capacity of doubly curved composite shells.

The results for dynamic elastic–plastic analysis indicate that the response is much more sensitive to the coupling between bending and stretching than for the corresponding static elastic–plastic situation [12]. The effects of lamina stack sequences on the dynamic elastic–plastic response are more significant than those of lamina angle sequences.

Comparison of the results of dynamic analysis shows that the bending stretching coupling effect in laminated spherical shells decreases with increasing lamina number as is the case for laminated plate situations but is not so pronounced as in laminated plate structures. A decrease of bending stretching coupling is also observed with increasing shell angle, φ, and this effect becomes more significant for shell angle, φ, varying between $0^{\cdot} - 20^{\cdot}$.

<u>References</u>

1. N. D. Phan and J. N. Reddy
    Analysis of laminated composite plates using a high–order shear deformation theory. Int. J. Num. Meth. Engng. **21**, 2201–2219, 1985.

2. J. M. Whiteney
    Stress analysis of thick laminated composite and sandwich plates. J. Comp. Mat. **6**, 426–441, 1972.

3. S. C. Panda and R. Natarajan
    Finite element analysis of laminated composite plates. Int. J. Num. Meth. Engng. **14**, 69–79, 1979.

4. N. J. Pagano
    Exact solutions for rectangular bidirectional composites and sandwich plates. J. Comp. Mat. **4**, 20–34, 1970.

5. N. J. Pagano and S. J. Hatfield
    Elastic behaviour of multilayered bidirectional composites. AIAA Jnl. **10**, 931–933, 1972.

6. S. T. Mau, P. Tong and H. H. Pian
    Finite element solutions for laminated thick plates. J. Comp. Mat. **6**, 304–311, 1972.

7. R. L. Spilker
    Hybrid–stress eight–node elements for thin and thick multilayer laminated plates. Int. J. Num. Meth. Engng. **18**, 801–828, 1982.

8. S. Srinivas
    A refined analysis of composite laminates. J. Sound Vibr. **30**, 495–507, 1973.

9. N. J. Pagano and S. R. Soni
    Global–local laminational mode. Int. J. Solids Struct. **19**, 207–228, 1983.

10. D. R. J. Owen and Z. H. Li
A refined analysis of laminated plates by finite element displacement methods I – Fundamentals and static analysis. Computers & Structures, **26**, 907–914, 1987.

11. D. R. J. Owen and Z .H. Li
A refined analysis of laminated plates by finite element displacement methods II – Vibration and stability. Computers & Structures, **26**, 915–923, 1987.

12. D. R. J. Owen and Z. H. Li
Elasto–plastic numerical analysis of anisotropic laminated plates by a refined finite element model. In Computational Plasticity: Models, Software and Applications. Pineridge Press, U.K. pp.749–775, 1987.

13. E.Hinton and D. R. J. Owen
Finite Element Software for Plates and Shells, Pineridge Press, U.K. 1984.

14. D. R. J. Owen and E. Hinton
Finite Elements in Plasticity: Theory and Practice, Pineridge Press, U.K. 1980.

15. O. C. Zienkiewicz
The Finite Element Method. McGraw–Hill, New York 1977.

16. T.K. Caughey
Classical normal modes in damped linear system. J. Appl. Math., **27**, 269–271, 1960.

17. T. J. R. Hughes and W. K. Liu
Implicit–explicit finite elements in transient analysis: Stability theory. J. Appl. Meth., **45**, 371–374, 1978.

18. N. S. Putcha and J. N. Reddy
Stability and vibration analysis of laminated plates by using a mixed element based on a refined plate theory. J. Sound & Vibration, **104**(2), 285–300, 1986.

19. M. D. Sciuva
Bending, vibration and buckling of simply supported thick multilayered orthotropic plates: An evaluation of a new displacement model, J. Sound & Vibration, **105**(3), 425–442, 1986.

20. N. J. Pagano and S. J. Hattfield
Elastic behaviour of multilayered bidirectional composites, AIAA Journal, 931–933, 1972.

On the Use of Bubble Modes in Mixed Plate and Shell Finite Elements for Laminated Composites

Peter M. Pinsky, Assistant Professor
Raja V. Jasti, Doctoral Research Assistant

Department of Civil Engineering
Stanford University
Stanford, California 94305-4020

ABSTRACT

Criteria for the initiation of damage and delamination in laminated composite plate and shell structures typically depend on the three-dimensional stress state. For some local failure processes, the initiation criteria may be quite sensitive to the value of the transverse shear stress. In this paper we consider the application of an extended form of the Hellinger-Reissner principle for plate and shell finite elements with improved accuracy in stresses. The variational basis includes additional equilibrium constraints on the stresses through introduction of Lagrange multiplier functions. In the finite element approximation, we consider the effects of taking these Lagrange multiplier functions to be element-based bubble modes. Two elements are developed: a four-node homogeneous shell and a four-node laminated composite plate. It is shown that these elements provide accurate stresses in generally distorted meshes while retaining numerical stability and freedom from shear locking.

1. INTRODUCTION

Progressive damage and delamination in multi-layered composite materials are important local failure modes. Initiation of local intralayer damage in the form of fiber breakage, matrix cracking and fiber-matrix debonding may occur due to critical stress levels, construction defects or impact. The progressive development of damage characterized by the initiation and growth of microcracks can precipitate the formation of macrocracks which in turn may lead to structural collapse. Interlayer damage in the form of delamination due to fiber buckling, important in compressive stress states, is an instability-driven phenomenon that can propagate rapidly leading to structure failure. Conditions for the initiation of this local failure mode depend on the three-dimensional stress state and are quite sensitive to the value of the transverse shear stress. Similarly, planar delamination is also possible in regions of high transverse shear stress.

If laminated composite shell finite elements are to be effective tools in the design process, it is critical that they provide optimally accurate stresses that can be used for evaluating criteria for the initiation of local damage and delamination. This requirement on the stresses must be satisfied for arbitrarily distorted meshes and must include accurate values for the transverse shear stresses. In this paper we consider the development of two new linear four-node elements: a homogeneous shell element and a laminated composite plate element. These elements are based on a mixed variational principle that provides for accurate stress approximations while retaining the properties of numerical stability and freedom from locking phenomena in the thin limit.

In principle, elements based on the assumed-stress hybrid approach can provide accurate stresses. Such elements can be derived from the Hellinger-Reissner principle in conjunction with the use of stress basis functions which are chosen so that the stress representations satisfy the stress equilibrium equations *a priori* [1,2]. This method has been successfully applied to Kirchhoff plate theory where its motivation derives from circumventing the C^1 continuity condition [3-5]. This application is known to provide accurate results for moment predictions [4,5]. More recently the assumed-stress hybrid method has been applied to Reissner-Mindlin plate theory with disappointing results [6-8]. This formulation is subject to a shear locking phenomena in the thin limit. Although this locking is not as severe as that found in the classical displacement-based elements, previous efforts to eliminate it have resulted in invariant or inaccurate elements [6-8].

The assumed-stress hybrid approach is also not well suited for general shell formulations since it is practically impossible to arrive at stress representations that satisfy the equilibrium conditions because of their complexity and, at the same time, satisfy the element coordinate invariance requirements. In addition, due to the coupling which will exist between the stress components, the element flexibility matrix, which must be inverted, will be full and unstructured thereby engendering significant computational overhead.

One of the Euler-Lagrange equations of the Hellinger-Reissner functional is the stress equilibrium equation which clearly suggests that stress representations used in conjunction with the Hellinger-Reissner functional need not satisfy any conditions *a priori*. Unfortunately, Fraejis de Veubeke's limitation principle [9] states that if the stress components are given *independent* representation and that if these are of a sufficient degree of completeness, then the mixed finite elements derived from Hellinger-Reissner functional will be equivalent to the corresponding displacement-based elements with full integration and will therefore inherit the shear locking phenomenon in the thin shell limit.

The hybrid method is one approach that "overrides" the equivalence principle by constraining the stress basis functions *a priori*. An alternative to the hybrid approach for overriding the limitation principle is provided by the use of an extended Hellinger-Reissner functional in which the uncoupled stress representations are subject to additional variational equilibrium constraints through the introduction of Lagrange multipiers [10]. It will be shown that this approach has a number of attributes in developing four-node C^0 shell elements including: complete elimination of transverse shear locking, improved in-plane shear behavior, accurate shear stresses even with distorted meshes, rank sufficient stiffness matrix and computational efficiency. The governing variational principle is described in Section 2, and two finite element applications, namely a four-node homogeneous shell element and a four-node laminated composite plate element, are then outlined in Sections 3 and 4, respectively. Section 5 addresses some issues concerning efficiency, accuracy and numerical stability. Some numerical examples are given in Section 6.

2. VARIATIONAL PRINCIPLE

For some finite element discretization, consider the following modified Hellinger-Reissner functional:

$$\Pi = \sum_{e=1}^{nel} \int_{V^e} \left\{ -\frac{1}{2}\, \boldsymbol{\sigma} : \boldsymbol{S} : \boldsymbol{\sigma} + \boldsymbol{\sigma} : \nabla^s \boldsymbol{u}^q \right\} dV^e - \int_{\partial V^e} \boldsymbol{\sigma} \cdot \boldsymbol{n} \cdot (\boldsymbol{u}^q - \bar{\boldsymbol{u}})\, d\Gamma^e$$

$$- \int_{\partial V_\sigma^e} \bar{\boldsymbol{t}} \cdot \bar{\boldsymbol{u}}\, d\Gamma^e - \int_{V^e} \boldsymbol{\lambda} \cdot (\nabla \cdot \boldsymbol{\sigma})\, dV^e \tag{1}$$

where $\boldsymbol{\sigma}$ is the stress field, $\boldsymbol{u}^q$ is a displacement field obtained from interpolation of nodal degrees of freedom, $\boldsymbol{\lambda} \in \boldsymbol{R}^3$ is a Lagrange multiplier field, $\bar{\boldsymbol{u}}$ is an independent displacement field defined on all element boundaries, $\boldsymbol{S}$ is the elastic compliance tensor and $\bar{\boldsymbol{t}}$ is the prescribed traction over the boundary ∂V_σ^e. Applying the divergence theorem to (1) yields:

$$\Pi = \sum_{e=1}^{nel} \int_{V^e} \left\{ -\frac{1}{2}\, \boldsymbol{\sigma} : \boldsymbol{S} : \boldsymbol{\sigma} + \boldsymbol{\sigma} : \nabla^s(\boldsymbol{u}^q + \boldsymbol{\lambda}) \right\} dV^e$$

$$- \int_{\partial V^e} \boldsymbol{\sigma} \cdot \boldsymbol{n} \cdot (\boldsymbol{u}^q + \boldsymbol{\lambda} - \bar{\boldsymbol{u}})\, d\Gamma^e - \int_{\partial V_\sigma^e} \bar{\boldsymbol{t}} \cdot \bar{\boldsymbol{u}}\, d\Gamma^e \tag{2}$$

From Eq. (2) the Lagrange multiplier field $\boldsymbol{\lambda}$ can be identified as a displacement field which, for emphasis, we will henceforth denote as $\boldsymbol{u}^\lambda$. In the present work we incorporate the following two restrictions: (1) the Lagrange multiplier field $\boldsymbol{u}^\lambda$ is assumed to vanish on all element boundaries, and (2) the Euler-Lagrange equation:

$$\bar{\boldsymbol{u}} = \boldsymbol{u}^q + \boldsymbol{u}^\lambda \quad \text{on } \partial V^e \tag{3}$$

is satisfied identically *a priori*. In this case, functional given by Eq. (2) specializes to:

$$\Pi = \sum_{e=1}^{nel} \int_{V^e} \left\{ -\frac{1}{2}\, \boldsymbol{\sigma} : \boldsymbol{S} : \boldsymbol{\sigma} + \boldsymbol{\sigma} : \nabla^s(\boldsymbol{u}^q + \boldsymbol{u}^\lambda) \right\} dV^e$$

$$- \int_{\partial V_\sigma^e} \bar{\boldsymbol{t}} \cdot (\boldsymbol{u}^q + \boldsymbol{u}^\lambda)\, d\Gamma^e \tag{4}$$

The Euler-Lagrange equations of the above functional are:

$$\nabla^s(\boldsymbol{u}^q + \boldsymbol{u}^\lambda) = \boldsymbol{S} : \boldsymbol{\sigma} \quad \text{in } V^e \tag{5}$$

$$\nabla \cdot \boldsymbol{\sigma} = 0 \quad \text{in } V^e \tag{6}$$

$$\boldsymbol{\sigma} \cdot \boldsymbol{n} = \bar{\boldsymbol{t}} \quad \text{on } \partial V_\sigma \tag{7}$$

Eqs. (6) and (7) are repeated, resulting from variations with respect to $\boldsymbol{u}^q$ and $\boldsymbol{u}^\lambda$.

Two applications of Eq. (4) are described in the sequel: a four-node homogeneous shell element, and a four-node laminated composite plate element.

3. Application I: Four-Node Homogeneous Shell Element

3.1 Geometry and Coordinate Systems

The coordinates of a typical point x in the shell element with half-thickness h are given as functions of the element natural coordinates ξ, η, ζ by the following relation:

$$x(\xi,\eta,\zeta) = \sum_{a=1}^{4} N_a(\xi,\eta)\,\bar{x}_a + \zeta h \sum_{a=1}^{4} N_a(\xi,\eta)\,\hat{x}_a \tag{8}$$

where $\bar{x}_a$ is the position vector of nodal point a in the shell reference surface and $\hat{x}_a$ is a unit vector in the fiber direction at node a and based at the point $\bar{x}_a$ in the reference surface, and $N_a(\xi,\eta)$ is the two-dimensional shape function for node a.

In addition to the fixed global Cartesian coordinate system we introduce two standard shell coordinate systems [11]: a lamina coordinate system which is a Cartesian coordinate system defined at every point on the element reference surface, with a set of orthogonal basis vectors denoted e_1^l, e_2^l, e_3^l, and a fiber coordinate system which is a Cartesian coordinate system defined at each node a, with a set of orthogonal basis vectors denoted $e_{a1}^f, e_{a2}^f, e_{a3}^f$.

3.2 Kinematic Assumption

The *lamina* components of u^q are given by the expression:

$$u^q(\xi,\eta,\zeta) = \bar{u}^q(\xi,\eta) + \zeta h\,\hat{u}^q(\xi,\eta) \tag{9}$$

where $\bar{u}^q$ is the displacement vector of the reference surface and $\zeta h \hat{u}^q$ is the relative displacement vector due to fiber rotation.

The displacement vectors $\bar{u}^q$ and $\hat{u}^q$ are parameterized in terms of nodal degrees of freedom. The nodal degree of freedom vector for node a is defined to be:

$$q_a = \left\{ \begin{array}{c} \bar{q}_a \\ \hat{q}_a \end{array} \right\}, \quad \bar{q}_a = \left\{ \begin{array}{c} \bar{u}_{a1}^g \\ \bar{u}_{a2}^g \\ \bar{u}_{a3}^g \end{array} \right\}, \quad \hat{q}_a = \left\{ \begin{array}{c} \theta_{a1}^f \\ \theta_{a2}^f \end{array} \right\} \tag{10}$$

where $\bar{u}_{ai}^g$ are the *global* components of $\bar{u}^q$ and θ_{a1}^f and θ_{a2}^f are positive rotations about the nodal fiber basis vectors e_{a1}^f and e_{a2}^f, respectively, at node a.

Using nodal interpolation and transforming the global nodal displacements and fiber rotations to the lamina basis, we obtain

$$\left\{ \begin{array}{c} \bar{u}^q(\xi,\eta) \\ \hat{u}^q(\xi,\eta) \end{array} \right\} = R(\xi,\eta) \sum_{a=1}^{4} N_a(\xi,\eta) T_a q_a \tag{11}$$

where

$$R = \begin{bmatrix} r & 0 \\ 0 & r \end{bmatrix}, \quad T_a = \begin{bmatrix} I & 0 \\ 0 & t_a \end{bmatrix} \tag{12}$$

$$r = [e_1^l \quad e_2^l \quad e_3^l]^T, \quad t_a = [-e_{a2}^f \quad e_{a1}^f] \tag{13}$$

3.3 Lagrange Multiplier Field

Having identified the Lagrange multiplier field u^λ as a displacement field, we set

$$u^\lambda(\xi,\eta,\zeta) = \bar{u}^\lambda(\xi,\eta) + \zeta h\, \hat{u}^\lambda(\xi,\eta) \tag{14}$$

where $\bar{u}^\lambda$ and $\hat{u}^\lambda$ are independent functions. We shall subsequently make the assumption that the transverse lamina normal stress is zero and as a result $\hat{u}_3^\lambda$ will not appear in the functional. Accordingly, we restrict $\hat{u}^\lambda$ such that:

$$\hat{u}_3^\lambda = 0 \tag{15}$$

Within an element, the Lagrange multiplier functions $\bar{u}^\lambda$ and $\hat{u}^\lambda$ can be expressed in terms of a set of internal displacement parameters λ and bubble basis functions as follows:

$$\left\{ \begin{array}{c} \bar{u}_i^\lambda \\ \hat{u}_\alpha^\lambda \end{array} \right\} = \sum_{J=1}^{NB} \left\{ \begin{array}{c} \bar{U}_i^J(\xi,\eta) \\ \hat{U}_\alpha^J(\xi,\eta) \end{array} \right\} \lambda^J \qquad i = 1,2,3 \text{ and } \alpha = 1,2 \tag{16}$$

where NB is the total number of bubble functions. Note that the Lagrange multiplier field has been defined directly in the lamina coordinate basis. No transformation from global to lamina basis is needed.

3.4 Stress Approximations

For consistency with the shell kinematics given by Eqs. (9) and (14), the *lamina* based stress components are assumed to have the following form:

$$\sigma_{\alpha\beta}(\xi,\eta,\zeta) = \sigma_{\alpha\beta}^m(\xi,\eta) + \zeta h\, \sigma_{\alpha\beta}^b(\xi,\eta) \quad \alpha,\beta = 1,2$$

$$\sigma_{\alpha3}(\xi,\eta,\zeta) = \frac{h^2}{2}(1-\zeta^2)\, \sigma_\alpha^s(\xi,\eta) \quad \alpha = 1,2 \tag{17}$$

$$\sigma_{33}(\xi,\eta,\zeta) = 0$$

where $\sigma_{\alpha\beta}^m$, $\sigma_{\alpha\beta}^b$ and σ_α^s are independent functions of the element natural coordinates ξ and η. The ζ-dependence given in Eq. (17.b) is found from the equilibrium equation:

$$\sigma_{\alpha3}(\xi,\eta,\zeta) = -h \int \left[\sigma_{\alpha\beta}^m(\xi,\eta) + \zeta h\, \sigma_{\alpha\beta}^b(\xi,\eta) \right]_{,\beta} d\zeta \tag{18}$$

and boundary conditions requiring $\sigma_{\alpha3}$ to vanish at the shell surfaces $\zeta = \pm 1$.

We introduce the following matrix notation for the *lamina* stresses:

$$\sigma^m = \left\{ \begin{array}{c} \sigma_{11}^m \\ \sigma_{22}^m \\ \sigma_{12}^m \end{array} \right\}, \quad \sigma^b = \left\{ \begin{array}{c} \sigma_{11}^b \\ \sigma_{22}^b \\ \sigma_{12}^b \end{array} \right\}, \quad \sigma^s = \left\{ \begin{array}{c} \sigma_1^s \\ \sigma_2^s \end{array} \right\} \tag{19}$$

and strains:

$$\varepsilon_m^q = \left\{ \begin{array}{c} \bar{u}_{(1,1)}^q \\ \bar{u}_{(2,2)}^q \\ 2\bar{u}_{(1,2)}^q \end{array} \right\}, \quad \varepsilon_b^q = \left\{ \begin{array}{c} \hat{u}_{(1,1)}^q \\ \hat{u}_{(2,2)}^q \\ 2\hat{u}_{(1,2)}^q \end{array} \right\}, \quad \varepsilon_s^q = \left\{ \begin{array}{c} \bar{u}_{3,1}^q + \hat{u}_1^q \\ \bar{u}_{3,2}^q + \hat{u}_2^q \end{array} \right\} \tag{20}$$

$$\varepsilon_m^\lambda = \left\{ \begin{array}{c} \bar{u}_{(1,1)}^\lambda \\ \bar{u}_{(2,2)}^\lambda \\ 2\bar{u}_{(1,2)}^\lambda \end{array} \right\}, \quad \varepsilon_b^\lambda = \left\{ \begin{array}{c} \hat{u}_{(1,1)}^\lambda \\ \hat{u}_{(2,2)}^\lambda \\ 2\hat{u}_{(1,2)}^\lambda \end{array} \right\}, \quad \varepsilon_s^\lambda = \left\{ \begin{array}{c} \bar{u}_{3,1}^\lambda + \hat{u}_1^\lambda \\ \bar{u}_{3,2}^\lambda + \hat{u}_2^\lambda \end{array} \right\} \tag{21}$$

where differentiation in the lamina strain definitions is performed with respect to the lamina coordinates.

The stresses σ^m, σ^b and σ^s may be expressed in terms of basis functions and generalized stress parameters as follows:

$$\sigma^m = P_m(\xi,\eta)\,\beta_m, \quad \sigma^b = P_b(\xi,\eta)\,\beta_b, \quad \sigma^s = P_s(\xi,\eta)\,\beta_s \tag{22}$$

Introducing Eqs. (19)-(21) into the functional given by Eq. (4), assuming isotropic linear elasticity and integrating through the shell thickness yields:

$$\Pi = \int_{V^e} \left\{ -\frac{1}{2}\,\sigma^{m^T} S_m \sigma^m - \frac{1}{2}\,\sigma^{b^T} S_b \sigma^b - \frac{1}{2}\,\sigma^{s^T} S_s \sigma^s \right\} dV^e$$

$$\cdot + \int_{V^e} \left\{ 2h\sigma^{m^T}(\varepsilon_m^q + \varepsilon_m^\lambda) + \frac{2h^3}{3}\sigma^{b^T}(\varepsilon_b^q + \varepsilon_b^\lambda) + \frac{2h^3}{3}\sigma^{s^T}(\varepsilon_s^q + \varepsilon_s^\lambda) \right\} dV^e \tag{23}$$

$$- \int_{\partial V^e} (u^q + u^\lambda)^T\,\bar{t}\,d\Gamma^e$$

where

$$S_m = \frac{2h}{E} \begin{bmatrix} 1 & -\nu & 0 \\ -\nu & 1 & 0 \\ 0 & 0 & 2(1+\nu) \end{bmatrix}, \quad S_b = \frac{h^2}{3}\,S_m, \quad S_s = \frac{8h^5(1+\nu)}{15E} \tag{24}$$

3.5 Lagrange Multiplier and Stress Basis Functions

For the homogeneous shell, the Lagrange multiplier and stress basis functions were taken to be:

$$\left\{ \begin{array}{c} \bar{u}_1^\lambda \\ \bar{u}_2^\lambda \\ \bar{u}_3^\lambda \\ \hat{u}_1^\lambda \\ \hat{u}_2^\lambda \end{array} \right\} = \begin{bmatrix} Q_1 & 0 & 0 & 0 & 0 \\ 0 & Q_1 & 0 & 0 & 0 \\ 0 & 0 & Q_2 & 0 & 0 \\ 0 & 0 & 0 & Q_2 & 0 \\ 0 & 0 & 0 & 0 & Q_2 \end{bmatrix} \{\lambda\} \tag{25}$$

$$Q_1 = (1-\xi^2)(1-\eta^2), \quad Q_2 = (1-\xi^2)(1-\eta^2)\langle 1,\xi,\eta \rangle \tag{26}$$

$$P_m = P_b = \begin{bmatrix} Q_3 & 0 & 0 \\ 0 & Q_3 & 0 \\ 0 & 0 & Q_3 \end{bmatrix}, \quad P_s = \begin{bmatrix} Q_3 & 0 \\ 0 & Q_3 \end{bmatrix} \tag{27}$$

$$Q_3 = \langle 1,\xi,\eta,\xi\eta \rangle \tag{28}$$

These particular basis functions have been selected on the basis of accuracy and numerical stability criteria which are reviewed in Section 5.

4. Application II: Laminated Composite Plate

4.1 Kinematics

Consider a plate of half-thickness h and consisting of l generally orthotropic perfectly bonded layers, and the reference surface of the plate Ω discretized into n finite elements, denoted by Ω^e, with $\mathbf{V}^e = \Omega^e \times [-h, h]$ and $\Omega = \bigcup\limits_{e=1}^{n} \Omega^e$, where $\mathbf{V}^e$ is the element domain. The functional given by Eq. (4) can be extended for a multilayered plate as follows:

$$\Pi = \sum_{e=1}^{n} \sum_{i=1}^{l} \left\{ \int_{V_e^i} \left[-\frac{1}{2}\, \boldsymbol{\sigma}^{(i)} : \mathbf{S}^{(i)} : \boldsymbol{\sigma}^{(i)} + \boldsymbol{\sigma}^{(i)} : (\nabla^s \mathbf{u}^q + \nabla^s \mathbf{u}^\lambda) \right] dV^i \right.$$
$$\left. - \int_{\partial_\sigma V_e} (\mathbf{u}^q + \mathbf{u}^\lambda) \cdot \bar{\mathbf{t}}\; d\Gamma^e \right\}$$

(29)

in which V_e^i denotes the element domain of the ith layer and $V_e = \sum\limits_{i=1}^{l} V_e^i$, $\boldsymbol{\sigma}^{(i)}$ and $\mathbf{S}^{(i)}$ denote the stress and the compliance tensors of ith layer respectively and $\partial_\sigma V_e$ represents the boundary of the element on which the tractions are prescribed. The displacement field $\mathbf{u}^q$ and Lagrange multiplier field $\mathbf{u}^\lambda$ are assumed to be *independent* of the layers and hence the number of nodal displacements are fixed irrespective of the number of layers. The kinematical description of the displacement field $\mathbf{u}^q$ is precisely that described for the shell in Section 3.2 but simply specialized for a plate. Similarly, the Lagrange multiplier field $\mathbf{u}^\lambda$ is characterized in terms of $\bar{\mathbf{u}}^\lambda$ and $\hat{\mathbf{u}}^\lambda$ exactly as described for the shell in Section 3.3.

4.2 Stress Approximations

In order to be consistent with the kinematic assumption, the stress approximations are given in terms of a set of parameters which are independent of the number of layers. However, it is important to be able to represent the jumps in in-plane stresses across layer interfaces when the bonded layers have different constitutive properties. This may be accomplished by assuming certain reasonable through-thickness distributions for the strains and obtaining the stress basis functions for each layer by using the constitutive relations of the corresponding layers. The assumed strain distributions, denoted by $\tilde{\boldsymbol{\varepsilon}}$, introduced solely to obtain the stress approximations, should not be confused with the strains derived as the symmetric gradients of the displacements. The following through thickness distributions was employed:

$$\tilde{\varepsilon}_{\alpha\beta} = \bar{\varepsilon}_{\alpha\beta}^m(\xi, \eta) + h\zeta\bar{\varepsilon}_{\alpha\beta}^b(\xi, \eta)$$
$$\tilde{\varepsilon}_{\alpha 3} = \frac{h^2}{2}(1 - \zeta^2)\bar{\varepsilon}_{\alpha 3}(\xi, \eta)$$

(30)

$$\tilde{\varepsilon}_{33} = 0.$$

Note that $\bar{\varepsilon}_{\alpha\beta}$ and $\bar{\varepsilon}_{\alpha 3}$ will be expressed using basis functions defined over the natural coordinates ξ and η, and also that they will be assumed to be uncoupled from each other. Once the expansions for $\bar{\varepsilon}_{\alpha\beta}$ and $\bar{\varepsilon}_{\alpha 3}$ are chosen, the stress basis functions for each layer can be obtained as:

$$\begin{Bmatrix} \sigma_{11}^{(i)} \\ \sigma_{22}^{(i)} \\ \sigma_{12}^{(i)} \end{Bmatrix} = \mathbf{Q}_i^{(i)} \begin{Bmatrix} \tilde{\varepsilon}_{11} \\ \tilde{\varepsilon}_{22} \\ 2\tilde{\varepsilon}_{12} \end{Bmatrix}$$

(31)

$$\left\{ \begin{array}{c} \sigma_{13}^{(i)} \\ \sigma_{23}^{(i)} \end{array} \right\} = \mathbf{Q}_o^{(i)} \left\{ \begin{array}{c} 2\tilde{\varepsilon}_{13} \\ 2\tilde{\varepsilon}_{23} \end{array} \right\} \tag{32}$$

where $\mathbf{Q}_i^{(i)}$ and $\mathbf{Q}_o^{(i)}$ are the inverses of the in-plane and out of plane parts of the compliance matrices corresponding to the ith layer, referred to previously as $\mathbf{S}^{(i)}$. It should be observed that, in general, the in-plane stresses in each layer are assumed to be linear in the thickness direction and all the stresses are allowed to be discontinuous at the inter-layer boundaries. This obviously violates the traction continuity condition in the x_3 direction, but appears to have no significant affects on the accuracy of the element for both displacement and stress predictions.

Introducing the approximation for the stresses, Eqs. (31) and (32), into the functional given by Eq. (29) results in:

$$\Pi = \sum_{e=1}^{n} \sum_{i=1}^{l} \left\{ \int_{V_e^i} \left[-\frac{1}{2} \tilde{\varepsilon}_i^{(i)^T} \mathbf{Q}_i^{(i)} \tilde{\varepsilon}_i^{(i)} - \frac{1}{2} \tilde{\varepsilon}_o^{(i)^T} \mathbf{Q}_o^{(i)} \tilde{\varepsilon}_o^{(i)} \right. \right.$$

$$\left. + (\mathbf{Q}_i^{(i)} \tilde{\varepsilon}_i^{(i)})^T (\varepsilon_i^q + \varepsilon_i^\lambda) + (\mathbf{Q}_o^{(i)} \tilde{\varepsilon}_o^{(i)})^T (\varepsilon_o^q + \varepsilon_o^\lambda) \right] dV_e^i \tag{33}$$

$$\left. - \int_{\partial V_e^i} (\mathbf{u}^q + \mathbf{u}^\lambda)^T \bar{\mathbf{t}} \; d\Gamma^e \right\}$$

where

$$\tilde{\varepsilon}_i = \left\{ \begin{array}{c} \tilde{\varepsilon}_{11} \\ \tilde{\varepsilon}_{22} \\ 2\tilde{\varepsilon}_{12} \end{array} \right\} \qquad \tilde{\varepsilon}_o = \left\{ \begin{array}{c} 2\tilde{\varepsilon}_{13} \\ 2\tilde{\varepsilon}_{23} \end{array} \right\} \tag{34}$$

$$\mathbf{Q}_i^{(i)} = \left[\begin{array}{ccc} C_{11} & C_{12} & C_{16} \\ C_{12} & C_{22} & C_{26} \\ C_{16} & C_{26} & C_{66} \end{array} \right]^{(i)} \quad \mathbf{Q}_o^{(i)} = \left[\begin{array}{cc} C_{44} & C_{45} \\ C_{45} & C_{55} \end{array} \right]^{(i)} \tag{35}$$

$$C_{11} = C^4 Q_{11} + 2C^2 S^2 (Q_{12} + 2G_{12}) + S^4 Q_{22}$$
$$C_{12} = C^2 S^2 (Q_{11} + Q_{22} - 4G_{12}) + (C^4 + S^4) Q_{12}$$
$$C_{16} = C^3 S (Q_{11} - Q_{12} - 2G_{12}) + S^3 C (Q_{12} - Q_{22} + 2G_{12})$$
$$C_{22} = S^4 Q_{11} + 2C^2 S^2 (Q_{12} + 2G_{12}) + C^4 Q_{22}$$
$$C_{26} = S^3 C (Q_{11} - Q_{12} - 2G_{12}) + C^3 S (Q_{12} - Q_{22} + 2G_{12}) \tag{36}$$
$$C_{66} = C^2 S^2 (Q_{11} + Q_{22} - 2Q_{12} - 2G_{12}) + (C^4 + S^4) G_{12}$$
$$C_{44} = C^2 G_{13} + S^2 G_{23}$$
$$C_{45} = CS (G_{13} - G_{23})$$
$$C_{55} = C^2 G_{23} + S^2 G_{13}$$

In Eq. (36),

$$Q_{11} = E_1 / (1 - \nu_{12} \nu_{21})$$
$$Q_{22} = E_2 / (1 - \nu_{12} \nu_{21}) \tag{37}$$
$$Q_{12} = \nu_{12} E_2 / (1 - \nu_{12} \nu_{21})$$

and $C = \cos\theta$, $S = \sin\theta$ where θ is the angle of fiber orientation measured from the x_1 axis. In the above expressions E_1, E_2 and ν denote the Young's moduli and Poisson's ratio respectively and G_{12}, G_{13} and G_{23} represent the shear moduli of the material.

4.3 Bubble and Stress Basis Functions:

The Lagrange multiplier fields $\bar{u}^\lambda$ and $\hat{u}^\lambda$ employ the bubble basis functions used for the shell and are given by Eqs. (25) and (26).

The assumed strain distribution given by Eq. (30) is expressed in terms of in-plane basis functions and generalized strain parameters as:

$$
\begin{aligned}
\tilde{\varepsilon}_{11} &= \mathbf{a}(\xi,\eta)\beta_1 + h\zeta\,\mathbf{a}\,(\xi,\eta)\beta_2 \\
\tilde{\varepsilon}_{22} &= \mathbf{a}(\xi,\eta)\beta_3 + h\zeta\,\mathbf{a}\,(\xi,\eta)\beta_4 \\
2\tilde{\varepsilon}_{12} &= \mathbf{a}(\xi,\eta)\beta_5 + h\zeta\,\mathbf{a}\,(\xi,\eta)\beta_6 \\
2\tilde{\varepsilon}_{13} &= \frac{h^2}{2}(1-\zeta^2)\,\mathbf{a}(\xi,\eta)\,\beta_7 \\
2\tilde{\varepsilon}_{23} &= \frac{h^2}{2}(1-\zeta^2)\,\mathbf{a}(\xi,\eta)\,\beta_8 \\
\tilde{\varepsilon}_{33} &= 0
\end{aligned}
\tag{38}
$$

where

$$
\mathbf{a} = \langle 1, \xi, \eta, \xi\eta \rangle
\tag{39}
$$

5. FINITE ELEMENT EQUATIONS

5.1 Element Stiffness Matrix

For both the plate and shell applications, the stress and the strain components may be expressed in terms of the stress and the displacement parameters using standard procedures:

$$
\begin{aligned}
\sigma(\xi,\eta) &= P(\xi,\eta)\,\beta &\tag{40} \\
\varepsilon^q(\xi,\eta) &= B^q(\xi,\eta)\,q &\tag{41} \\
\varepsilon^\lambda(\xi,\eta) &= B^\lambda(\xi,\eta)\,\lambda &\tag{42}
\end{aligned}
$$

where

$$
\begin{aligned}
\varepsilon &= \langle \varepsilon_{11}, \varepsilon_{22}, \varepsilon_{33}, 2\varepsilon_{12}, 2\varepsilon_{13}, 2\varepsilon_{23} \rangle^T \\
\sigma &= \langle \sigma_{11}, \sigma_{22}, \sigma_{12}, \sigma_{13}, \sigma_{23} \rangle^T
\end{aligned}
$$

Introducing Eqs. (40)-(42) into either the functional given by Eq. (23) or (33), and integrating through the thickness, leads to the following general matrix form:

$$
\Pi = \sum_{e=1}^{n} \left[-\frac{1}{2}\beta^T H\beta + \beta^T Gq + \beta^T R\lambda - F_1^T q - F_2^T \lambda \right]
\tag{43}
$$

where

$$H = \int_{V^e} P^T \, S \, P \, dV^e \qquad G = \int_{V^e} P^T \, B^q \, dV^e \qquad R = \int_{V^e} P^T \, B^\lambda \, dV^e \qquad (44)$$

$$F_1^T q = \int_{\partial_\sigma V^e} u^q \cdot \bar{t} \, d\Gamma^e \qquad F_2^T \lambda = \int_{\partial_\sigma V^e} u^\lambda \cdot \bar{t} \, d\Gamma^e \qquad (45)$$

The stationarity of Π with respect to the variations in β and λ yields

$$\beta = H^{-1}(Gq + R\lambda) \qquad (46)$$

$$R^T \beta = F_2 \qquad (47)$$

Using Eqs. (46) and (47) to eliminate β and λ in Eq. (43), we obtain:

$$\Pi = \sum_{e=1}^{n} \left[\frac{1}{2} q^T k q - F q \right] \qquad (48)$$

$$k = \tilde{G}^T H^{-1} \tilde{G} \qquad (49)$$

$$\tilde{G} = G - R(R^T H^{-1} R)^{-1} R^T H^{-1} G \qquad (50)$$

$$F = F_1 - G^T H^{-1} R (R^T H^{-1} R)^{-1} F_2 \qquad (51)$$

where k is the element stiffness matrix.

5.2 Inversion of the Element Flexibility Matrix

As observed above, one of the disadvantages of the hybrid approach results from the necessity to invert the element flexibility matrix which will always be full and will have dimension equal to the number of stress parameters. Since the present approach uses uncoupled stress components leading to many stress parameters one may be tempted to conclude that the element may be even more expensive than the hybrid-type elements. However, the use of uncoupled stress components provides sparseness and a special structure in the flexibility matrix that is easily exploited to provide economy in forming the element stiffness matrix [2]. It is only necessary to invert a small submatrix of the flexibility matrix. Consider the homogeneous shell. It may be shown that the inverse of the element flexibility matrix has the form:

$$H^{-1} = \begin{bmatrix} H_m^{-1} & 0 & 0 \\ 0 & H_b^{-1} & 0 \\ 0 & 0 & H_s^{-1} \end{bmatrix} \qquad (52)$$

where

$$H_m^{-1} = \frac{E}{2h(1-\nu^2)} \begin{bmatrix} V^{-1} & \nu V^{-1} & 0 \\ \nu V^{-1} & V^{-1} & 0 \\ 0 & 0 & \frac{(1-\nu)}{2} V^{-1} \end{bmatrix}$$

$$H_b^{-1} = \frac{3}{2h^2} H_m^{-1} \qquad (53)$$

$$H_s^{-1} = \frac{15E}{8h^5(1+\nu)} \begin{bmatrix} V^{-1} & 0 \\ 0 & V^{-1} \end{bmatrix}$$

$$V = \int_{\Omega^e} Q_3^T Q_3 \, d\Omega^e$$

in which Q_3 is defined by Eq. (28).

Observing that $dim(V) = dim(Q_3) = 4$, it is only necessary to invert a matrix of dimension four for computing the inverse of the flexibility matrix, which is of dimension 32. Similarly for the composite plate, it is only necessary to invert a matrix of dimension four. Note, however, from Eq. (50) that in addition to H^{-1}, it is also necessary to compute the inverse of $R^T H^{-1} R$. This matrix has dimension eleven corresponding to the number of Lagrange multiplier parameters. Nevertheless, the proposed shell formulation offers significant computational efficiency when compared to the hybrid approach, where the typical size of the flexibility matrix will be much larger than eleven.

5.3 Equilibrium Constraints

In mixed methods, the number of parameters used to define each independent field must be selected to satisfy stability and accuracy requirements. In this section we use a heuristic argument to obtain the optimal relation between the number of field parameters on the basis of accuracy considerations. The numerical stability of the elements based on the selected parameterization of the fields is then checked in Section 5.5. Consider the assembled finite element equations derived from the matrix form of the functional given by Eq. (23) or (33):

$$
\begin{bmatrix} -\mathbf{H} & \mathbf{G} & \mathbf{R} \\ \mathbf{G}^T & \mathbf{0} & \mathbf{0} \\ \mathbf{R}^T & \mathbf{0} & \mathbf{0} \end{bmatrix} \begin{Bmatrix} \beta \\ \mathbf{q} \\ \lambda \end{Bmatrix} = \begin{Bmatrix} \mathbf{0} \\ \mathbf{f}_q \\ \mathbf{f}_\lambda \end{Bmatrix} \tag{54}
$$

where the submatrices are defined in Section 5.1. The first row partition of (54) is the discrete approximation of the constitutive equations and has N_β rows, equal to the total number of stress parameters. The second and third row partitions are the discrete approximation of the equilibrium equations and have a total of $N_q + N_\lambda$ rows, equal to the total number of nodal degrees of freedom N_q and bubble functions N_λ. The ratio of the number of constitutive equations to the number of equilibrium equations in the finite element model is given by $r_c = N_\beta/(N_q + N_\lambda)$. We argue that the optimal value of r_c should be close to the corresponding ratio for the continuum boundary value problem. Furthermore, this ratio should be considered for in-plane behavior, bending behavior and the combination of these two cases.

An estimate for r_c can be obtained by considering an infinite number of elements, to which is added one more element. Then we have

$$
r_c = \frac{n_\beta}{(n_q + n_\lambda)} \tag{55}
$$

in which n_β is the number of stress parameters associated with the element, n_q is the number of incremental nodal degrees of freedom added to the mesh, and n_λ is the number of bubble functions associated with the element.

If r_c of an element is much smaller than the corresponding continuum ratio, we argue that there are too many equilibrium constraints (the resulting element will probably also suffer from numerical stability problems). On the other hand, if r_c is much larger than the continuum ratio, we argue that there are not enough equilibrium constraints and, consequently, the finite element solution will be excessively stiff.

For both the plate and shell, the constraint ratio for the continuum boundary value problem has a value of 1.5 for in-plane behavior, 1.67 for bending behavior and 1.67 for the combined case. If the bubble modes are not included in the proposed plate and shell finite elements, the r_c ratio has values of 6.0, 6.67 and 6.4 for in-plane, bending and combined behaviors respectively. These elements, by de Veubeke's limitation principle [9], are equivalent to the corresponding fully integrated displacement-based elements. Inclusion of the bubble modes according to Eqs. (25) and (26) improves the r_c ratios to 3.0, 1.67 and 2.0 for in-plane, bending and combined behaviors respectively. It can be seen that the elements have a respectable count for all three cases and therefore can be expected to show optimal accuracy for all classes of problems.

5.4 Shear Locking

In addition to providing variational constraints on the stresses, the Lagrange multiplier bubble modes also provide another very important function; they precisely eliminate the source of shear locking observed in Mindlin-type elements in the thin limit. This result is briefly demonstrated for the shell element, but the arguments for the plate element are similar.

Taking the variation of the functional given by Eq. (23) with respect to the shear stress parameters β_s defined by Eq. (22) and setting the result to zero gives:

$$- \beta_s^T H_s \beta_s + \frac{2h^3}{3} \int_{\Omega^e} \beta_s^T P_s^T \left(B_s^q q + B_s^\lambda \lambda \right) \, d\Omega^e = 0 \tag{56}$$

$$H_s = \frac{8h^5(1+\nu)}{15E} \int_{\Omega}^{e} P_s^T P_s \, d\Omega^e \tag{57}$$

where B_s^q and B_s^λ are the strain-displacement matrices associated with the transverse shear strain and P_s is given by Eq. (27). Eq. (56) must hold for all arbitrary β_s. In the limiting case of a thin shell $H_s \approx 0$ and Eq. (56) takes the form

$$\int_{\Omega^e} \beta_s^T P_s^T \left(B_s^q q + B_s^\lambda \lambda \right) \, d\Omega^e = 0. \tag{58}$$

Eq. (58) is the weak form of the Kirchhoff condition that enforces the vanishing of the transverse shear strains. In the absence of the bubble modes Eq. (58) results in spurious constraints on the nodal degrees of freedom that forces the element to lock. However, the inclusion of the bubble modes removes these spurious constraints. Consider a square plate of dimension 2. Introducing P_s given by Eq. (27) into Eq. (58), yields, after integration through the thickness:

$$\begin{aligned}
&\delta\beta_{25}[4(\alpha_{13} + \alpha_{10} + \tfrac{4}{9}\lambda_6)] + \delta\beta_{26}[\tfrac{4}{3}(\alpha_{14} + \tfrac{4}{3}\lambda_3 + \tfrac{4}{15}\lambda_7)] \\
&+\delta\beta_{27}[\tfrac{4}{3}(\alpha_{15} + \alpha_{12} + \tfrac{4}{15}\lambda_8)] + \delta\beta_{28}[\tfrac{4}{9}(\alpha_{16} + \tfrac{4}{5}\lambda_5)] \\
&+\delta\beta_{29}[4(\alpha_{17} + \alpha_{11} + \tfrac{4}{9}\lambda_9)] + \delta\beta_{31}[\tfrac{4}{3}(\alpha_{19} + \tfrac{4}{3}\lambda_3 + \tfrac{4}{15}\lambda_{11})] \\
&+\delta\beta_{30}[\tfrac{4}{3}(\alpha_{18} + \alpha_{12} + \tfrac{4}{15}\lambda_{10})] + \delta\beta_{32}[\tfrac{4}{9}(\alpha_{20} + \tfrac{4}{5}\lambda_4)] = 0.
\end{aligned} \tag{59}$$

which must hold for all arbitrary β_i and where the α_i are linear combinations of the nodal degrees of freedom. In the absence of the λ_i parameters (i.e without the bubble functions) equation (59) implies

$$\alpha_{14} = \alpha_{16} = \alpha_{19} = \alpha_{20} = 0. \tag{60}$$

It may be shown that the only solution to Eq. (60), besides the rigid body modes, is that all the rotational degrees of freedom should vanish, leading to shear locking. On the other hand, the bubble function parameters λ_i remove the spurious constraints, hence eliminating the shear locking.

5.5 Numerical Stability

The variational problem of a mixed functional is a saddle point problem. The existence of a saddle point solution to a mixed variational problem depends on a coerciveness condition on a bilinear form in the functional as well as the satisfaction of the Babuška-Brezzi condition [12,13]. Hartmann [14] has described the discrete analog of the coerciveness condition and the Babuška-Brezzi condition. These analogous conditions are applicable to the discrete mixed variational problem based on the use of finite element trial functions. When these conditions on the discrete problem are satisfied, it may be shown [14] that the resulting system of equations has a unique solution which is a saddle point solution of the discrete mixed functional and, furthermore, the "stiffness" matrix of the problem is positive definite.

In the present formulation based on $\sigma \in L_2(\Omega)$ and $\lambda \in L_2(\Omega)$, it is possible to consider local condensation of the element stress variables β and element Lagrange multiplier variables λ resulting in an element stiffness matrix based on Eq. (49) which may then be assembled in the standard manner. The elimination of these variables at the element level is based on a local or element form of Eq. (54) and in this case a discrete element level Babuška-Brezzi condition can be stated . Satisfaction of this condition will ensure correct rank of the element stiffness matrix and eliminate the possibility of zero energy modes other than the rigid body modes.

The requirements of the discrete coerciveness condition and Babuška-Brezzi condition, written at the element level, are :

1. $N_\beta \geq N_q + N_\lambda - N_{rb}$, where N_{rb} is the number of rigid body modes associated with the element.

2. The flexibility matrix $\mathbf{H}$ must be positive definite with respect to the kernel vectors of the $\mathbf{G}$ and the $\mathbf{R}$ matrices.

3. The column rank of the $[\mathbf{G}\ \mathbf{R}]$ matrix must be equal to the total number of its columns minus the number of rigid body modes N_{rb}.

The four-node elements described above clearly satisfy condition (1). The satisfaction of condition (2) follows immediately from the positive definiteness of the elastic compliance matrix. Condition (3) is a crucial requirement. It can be shown that for the case of a square flat element, the proposed formulation identically satisfies condition (3) and, therefore, is stable in the sense of the Babuška-Brezzi condition. Since curvature of the element will induce coupling between the membrane and bending modes, the curved element will be less susceptible to spurious modes than the flat element in which the modes are uncoupled. Numerical stability of the flat element implies stability of the curved element. Numerical rank checks based on an eigen-value analysis of single element stiffness matrices for elements used in non-planar trial meshes including various element shapes, verify this conclusion. This element exactly represents all rigid body modes.

6. Numerical Examples

6.1 Homogeneous Shell Element

In this section, the performance of the homogeneous shell element is evaluated using two numerical examples. The following notation is used to identify elements:

4-Bubble—Proposed four-node mixed shell element.

4-SRI　—four-node displacement based shell element with selective/reduced integration on in-plane and out-of-plane shear terms. This element possesses two zero-energy modes [15].

Square Plate

Lasry and Belytschko [16] observed that transverse shear stresses may oscillate when obtained using 4-node quadrilateral Mindlin elements based on selective/reduced integration. In particular, these oscillations in the transverse shear stress, which may be quite severe, are induced by even very small perturbations in a regular mesh. The 4-Bubble element is tested to observe its behavior under such mesh perturbations by using the problem discussed by Lasry and Belytschko. This problem involves a clamped square plate of dimension 10 and thickness of 0.1. The plate is subjected to a uniformly distributed loading of magnitude 1.0. Young's modulus of the material is 1.092×10^6 and Poisson's ratio is 0.3. Only a quarter of the plate was analyzed using symmetry. Six nodes per side were employed in the analysis and the mesh is presented in Fig. 1a. An analytical solution for the problem may be found in [17]. Both 4-Bubble and 4-SRI elements show excellent accuracy for the transverse shear forces when a regular mesh is employed and therefore the results are not presented. A perturbation was introduced at node A, which has coordinates $(3.0, 3.0)$ in the regular mesh and $(3.1, 2.9)$ in the perturbed mesh. Results for transverse shear forces Q_{13} and Q_{23} along the line $x_2 = 1.5$ are plotted in Figs. 1b and 1c, respectively. These figures verify the severe oscillations in transverse shear forces found in the 4-SRI element and show that the 4-Bubble element is unaffected by the perturbation.

Pinched Cylinder with End Diaphragms

This example tests the elements ability to represent inextensional bending strain and complex membrane strain states. The cylinder has a length of 600, a radius of 300 and a thickness of 3. The geometry of the problem is presented in Fig. 2a. Young's modulus of the material is assumed to be 3.0×10^6 and the Poisson's ratio is taken to be 0.3. The cylinder is subjected to two concentrated loads of magnitude 1.0 applied radially at two diametrically opposite points. Both the ends of the cylinder are restrained by rigid diaphragms. An analytical solution for the problem is given by Flugge [18] and the solution for the transverse displacement under the load is $0.18248 \times$

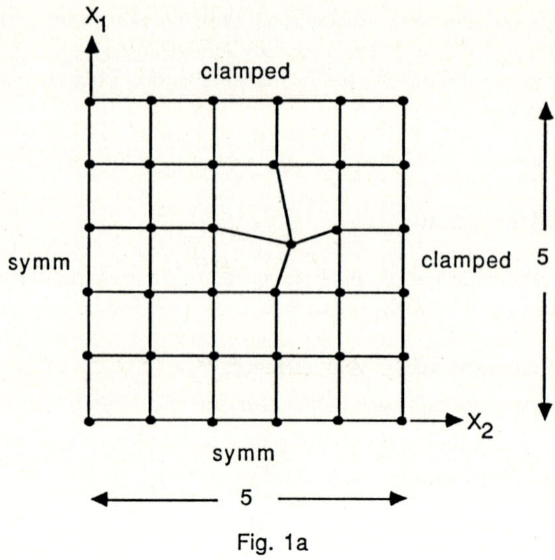

Fig. 1a

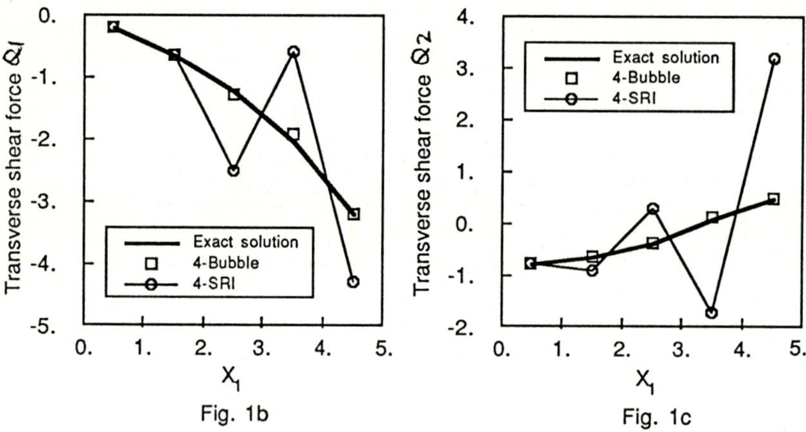

Fig. 1b

Fig. 1c

Fig. 1. - Square homogeneous plate.

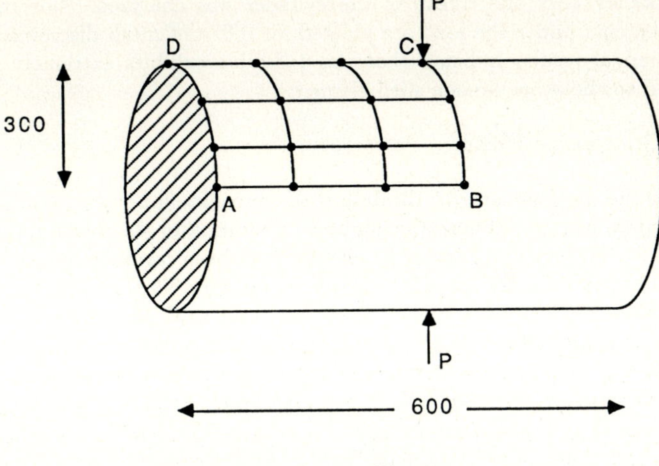

Fig. 2a

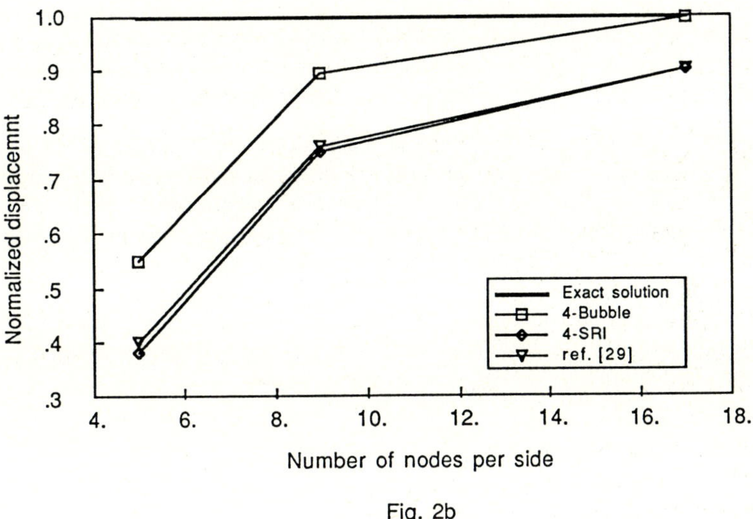

Fig. 2b

Fig. 2. - Pinched cylindrical shell with closed ends.

10^{-4}. Using symmetry only one eighth of the cylinder was analyzed. Normalized transverse displacements under the load are plotted for different mesh discretizations in Fig. 2b. Numerical results indicate that the 4-Bubble exhibits extremely good accuracy compared to the other 4-node shell elements.

6.2 Laminated Composite Plate

In this section the performance of the proposed four-node laminated composite element is documented using a numerical example. In addition to the four-node plate element described in the paper, a nine-node element has been formulated using the same principles and results for this element are also presented. The example problem involves a square plate and only a quarter of the plate was analyzed. All the values presented were normalized with respect to the corresponding exact solution. The following notations are employed in this section: SS - simply supported boundary conditions, U - uniform load. Element notation is as follows: MQH3 - 8-node hybrid stress element proposed by Spilker and Jacobs [19], MQH3TS - 8-node hybrid stress element proposed by Spilker and Engelmann [20], 9-Bubble - 9-node mixed element based on the present formulation.

Uniformly Loaded Square Plate (2 layer angle-ply $\pm\theta$)

In this study a two layered square plate with angle-ply construction ($\pm\theta$) with the following material properties is considered:

$$E_{11} = 40 \times 10^6, E_{22} = 10^6, \nu_{12} = \nu_{23} = 0.25, G_{12} = G_{23} = 0.5 \times 10^6$$

The thickness of the plate is 0.02. The series solution developed by Whitney [21] was used for evaluating the performance of the elements. The boundary conditions and fiber orientations are given in Fig. 3. The convergence behavior of the elements for the center displacement and center moments are also presented in Fig. 3., for a square plate of length 10. The bubble elements are compared to the hybrid stress composite elements developed by Spilker et al. [19, 20]. It was found that seven nodes per side were sufficient to converge to the exact solution in both displacements and stresses. Results show that 4-Bubble element displays improved performance over the hybrid-stress elements.

CONCLUSIONS

The design of laminated composite plate and shell structures requires accurate knowledge of all stress components for input into local failure criteria. We have proposed an extended version of the Hellinger-Reissner principle which includes Lagrange multiplier constraints on the stresses. In particular, we have considered the effects of choosing the Lagrange multiplier functions to be element-based bubble functions in the finite element approximation. In the two elements developed on this basis, we obtained accurate stresses, including transverse shear stresses, for arbitrarily distorted meshes. In addition, the elements are numerically stable and have no shear locking in the thin limit.

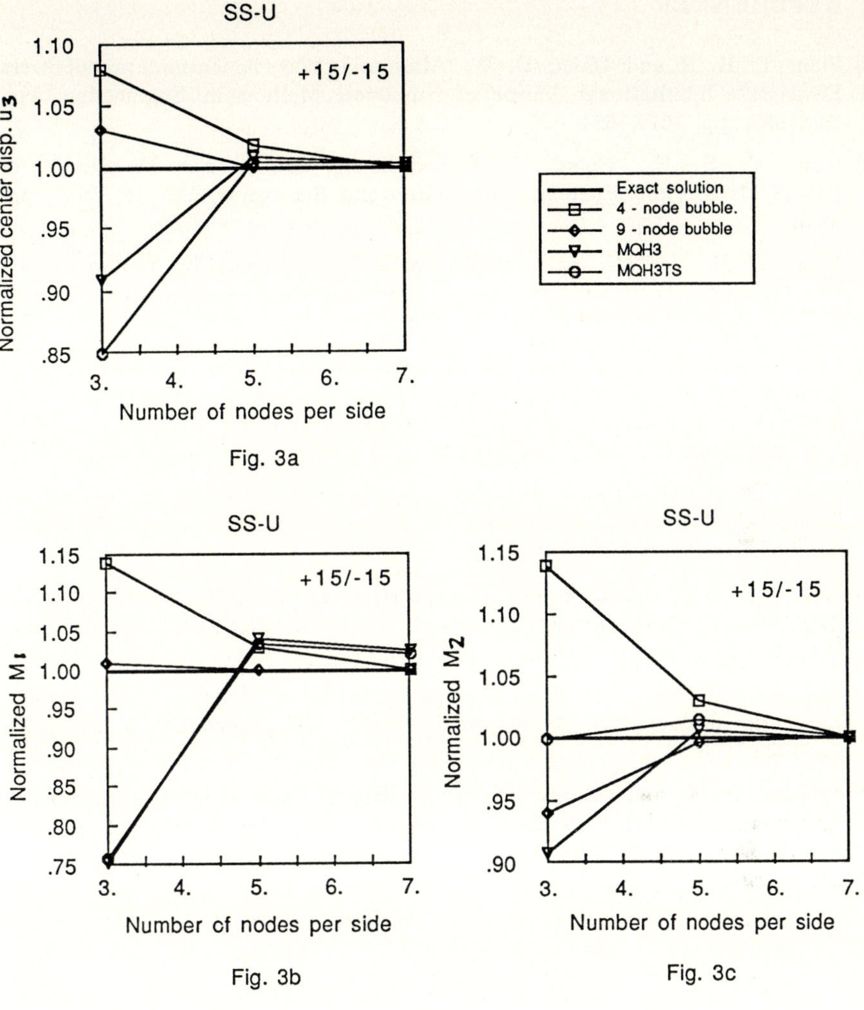

Fig. 3. - Square composite plate example.

REFERENCES

[1] Pian, T. H. H. and Chen, D. P., "Alternative Ways for Formulation of Stress Elements", International Journal of Numerical Methods in Engineering, Vol. 18, 1982, pp. 1679-1684.

[2] Pian, T. H. H., Chen, D. P. and Kang, D., "A New Formulation of Hybrid/Mixed Finite Element", Computers and Structures, Vol. 16, 1983, pp. 81-87.

[3] Pian, T. H. H., "Derivation of Element Stiffness Matrices by Assumed Stress Distribution", American Institute of Aeronautics and Astronautics Journal, Vol. 2, 1964, pp. 1333-1336.

[4] Pian, T. H. H., "Element Stiffness Matrices for Boundary Compatibility and for Prescribed Boundary Stresses", Proceedings of First Conference on Matrix Methods in Structural Mechanics, Wright-Patterson Air Force Base, Ohio, A-FFDL-TR-66-80, 1966, pp. 457-477.

[5] Pian, T. H. H., "Hybrid Models", Numerical and Computer Methods in Structural Mechanics, eds. Fenves, S. J. et al., Academic Press, NY, 1971, pp. 59-78.

[6] Spilker, R. L., "Invariant 8-node Hybrid Stress Elements for Thin and Moderately Thick Plates", International Journal of Numerical Methods in Engineering, Vol. 18, 1972, pp. 1153-1178.

[7] Spilker, R. L. and Munir, N. I., "The Hybrid-Stress Models for Thin Plates", International Journal of Numerical Methods in Engineering, Vol. 15, 1980, pp. 1239-1260.

[8] Spilker, R. L. and Munir, N. I., "A Hybrid-Stress Quadratic Serendipity Displacement Mindlin Plate Bending Element", Computers and Structures, Vol. 12, 1980, pp. 11-21.

[9] Fraejis de Veubeke, B., "Displacement and Equilibrium Models in the Finite Element Method", Stress Analysis, eds. Zienkiewicz, O. C. and Holister, G. S.), Wiley, London, 1965, pp. 145-197.

[10] Pinsky, P. M. and Jasti, R. V., "A Mixed Finite Element Formulation for Reissner-Mindlin Plates Based on the Use of Bubble Functions", International Journal of Numerical Methods in Engineering, Vol. 28, 1989, pp. 1677-1702.

[11] Babuška, I., "The Finite Element Method with Lagrangian Multipliers", Numerical Mathematics, Vol. 20, 1973, pp. 179-192.

[12] Brezzi, F., "On the Existence, Uniqueness and Approximation of Saddle-Point Problems Arising from Lagrangian Multipliers", Revue Francaise de Automatique de Informatique de Recherche Operationelle, Vol. R2., 1974, pp. 129-151.

[14] Hartmann, F., "The Discrete Babuška-Brezzi Condition", Ingenieur-Archiv, Vol. 56, 1986, pp. 221-228.

[15] Stanley, G. M., "Continuum Based Shell Elements", Doctoral Thesis, Division of Applied Mechanics, Stanford University, 1985.

[16] Lasry, D. and Belytschko, T., "Transverse Shear Oscillations in Four-Node Quadrilateral Plate Elements", Computers and Structures, Vol. 27, 1987, pp. 393-398.

[17] Timoshenko, S. and Woinowsky-Krieger, S., "Theory of Plates and Shells", 2nd edition, McGraw-Hill, NY, 1959.

[18] Flugge, W., "Stresses in Shells", 2nd edition, Springer, Berlin, 1973.

[19] Spilker, R. L. and Jacobs, D. M., "Hybrid Stress Reduced-Mindlin Elements for Thin Multilayered Plates", International Journal of Numerical Methods in Engineering, Vol. 23, 1986, pp. 555-578.

[20] Spilker, R. L. and Engelmann, B. E., "Hybrid-Stress Isoparametric Elements for Moderately Thick and Thin Multilayered Plates", Computer Methods in Applied Mechanics and Engineering, Vol. 56, 1986, pp. 339-361.

[21] Whitney, J. M., "Bending-Extensional Coupling in Laminated Composite Plates Under Transverse Load", Journal of Composite Materials, Vol. 3, 1969, pp. 20-28.

**Part 4
Specific Applications
of Nonlinear Shell Concepts**

Nonlinear Shell Analysis of the Space Shuttle Solid Rocket Boosters

N. F. KNIGHT, JR.†, R. E. GILLIAN‡, and M. P. NEMETH†

NASA Langley Research Center, Hampton, Virginia 23665-5225, USA

Abstract

A variety of structural analyses have been performed on the Solid Rocket Boosters (SRB's) to provide information that would contribute to the understanding of the failure which destroyed the Space Shuttle Challenger. This paper describes nonlinear shell analyses that were performed to characterize the behavior of an overall SRB structure and a segment of the SRB in the vicinity of the External Tank Attachment (ETA) ring. Shell finite element models were used that would accurately reflect the global load transfer in an SRB in a manner such that nonlinear shell collapse and ovalization could be assessed. The purpose of these analyses was to calculate the overall deflection and stress distributions for these SRB models when subjected to mechanical loads corresponding to critical times during the launch sequence. Static analyses of these SRB models were performed using a "snapshot picture" of the loads. Analytical results obtained using these models show no evidence of nonlinear shell collapse for the pre-liftoff loading cases considered.

Introduction

The basic elements of the Space Shuttle system are the Orbiter, the External Tank (ET), and the two reusable Solid Rocket Boosters (SRB's) as shown in figure 1. The SRB's provide the primary Shuttle ascent boost for the first two minutes of flight with an assist from the three Space Shuttle Main Engines (SSME's) on the Orbiter. The SRB structural subsystems include the Solid Rocket Motor (SRM) which consists of four lined, insulated rocket motor segments. These segments are connected using pinned tang-clevis joints (see figure 2). The upper end of the lower cylindrical, motor segment forms the clevis. The lower end of the upper cylindrical, motor segment forms the tang which mates with the lower clevis. Around the circumference of both tang and clevis ends are 180 holes into which one-inch-diameter connecting pins are inserted and are held in place by retainer bands. The seal between two motor segments is provided by two O-rings in the "inner arm" of the clevis. The O-rings are compressed upon assembly of the SRM segments by a flat sealing surface on the tang.

The accident which destroyed the Space Shuttle Challenger is believed to have been caused by the failure of a case joint in the right solid rocket motor.[1] Several characteristics of the original SRM joint design have been identified as potential contributors to the failure. One characteristic is the behavior of the joint under internal pressure load. For this loading, the motor case expands radially outward. Because the joint has a higher hoop stiffness than the case wall on either side of the joint, its radial expansion is less than that of the case wall. In addition to this nonuniform stiffness in the longitudinal direction associated with the case joints, a nonuniform stiffness in the circumferential direction exists in the aft attachment segment due to the external tank attachment (ETA) ring. Nonuniform radial expansion is the primary cause of relative motion

---

† *Aerospace Engineers, Structural Mechanics Branch, Structural Mechanics Division.*
‡ *Mathematician, Structural Mechanics Branch, Structural Mechanics Division.*

between the inner clevis arm and the sealing surface on the tang. This relative motion can cause the O-rings to become unseated and therefore lose their sealing capability.

Structural analyses at various levels of detail have been performed in support of the SRB structural design assessment and redesign efforts. These levels vary from local three-dimensional solid models of a one-degree segment of the tang-clevis joint reported in reference 2 to global models of the SRB reported in references 3 through 5. Axisymmetric shell-of-revolution and axisymmetric solid models of the local joint have been analyzed wherein the overall shell response characteristics are represented well. However, the use of these models to determine the structural response locally in the joint requires substantial engineering judgement due to inherent asymmetries associated with the pins, friction, and contact. In reference 2, the three-dimensional solid finite element models of a one-degree segment (centerline of a pin to midway between pins) of the field joint were used to determine the local joint response to pressure loading including the induced axial load. In reference 3, a linear analysis of the entire SRB, including the propellant, was performed with the MSC/NASTRAN finite element computer code[6] using solid elements throughout the model. As such, the local joint behavior could be assessed and data then used in more detailed, local joint models. In reference 4, nonlinear analyses of the entire SRB were performed using shell elements (see figure 3) wherein equivalent stiffness joint models were used, and as such this model cannot predict the local joint behavior. In reference 5, nonlinear analyses of the SRB/ETA ring interface region were performed using a shell finite element model also (see figure 4). These two-dimensional shell models of the SRB have been developed using the STAGSC-1 computer code[7,8]. These 2-D shell models have been used to calculate the overall deflection distributions for the SRB when subjected to mechanical loads corresponding to selected times during the launch sequence. The mechanical loading conditions for the full SRB arise from the ET attachment points, the SRM pressure load, and the SRB aft skirt hold down posts. Static analyses of the full SRB were performed using a "snapshot picture" of the loads. The purpose of this paper is to provide a summary of the nonlinear analyses performed in references 4 and 5 including a description of the computational approach and requirements, and a presentation of results from the analyses of the Space Shuttle solid rocket boosters subjected to selected pre-liftoff loads corresponding to Space Transportation System (STS) Flight 51-L.

Method of Analysis

The STAGSC-1 computer code[7,8] has been under development for over 15 years and its development was initiated to support the design and analysis of the space shuttle system. STAGSC-1 is a 2-D shell finite element analysis code based on the displacement formulation. The element library includes nonlinear spring (or mount) elements, 1-D beam elements, and 2-D plate/shell elements. Analysis options are provided for including geometric and material nonlinearities for buckling, collapse, vibration, or transient dynamic analysis. STAGSC-1 is supported on CDC, VAX, and Cray computers and is available through COSMIC[†].

The STAGSC-1 computer code is comprised of six modules: STAGS1; STAGS2; UNFFMT; FMTUNF; POSTP; and STAPL. The STAGS1 module is a preprocessing module which handles model generation, degree-of-freedom tables, element shape functions, element constitutive matrices, and so forth. Execution of this module typically precedes the execution of STAGS2, POSTP, or STAPL. The STAGS2 module is the computational module which performs matrix decomposition, linear and nonlinear stress analyses, eigenvalue (buckling or vibration) analyses, transient dynamic response predictions, and so forth. The modules UNFFMT and FMTUNF are used to translate the STAGS restart file (TAPE22) to an ASCII file for transfer to another computer type and then to convert that ASCII file back to a STAGS restart file for further processing. For example, the analysis could be performed on a Cray computer and the restart file translated into ASCII format. This ASCII file is transferred to a different computer system, say a VAX, and then converted back to a restart file for postprocessing. The POSTP module is a postprocessing module for printing primary

---

† *COSMIC is a non-profit agency, established by NASA as the one central office to collect, evaluate, and distribute software that is developed with NASA funding. COSMIC, The University of Georgia, 382 East Broad Street, Athens, GA 30602, USA.*

solutions or recovering secondary solutions (such as stresses, strains, stress resultants) from previously calculated displacement solutions which have been written to the restart file. The STAPL module is also a postprocessing module for plotting undeformed and deformed geometries, as well as contour plots of solution vectors (primary or secondary solutions).

The modeling strategy used in the STAGSC-1 computer code involves the concepts of a shell unit and an element unit. A shell unit may be viewed as a substructure or superelement for the purpose of modeling convenience only. A shell unit may be composed of hundreds of nodes and elements, and automatic mesh generation facilities are provided for several common geometries for plate and shell structures. Mesh generation for a shell unit is accomplished by specifying the number of rows and columns of grid lines in each coordinate direction, not the number of elements. For example, a mesh with two rows and two columns represents one quadrilateral finite element. An element unit is perhaps more like conventional finite element codes in terms of required input data (*e.g.*, node and element numbers, nodal coordinates, nodal connectivities) and provides the flexibility to model general shell-type structures. User-written subroutines may be input by the user to utilize mesh generation utilities that meet specific needs. For example, the user-written subroutine WALL is used in the analyses reported herein to vary the shell wall properties longitudinally along the SRB.

Finite Element Modeling

The modeling philosophy adopted in this study was substantially influenced by the size of the structure to be analyzed and the resulting number of the degrees-of-freedom in the equations to be solved. The underlying philosophy was to construct a finite element model that would accurately reflect the global load transfer in the SRB in a manner such that nonlinear shell collapse and shell ovalization under pre-launch loads could be assessed. The STAGSC-1 computer code was used to perform nonlinear shell analyses to characterize the behavior of an overall SRB structure and a segment of the SRB in the vicinity of the ETA ring. The overall SRB model and the segment model, referred to as the SRB/ETA ring interface model, are described in this section. Although, the resulting finite element model of the entire SRB involved nearly 85,000 degrees-of-freedom, it does not have the necessary fidelity to determine detailed stress distributions in particular SRB subsystems. Details of the finite element modeling for the entire SRB shell structure are given in references 4 and 5.

Equivalent Joint Modeling

The factory and field joints of the SRB are complicated structural assemblies that behave nonlinearly due to contact, friction, and local material yielding in the joints. In the present analysis, the field and factory joints are modeled by using equivalent stiffness joints instead of detailed models of the joint. As such, the influence of the joints on the global shell response is included; however, local joint behavior (*i.e.*, gap motion) cannot be recovered from these global models. Local structural behavior of the joints is described in detail in reference 2. Global shell behavior of the SRB can be obtained using equivalent stiffness joints for the field and factory joints, and an evaluation of nonlinear effects such as shell collapse and ovalization can be performed.

In keeping with the underlying modeling philosophy of the SRB shell structure, the assembly joints of the SRB were modeled by 2-D shell elements that were assigned stiffnesses that reflected the membrane, bending, and shear load transfer through the joints in a statically-equivalent manner. The properties for the equivalent stiffness joint are determined through parametric studies and comparisons with the referee test data[9]. In these studies, the SRM shell wall thickness was varied in the vicinity of the joints. When a combination of thickness and effective length for the joints yields analytical results which agree with the measured radial deflections from the referee test girth gages, the equivalent stiffness joint properties are determined. In these analyses, the equivalent joint is modeled as a 6-inch-long portion of the shell with an 0.8-inch thickness.

Overall SRB Model

Each SRB is approximately 144 feet long and 12 feet in diameter. The rocket consists of several segments including the forward nose cone assembly, the forward motor case, the forward center motor case, the aft center motor case, the aft attach motor case, and the aft skirt and nozzle assembly. The motor cases connected together by a tang-clevis joint are assembled at the launch site. These joints are referred to as "field joints". Each of the upper three motor segments contain an additional tang-clevis joint that is assembled at the factory. These joints are referred to as "factory joints". The aft attach motor case has two of these factory joints. In addition, the forward motor case and the aft attach case each have what is referred to as a "Y-shaped factory joint" that connects the pressure domes to the SRM. These Y-shaped factory joints have an appendage that is used to connect the forward nose assembly and the aft skirt to the SRM. The SRB structural subsystem provides the necessary structural support for the Shuttle vehicle on the launch pad, transfers thrust loads to the Orbiter and ET, and provides the housing, structural support and bracketry needed for the recovery system, the electrical components, the separation motors, and the thrust vector control system. This subsystem consists of the nose cone assembly, the forward skirt including the forward SRB/ET attach fitting, the aft SRB/ET attach ring and attach struts, the aft skirt including the heat shield, the systems tunnel, and structure for mounting other SRB subsystems components.

The STAGSC-1 two-dimensional shell finite element model of the entire SRB shown in figure 3 involves 9205 nodes with 1273 two-node 211-beam elements, 90 three-node 321-triangular elements, and 9156 four-node 411-quadrilateral elements. The elements denoted as 211, 321, and 411 are elements in the STAGSC-1 element library. Although the resulting finite element model involves nearly 85,000 degrees-of-freedom, it does not have the fidelity necessary to determine detailed stress distributions in particular SRB subsystems. In this global shell model, the field and factory joints are modeled by using equivalent stiffness joints instead of detailed models of the joint. Additional details of the finite element modeling for the entire SRB shell structure are described in reference 4.

SRB/ETA Ring Interface Model

The definition of the SRB/ETA ring interface region for this paper includes both of the ETA rings (ring webs are approximately 12 inches apart), a portion of the SRM aft attachment segment including the factory joint at station 1577 (approximately sixty inches of shell), and a portion of the aft center segment including the field joint at station 1491 (approximately 64 inches of shell). Failure of the O-rings to seal at this field joint is believed to have caused the Challenger disaster. The total length of the SRB/ETA ring interface region considered in this paper is 136 inches. This length corresponds roughly to one shell radius on either side of the ETA rings.

The aft ETA ring assembly is shown in figure 1 and its center is located at station 1511, approximately twenty inches below the aft attachment segment field joint. The ETA ring assembly is comprised of two tapered, partial rings(ring webs are approximately 12 inches apart), H-fittings to attach the ET struts, cover plates, and various other intercostals and brackets. The ETA rings are bolted every 2-degrees around the circumference to two stub rings which are integral parts of the SRM aft attachment segment. The ETA ring assembly extends only 270-degrees circumferentially around the SRM segment. Three struts attach the aft end of the SRB with the ET as shown in figure 4. These three attachment struts are designated the lower strut (P9), the diagonal strut (P10), and the upper strut (P8).

The finite element model of the SRB/ETA ring interface has 45 elements uniformly spaced around the shell circumference and 26 elements along its length as shown in figure 4. The finite element used in these analyses is designated as 411 in the STAGSC - 1 element library. Symmetry boundary conditions are imposed at the forward end of the model with the exception that the longitudinal direction is unrestrained. Simple-support boundary conditions are imposed at the aft end of the model with the exception that the radial direction is unrestrained. This set of boundary conditions requires the aft end of the shell to remain circular.

The finite element model of the ETA ring is shown in figure 5. User-written subroutine USRPT and USRELT are used to generate the geometry and finite element discretization of the ETA ring. The ETA ring webs are modeled with one element through the depth of the web and has a uniform thickness of 0.25 inches. The ETA ring cap is modeled as a discrete stiffener with a rectangular cross section of 1.0 inches by 1.79 inches. The ETA ring cover plates, intercostals, H-fittings, and various other brackets are not included in these finite element models. Further discussion of the SRB/ETA ring interface is provided in reference 5.

SRB Loading

SRM Pressure Loading

The SRM pressure loading results from the burning of the solid propellant. Only the SRM components of the SRB are directly loaded by the internal pressure distribution. SRM ignition occurs approximately 6.6 seconds after SSME ignition and require 600 milliseconds to reach full pressurization. The nominal SRM internal pressure is approximately 1000 psi. The SRM longitudinal pressure distribution varies by approximately 100 psi over the length of the SRM. User-written subroutine UPRESS is used to model this pressure variation. The internal pressure elongates the SRM case before liftoff and imparts a significant load on the forward SRB/ET attach point. The SRM axial tension loads from SRM ignition to SRB separation are the result of internal pressure, thrust, and inertial loads.

SRB/ET Interface Loads

The reconstructed flight loads for the Space Shuttle Challenger STS Flight 51-L were obtained from NASA Johnson Space Center (JSC) initially in the form of strip charts and later in the form of data stored on a magnetic tape. The loads data consist of equivalent beam forces and moments and vehicle interface loads for the first ten seconds of the flight. The equivalent beam forces and moments are given at 19 locations along the SRB and include inertial effects. The interface loads include components of the loads in the forward and aft struts connecting the SRB to the external tank as shown in figure 6. A computer program was written to extract the interface loads from the JSC database, compute the SRB/ET strut loads, and print the desired data.

An overview of the pre-liftoff loads variations after SSME ignition is shown in figure 7. The loading cases considered in these analyses correspond to t = 0, 5.3, 6.6, and 7.2 seconds after SSME ignition. The time-consistent SRB/ET interface loads for these four "snapshots" in time are given in Table 1. At t = 0, the loads induced into the SRB are due to the eccentric weight of the orbiter and the external fuel tank and due to cryogenic shrinkage of the external tank during fueling. At SSME ignition, an eccentric thrust of approximately one million pounds is produced that causes the space shuttle to bend over to a maximum deflection (referred to as "max twang") and then spring back towards its original static configuration. For STS 51-L, the maximum bending occurred at t = 5.3 seconds. The SSME reached full thrust approximately 6.6 seconds after ignition. At t = 6.6 seconds, the space shuttle system has fully rebounded and the signal is issued to ignite the solid rocket motor propellant. The SRM ignition pressure transient has a duration of approximately 600 milliseconds for the pressure to build up inside the solid rocket boosters.

The reconstructed beam forces and moments for the SRB/ETA ring interface shell analysis model are given in Table 2. The aft ET attachment loads are equally divided between the two ETA ring webs and applied as point forces. The reconstructed beam forces and moments are used to generate a statically equivalent set of shell stress resultants for inplane compression and shear.

Computational Approach

The SRB analyses were performed using various NASA computer systems. The operational aspects of using various computer systems including remote access to the Numerical Aerodynamic Simulator (or NAS)[10] facility at the NASA Ames Research Center are described on figure 8. A Langley VAX computer is used for model preparation and verification. The datasets are then transmitted to the NAS computers for execution. The output files and restart files are returned to a Langley VAX computer for postprocessing. The output files and plot vector files are transmitted to the Langley central computer system for printing, microfiche generation, plotting, and archival storage. For example, a nonlinear analysis of the SRB is performed on the NAS Cray-2 computer and a restart file containing calculated displacement solution vectors is generated. Then the restart file is translated into ASCII format and transferred using the NASnet wide-area network to a Langley VAX computer. This ASCII file is converted back to a restart file for postprocessing. The output file (5 to 50 megabytes in size) generated on NAS is also returned to Langley and then transferred using the LaRCnet local-area network to the central computer site for making microfiche copies (7 to 25 microfiche). The postprocessing of these results and the model verification task are performed on a Langley VAX computer system instead of the Langley central site CYBER computers because of the large memory requirements.

Computer Systems

In building the finite element models, calculating results, and evaluating the output, three different classes of computer systems are used (*i.e.*, workstations, minicomputers and supercomputers). A fourth class of computer system, the mainframe system, is used to store data, process microfiche, and produce report-quality graphics output.

The first class of computer system used in building and verifying the finite element models is a minicomputer. A VAX 11/785 minicomputer running the VMS operating system is used. Since the STAGSC-1 computer code is developed and enhanced under the VMS operating system, this computer system is a natural part of the overall computing environment. The VAX 11/785 computer system allowed the finite element models to be generated, solutions using coarse grids to be computed, and "quick-look" graphics output to be evaluated prior to submitting large-scale analyses to the supercomputer.

The second class of computer system used to perform these analyses is a VAXstation II GPX workstation running the ULTRIX operating system. This computer system provides basic communication between the VAX/VMS DECnet environment of a minicomputer and the UNIX TCP/IP environment of a supercomputer. The workstation capability also provides the "quick-look" graphics display to verify the model generation and to evaluate results.

The third class of computer systems used consists of supercomputers. The computational portion of these analyses is performed on the NAS facility at the NASA Ames Research Center. The goals and objectives of the NAS require that the computer system hardware and software change. At the beginning of this project, the NAS computer environment was in transition. Cray Research, Inc. had delivered the first commercial Cray-2 supercomputer, and an interim Cray X-MP/12 was still on-site at Ames. Work at NAS began using the Cray X-MP/12 supercomputer running the COS operating system. This Cray X-MP/12 was made available for the initial phase of the work. When the initial phase of the work was completed, the Cray X-MP/12 was removed, and the STAGSC-1 code was ported to the Cray-2 to complete the preliminary analyses reported in this paper. The NAS Cray-2 supercomputer uses the UNICOS[†] operating system. Various upgrades occurred during the remainder of this activity as NAS transitioned through a Cray-2 running UNICOS 1.0, a Cray-2 running UNICOS 2.0, and finally to a Cray-2S running UNICOS 3.0.

Finally, the fourth class of computer system is the mainframe system. The Langley Central Site mainframe environment, consisting of several CDC Cyber computers running the NOS operating system, provided a capability for producing printed output, mass storage for the large output files produced on the supercomputer, archival output in the form of microfiche, and report quality graphics. This local capability provided the labor intensive functions required throughout the distributed environment and complements the other distributed capabilities required for this project.

[†] *The UNICOS operating system is derived from the AT&T UNIX System V operating system. UNICOS is also based in part on the Fourth Berkeley Software Distribution under license from The Regents of the University of California*

Network Access

The changes occurring in the field of computer networking represent probably the most dramatic changes affecting structural analysts over the period of this project. Networking removed the constraint of physical distance. Working remotely from the supercomputers used in the computational phase of this project presented a new set of problems that, once solved, resulted in a unique new capability for the structural analysts.

The network at Langley uses Ethernet within buildings and a fiber optic Pronet 10 token-passing ring network called LaRCnet between Langley buildings. Initially the gateways between buildings would route only a Xerox XNS-based protocol developed at Langley. Connected to one of the Ethernets is a Vitalink Bridge that would route both TCP/IP and DECnet to the NAS facility at Ames over a 256 kilobits per second satellite link. The evolution of this network over the course of the project followed the networks being developed in industry over that time. Much of the communication was done manually at first. LaRCnet was used to copy files to a computer in the building that contained the Vitalink Bridge, and DECnet was used to cross the country to a staging computer at Ames, and then the Cray Station software on the Ames VMS VAX computer was used to complete the link. The steps would be reversed to bring data back to Langley. This communication path was simplified over the course of the project to the network now in place. Workstations in separate buildings are supported with routing gateways through the LaRCnet fiber optic system, which uses a Pronet P4200 gateway connected to the Vitalink directly. The communication link with Ames has been upgraded to a one megabit per second transfer rate (*i.e.*, T1 link) and it was discovered that a land line is preferable to a satellite link for interactive use. The result is that the miles between Ames and Langley are no longer a problem; researchers can use the NAS system at Ames as if it were located at Langley. The Cray-2 appears to the structural analyst as if it were embedded in the local workstation.

Performance

The performance of the STAGSC-1 program on the Cray X-MP/12 and Cray-2 computer is shown in Table 3. Much of the early work was done without the advantage of FORTRAN optimization from the Cray-2 UNICOS 1.0 compiler due to the newness of the compiler and the errors in the compiler optimization. The Cray-2 flowtrace capability was used to identify the routines that required the most CPU time and efforts were directed at optimizing those routines. When the FORTRAN compiler optimization had been completed, the Cray-2 still took twice as long as the Cray X-MP/12 to perform a linear stress analysis of the SRB finite element model described in this paper. With the use of the Cray-2 vector library routine "sdot", the run time was decreased by a factor of two, achieving the same overall rate as the Cray X-MP/12. Increasing the amount of memory managed within the program itself also resulted in significant I/O savings. Since STAGSC-1 was designed on static memory machines, the program made use of blank common to provide out-of-core solutions. By controlling the amount of managed memory within the STAGSC-1 program (*i.e.*, changing the size of the blank common), the I/O rate required for efficient execution could be balanced with the restrictions of the specific implementation of UNICOS job processing.

The NAS Cray-2 supercomputer has four processors, each with a clock cycle time of 4.1 billionths of a second and a total memory size of 256 million 64-bit words. This Cray-2 is a supercomputer capable of over one hundred times the computational capability of a VAX 11/785 computer. In addition, the Cray-2 is a native 64-bit wordsize machine, and roundoff problems that are a problem on 32-bit machines are eliminated. The STAGSC-1 computer code, designed nearly fifteen years ago, uses basic algorithms that provide out-of-core solution methods that also work well on the Cray-2. Even with 256 million words of main memory, the larger matrices could not be held in memory. This application program made use of 60 million words of main memory (*e.g.*, blank common is dimensioned to 32 million words) to avoid excessive I/O and to fit execution runs into the normal processing queues eliminating the need for special priority. Auxiliary data storage requirements for these analyses is another concern. During the large SRB runs, a single temporary file requires in excess of 800 megabytes of storage. Hence, coordination or scheduling of these runs by the analyst is necessary to avoid exceeding the available auxiliary storage.

A comparison of the NAS Cray-2 performance with the other classes of computers used in this study is given in Table 4. The problem solved in this case is a linear stress analysis of the finite element model of the SRB/ETA ring interface (see ref. 5). The finite element model has approximately 12,000 active degrees-of-freedom with an average semi-bandwidth of 510 in the global stiffness matrix. This problem is selected for comparison since it is the largest reasonable problem that could be expected to run on VAX 11/785 computers (CPU speed limitation) and CDC NOS computers (fixed memory limitation). The times presented in Table 4 are in CPU seconds and demonstrate clearly why a supercomputer is needed for these calculations, and it justifies the time spent in applying vector optimization techniques.

Results and Discussion

The purpose of the 2-D shell analyses of the SRB/ETA ring interface region and the entire SRB is to calculate the overall deflection distributions for the SRB when subjected to mechanical loads corresponding to critical times during the launch sequence. The field and factory joints are modeled by using equivalent stiffness joints instead of detailed models of the joint. As such, local joint behavior cannot be obtained from this global model. However, global shell behavior can be obtained and an assessment of nonlinear effects such as shell collapse and ovalization can be performed for selected times during the launch sequence.

SRB/ETA Ring Interface Shell Model

The SRB/ETA ring interface model was developed using the original (*i.e.*, STS 51-L) geometry configuration. Approximately sixty inches of the SRM motor case are modeled on either side of the SRB/ETA ring interface. Two loading conditions are considered. The first condition corresponds to internal pressure only. The second condition corresponds to selected pre-liftoff, time-consistent loads for flight STS 51-L. Two sets of STS 51-L flight loads are considered; namely, those at maximum bending moment prior to SRM ignition ("max twang") and those just prior to liftoff but after the SRM ignition pressure transient. The applied loads for the second conditions include SRM internal pressure, ET strut loads, and inplane shell loads.

Approximately 12,000 active degrees-of-freedom are in the finite element model of the SRB/ETA ring interface region shown in figure 4. The average semi-bandwidth of the global stiffness matrix is 510. To form the elemental stiffness matrices and then assemble the global stiffness matrix required 19 CPU seconds on the NAS Cray-2 computer. A single decomposition of the global stiffness matrix required an additional 100 CPU seconds. One forward-reduction/back-substitution cycle either to obtain the linear stress solution or to perform one nonlinear iteration required an additional 7 CPU seconds. The complete nonlinear shell analysis of this model required a total of three decompositions of the global stiffness matrix and 16 nonlinear iterations.

The linear and nonlinear response of the radial deflection normalized by the nominal shell thickness (*i.e.*, 0.479 inches) of two points diametrically opposite are shown in figure 9 as a function of internal pressure. The point labeled A is located midway between the ends of the ETA ring. The point labeled B is located approximately 180-degrees away and located on the ETA ring. The longitudinal location of these points is midway between the ETA ring webs (*i.e.*, station 1511). The shell response of point A exhibits significant nonlinearity. The radial deflection from the nonlinear solution for point A is nearly twice as large as the linear solution. Conversely, the shell response of point B is only mildly influenced by including the geometrically nonlinear effects and exhibits a stiffening trend.

The radial deflection patterns of the SRM stub ring are shown in figure 10 for both the linear and nonlinear solutions for a 1000 psi internal pressure load. The radial deflections are normalized by the nominal shell thickness. At the ends of the ETA ring, the radial deflections from the linear analysis are equal to the shell thickness. This result indicates that a nonlinear analysis is required to predict accurately the structural response. The radial deflection pattern from the nonlinear analysis indicates a stiffening of the shell response due to the inclusion of geometric nonlinearities. These deflection patterns exhibit large changes in amplitude near the ends of the ETA ring.

The hoop stress distributions of the SRM stub ring are shown in figure 11 for the linear and nonlinear solutions for a 1000 psi internal pressure load. At the ends of the ETA ring, the hoop stress peaks due to the discontinuity in stiffness resulting from a partial ETA ring. The linear and nonlinear hoop stress distributions are similar, and their magnitudes are nearly the same except at the ends of the ETA ring. The nonlinear hoop stress at the end of the ETA ring is approximately 10% less than the linear hoop stress.

The linear and nonlinear solutions for the model of the SRB/ETA ring interface subjected to an internal pressure of 1000 psi are obtained. Deformed geometries with exaggerated deflections corresponding to the linear and nonlinear solutions are shown in figure 12. Deformed geometries of the entire model are shown in the upper half of the figure and those of the SRM stub ring alone are shown in the lower half. Both the linear and nonlinear solutions exhibit an abrupt change in deflections near the ends of the ETA rings. This high local bending causes large tangential shearing and normal forces to develop between the SRM stub rings and the ETA rings.

The axial distribution of the nonlinear radial deflections for the SRB/ETA ring interface model subjected to 1000 psi internal pressure only is shown in figure 13 for three circumferential locations. The radial deflections are normalized by the nominal shell thickness and are shown as a function of SRB station number. Station numbers corresponding to tang-clevis joint (field and factory) locations and the upper and lower ETA ring webs are also noted on the figure. The first circumferential location (point A) corresponds to midway between the ends of the ETA ring. The radial deflection pattern is denoted as the solid curve. This pattern exhibits a marked change in radial deflections near the field joint at station 1492 and is such that a tang-clevis joint would tend to open. The second location corresponds to point B which is approximately 180 degrees opposite to point A. The radial deflection pattern at this location is denoted by a dashed line. The pattern near the field joint is again similar to the patterns at the other locations. However, near the ETA ring webs the radial deflection pattern is different and the stiffening influence of the ETA ring on the shell response can be seen. The third location (point C) corresponds to an end of the ETA ring. The radial deflection at this location is denoted by a line with filled symbols. This pattern is similar to that of point A with the exception being an increase in amplitude of the radial deflections. For comparison, two additional curves are shown on figure 13. One curve represents the membrane solution, including the biaxial effect, for a uniform thickness (0.479 inches) cylindrical shell with an internal pressure of 1000 psi. The other curve represents the nonlinear solution for the same finite element model used to generate the other results except without the partial ETA ring. These radial deflection patterns from the nonlinear solution also indicate that end effects due to imposed boundary conditions appear to be localized near the ends of the model. Also, these deflection patterns indicate that the STS 51-L tang-clevis field joint at station 1492 would tend to open. However, these models only reflect STS 51-L geometry and not the new SRM joint redesign with an interference-fit capture feature that is designed to restrict the motion between the inner clevis arm and the tang. Extensive analytical studies are reported in reference 4 which describe the structural behavior of the original and modified tang-clevis joint designs. These studies included three-dimensional stress analysis, nonlinear contact, and correlation between test and analytical results.

Nonlinear analyses of the SRB/ETA ring interface model using the reconstructed loads for time t=5.3 and 7.2 seconds after SSME ignition have been performed. The nonlinear radial deflections at the ETA ring normalized by the nominal SRB case thickness (0.479 inches) are shown in figure 14 as a function of circumferential location around the SRB. Time t=5.3 seconds corresponds to the time at which maximum bending ("max twang") occurs. At this time, the SRM is unpressurized since it has not yet been ignited. The ET strut loads at maximum bending result in an asymmetric radial deflection pattern; however, the amplitudes of these deflections are small compared to either the nominal shell thickness or the radial deflections caused by internal pressure loading only. Time t=7.2 seconds corresponds to the time at which the SRM reaches full pressure and liftoff occurs. At SRM pressurization, the overall shell response is dominated by the effects of the internal pressure, and the effect of the ET strut loads is secondary.

These analytical results for the SRB/ETA ring interface indicate significant differences between the linear and nonlinear deflection patterns. The loading component which has been shown to significantly affect the shell deflection patterns is the SRM internal pressure. Both the deflection pattern and the tangential shearing force distribution changed only slightly when an axial force was combined with the internal pressure loading case. Two STS 51-L pre-liftoff, time-consistent loading cases have been considered. The ET strut loads at maximum bending result in an asymmetric radial deflection pattern; however, the amplitudes of these deflections are small compared to either the nominal shell thickness or the radial deflections caused by internal pressure loading only. At SRM pressurization, the overall shell response has been shown to be dominated by the effects of the internal pressure.

Global SRB Shell Model

The SRB finite element model was also developed using the original (*i.e.*, STS 51-L) SRB geometry configuration. The loading cases considered in these analyses correspond to the loadings at t = 0, 5.3, 6.6, and 7.2 seconds after SSME ignition. These four loading cases are used to examine the extremes of the actual loading expected prior to liftoff. The time-consistent SRB/ET interface loads for these four points in time are given in Table 1.

Approximately 85,000 active degrees-of-freedom are in the finite element model of the entire SRB shown in figure 3. The average semi-bandwidth of the global stiffness matrix is 510. To form the elemental stiffness matrices and then assemble the global stiffness matrix required 101 CPU seconds on the NAS Cray-2 computer. A single decomposition of the global stiffness matrix required an additional 685 CPU seconds. One forward-reduction/back-substitution cycle either to obtain the linear stress solution or to perform one nonlinear iteration required an additional 45 CPU seconds. The complete nonlinear shell analysis of the model required a total of two decompositions of the global stiffness matrix and 30 nonlinear iterations. The complete nonlinear shell analysis of the SRB for each pre-liftoff loading case required a total of nearly 3900 CPU seconds on the Cray-2. The same computation on a VAX 11/785 computer is estimated to take 7.4 CPU days.

At t = 0, the loads induced into the SRB's are due to the eccentric weight of the Orbiter and ET and also due to cryogenic shrinkage of the ET during fueling. The t = 0 load case is included to investigate the initial static configuration prior to the ignition of the SSME's. The SRB finite element model with exaggerated deflections from a nonlinear analysis for the t = 0 load case is shown in figure 15a.

When the SSME's are ignited, an eccentric thrust of approximately 1,000,000 pounds is produced that causes the Space Shuttle to bend over to a maximum deflection ("max twang"). Maximum bending occurs at t = 5.3 seconds after SSME ignition. The SRB finite element model with exaggerated deflections from a nonlinear analysis for the t = 5.3 seconds load case is shown in figure 15b. Although the structural response of the SRB to the "max twang" loads at t = 5.3 seconds exhibits large deflections, the overall response as characterized by the SRB tip deflection is nearly linear as indicated in reference 4.

The SSME's reach full thrust approximately 6.6 seconds after their ignition. At t = 6.6 seconds, the SRB has rebounded, and the signal is given to ignite the SRM propellant and to release the bolts in the hold-down post of the aft skirt. The SRB finite element model with exaggerated deflections from a nonlinear analysis for the t = 6.6 seconds load case is shown in figure 15c. Comparing the deformed geometries given in figure 15b for the "max twang" condition with that given in figure 15c for the SRM ignition condition indicates that the SRB is rebounding to its original position.

After approximately another 600 milliseconds (*i.e.*, t = 7.2 seconds), the pressure inside the SRM has built up to nearly 1000 psi. The effect of the variation of the internal pressure distribution along the length of the SRM on the structural response is minimal. Since the SRB is not restrained to the launch pad at this time, liftoff occurs. However, the finite element model of the SRB for this load case assumes that the entire edge of the aft skirt is clamped so that the structural response for the pressurized SRM could be analyzed with the head pressure on the forward dome also applied. The SRB finite element model with exaggerated deflections from a linear analysis for the t = 7.2 seconds load case is shown in figure 15d. Comparing the deformed geometries given in figures 15a, 15b, and 15c for the unpressurized SRM condition with that given in figure 15d for the pressurized SRM condition indicates that the SRM field and factory joints influence the global structural response much like frames on an aircraft fuselage (*i.e.*, they cause "pressure pillowing").

The global structural response of the SRB to these four pre-liftoff loading cases (time-consistent SRB/ET interface loads for t = 0, 5.3, 6.6, and 7.2 seconds after SSME ignition) is given in figure 15. These four loading cases are used to examine the extremes of the actual loading expected prior to liftoff. At t = 0, the SRB's deflect toward the Orbiter due to the eccentric weight of the Orbiter and ET. At t = 5.3 seconds after SSME ignition, the SRB's deflect to a maximum value due to the eccentric thrust of the SSME's. At t = 6.6 seconds after SSME ignition, the SSME's reach for thrust, the Space Shuttle system is rebounding to its vertical position, and the SRM's are ignited. At t = 7.2 seconds after SSME ignition, the SRM's reach maximum operating pressure and liftoff occurs. No evidence of nonlinear shell collapse was observed in these preliminary 2-D shell analyses of the SRB for the pre-liftoff loading cases considered.

Summary

The SRB analyses reported in this paper utilized various NASA computer systems. In building the finite element models, calculating results, evaluating the output, generating report-quality graphics output, and providing archival storage of datasets and results, four different classes of computer systems are used (*i.e.*, workstations, minicomputers, mainframes, and supercomputers). The computational approach using this variety of computer systems is described.

The results of these analyses represent a preliminary assessment of the overall structural response of the SRB to selected pre-liftoff loads for STS 51-L. The present analyses neglect the effects of the SRM propellant and any dynamics. The field and factory tang-clevis joints are modeled as equivalent stiffness joints, and bolted connections (*e.g.*, ETA rings) are modeled as "welded" sections. The overall structural response predicted by these analyses characterizes the global shell behavior of the Space Shuttle SRB.

References

1. *Report of the Presidential Commission on the Space Shuttle Challenger Accident*, Washington, D.C., June 6, 1986.

2. Greene, William H.; Knight, Norman F., Jr.; and Stockwell, Alan E.: Structural Behavior of the Space Shuttle SRM Tang-Clevis Joint. *Journal of Propulsion and Power*, Vol. 4, No. 4, July-August 1988, pp. 317-327. (Also NASA TM-89018, September 1986.)

3. Christensen, N. G.: Supercomputing Gives a Boost to Shuttle Solid Rocket Motor Redesign. *Science and Engineering on Cray Supercomputers*, Eric J. Pitcher (editor), Cray Research, Inc., pp. 109-125, 1988.

4. Knight, Norman F., Jr.; Gillian, Ronnie E.; and Nemeth, Michael P.: *Preliminary 2-D Shell Analysis of the Space Shuttle Solid Rocket Boosters*. NASA TM-100515, March 1988.

5. Knight, Norman F., Jr.: *Nonlinear Shell Analyses of the SRB/ETA Ring Interface*, NASA TM-89164, July 1987.

6. Anon., *MSC/NASTRAN User's Manual*. MacNeal-Schwendler Corporation, Los Angeles, CA, 1985.

7. Almroth, B. O.; Brogan, F. A.; and Stanley, G. M.: *Structural Analysis of General Shells, Vol. II, User Instructions for STAGSC-1*, Report No. LMSC-D633873, Lockheed Palo Alto Research Laboratory, Palo Alto, CA, December 1982.

8. Rankin, C. C.; Stehlin, P.; and Brogan, F. A.: *Enhancements to the STAGS Computer Code*, NASA CR-4000, November 1986.

9. Oostyen, J. E.; Bright, D. D.; Hawkins, G. F.; McCluskey, P. M.; and Larsen, G. L.: *SRM Joint Deflection Referee Test: Phase 2 Final Report*, Morton Thiokol, Inc., Wasatch Operations, Document Number TWR-300149, April 3, 1986.

10. Bailey, F. R.: NAS — Current Status and Future Plans. *Supercomputing In Aerospace*, NASA CP-2454, pp. 13-21, 1987.

Table 1. Time-Consistent SRB/ET Interface Loads.

ET Strut Loads, lbs.	Time after SSME Ignition, sec.			
	0.0	5.3	6.6	7.2
P8	98,394	41,188	53,639	109,626
P9	80,397	93,003	88,198	65,442
P10	47,834	-10,042	14,525	-22,958
P14	-971,398	-424,394	-422,609	-946,084
P15	-104,089	-59,770	-59,065	-33,267
P16	-42,618	174,856	80,639	38,454

Table 2. Time-Consistent Reconstructured Beam
Forces and Moments.

	Time After SSME Ignition, sec.	
	$t = 5.3$	$t = 7.2$
Equivalent <u>Beam Loads</u>		
F_x, lb.	1,252,029	-12,906,980
F_y, lb.	61,859	109,565
F_z, lb.	-200,345	1,185
M_x, in.-lb.	13,879,510	313,331
M_y, in.-lb.	-194,578,500	-2,544,368
M_z, in.-lb.	-28,056,290	-4,455,716

Table 3. STAGSC-1 Performance Comparison on NAS Computers.

	Cray X-MP/12 COS, CPU sec.	Cray-2 UNICOS CPU sec.	
		No Vector Library	Vector Library
STAGS1	107	137	142
STAGS2	781	1577	853
TOTAL	888	1714	995

**Table 4. CPU Performance of Various Classes of Computer Systems
Using the STAGSC-1 Computer Program**

	VAX 11/785 VMS 4.5, CPU sec.	CDC 173 NOS 2.4, CPU sec.	CDC 855 NOS 2.4, CPU sec.	CRAY-2 UNICOS 2.0, CPU sec.
Form and Assemble Global Stiffness Matrix	901	478	96	18
Decompose the Global Stiffness Matrix	18536	9917	1983	96
Forward/Backward Substitution and Stress Recovery	514	236	47	7
Total	19951	10631	2126	121

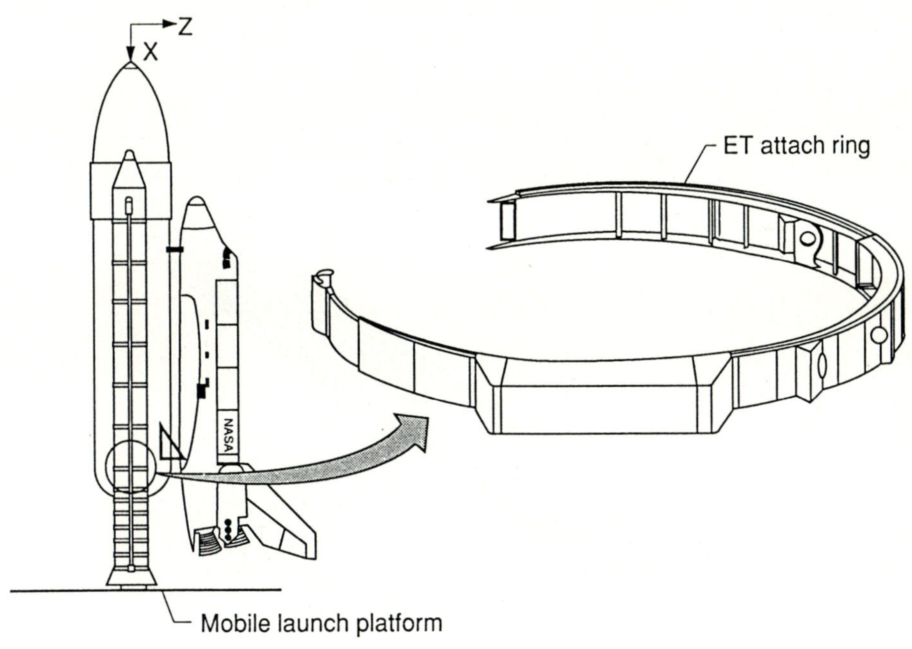

Fig. 1 Space Shuttle system.

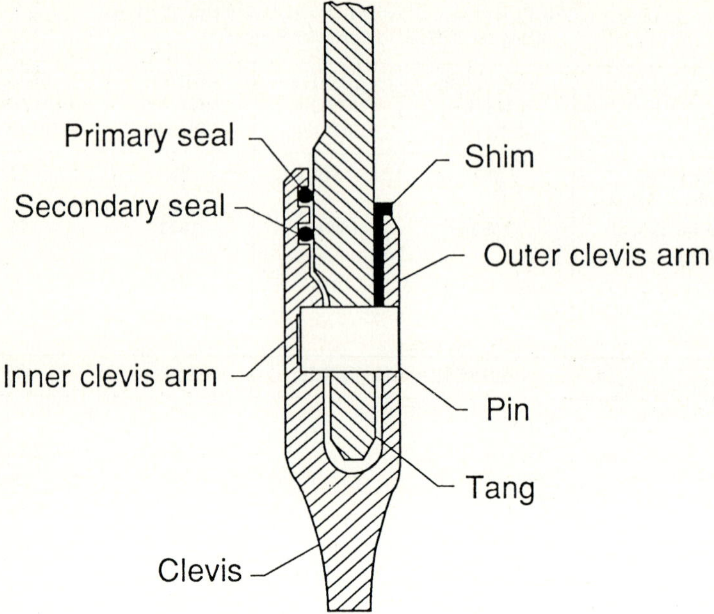

Fig. 2 Solid rocket motor case joint cross section.

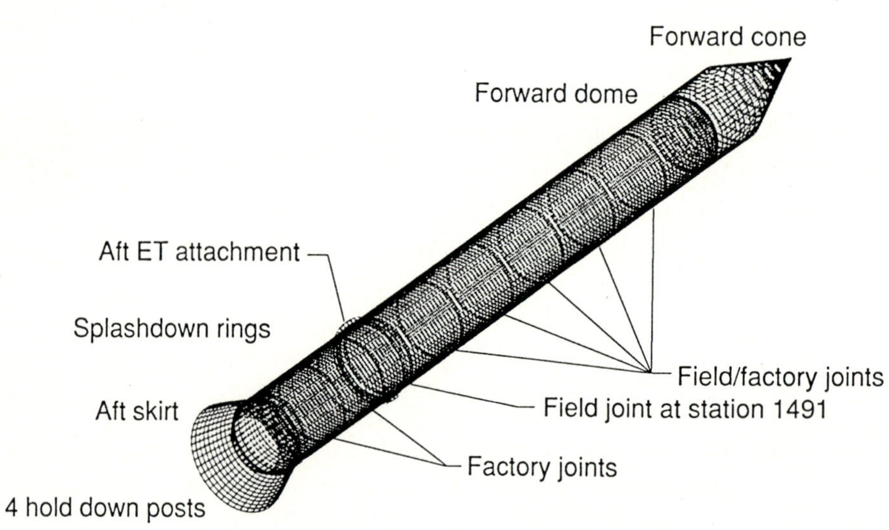

Fig. 3 Finite element model of entire SRB.

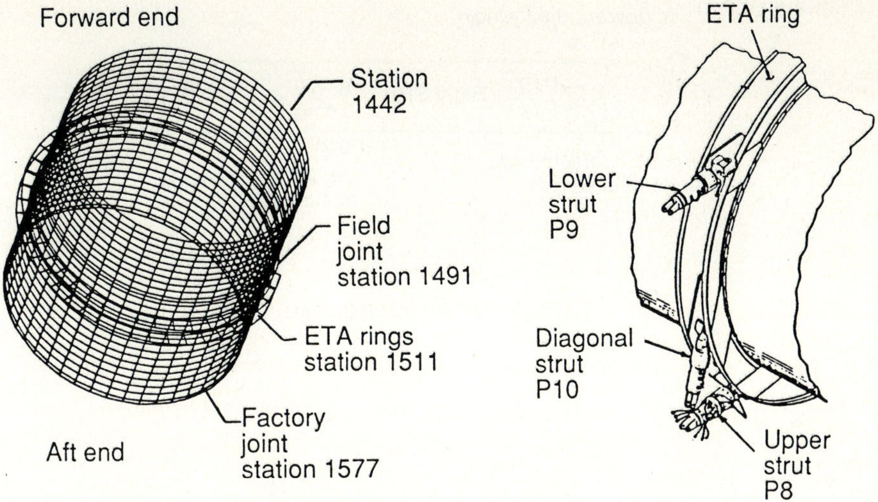

Fig. 4   Finite element model of the SRB/ETA ring interface region.

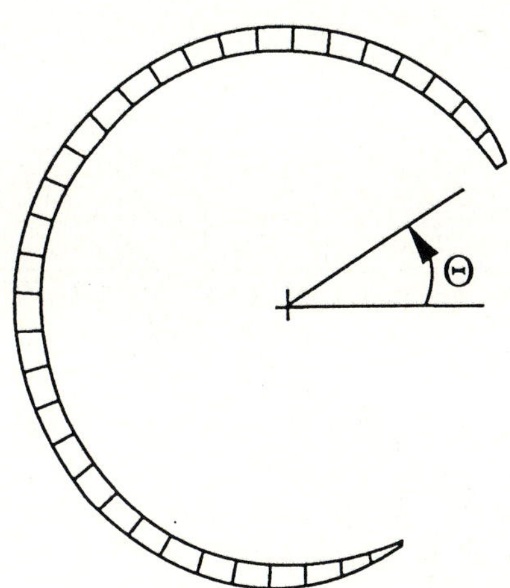

Fig. 5   Finite element model of ETA ring.

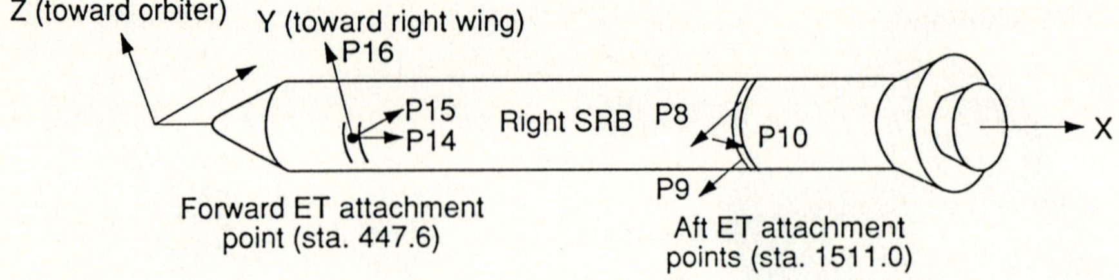

Fig. 6   SRB/ET interface loads.

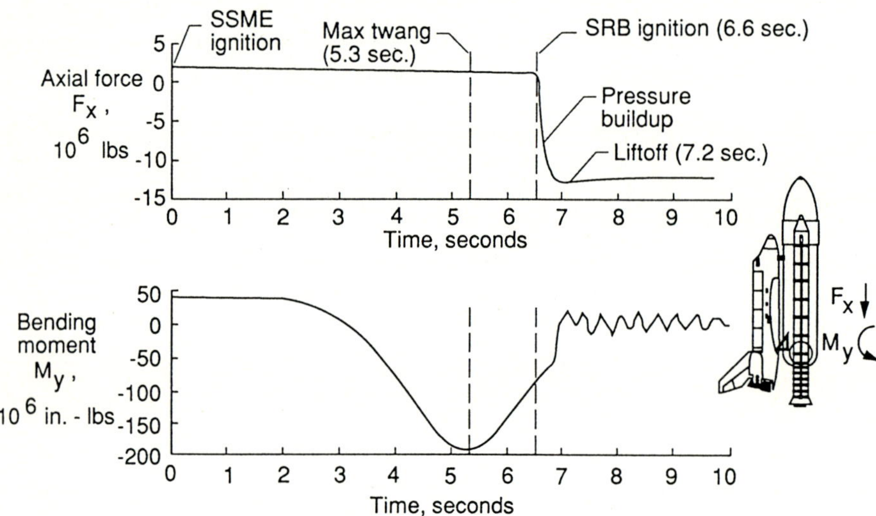

Fig. 7   Overview of pre-liftoff loads transient.

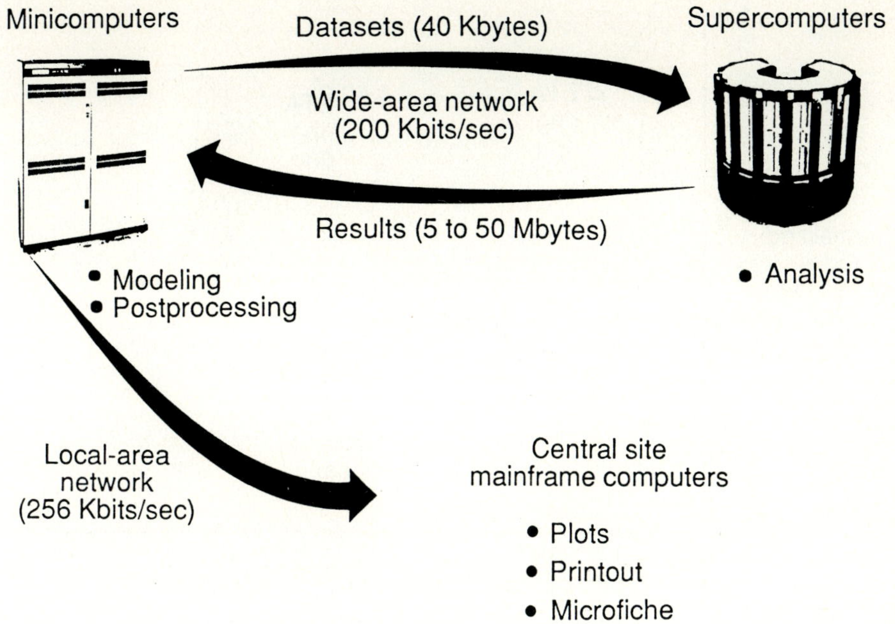

Fig. 8   Remote access to NAS computers.

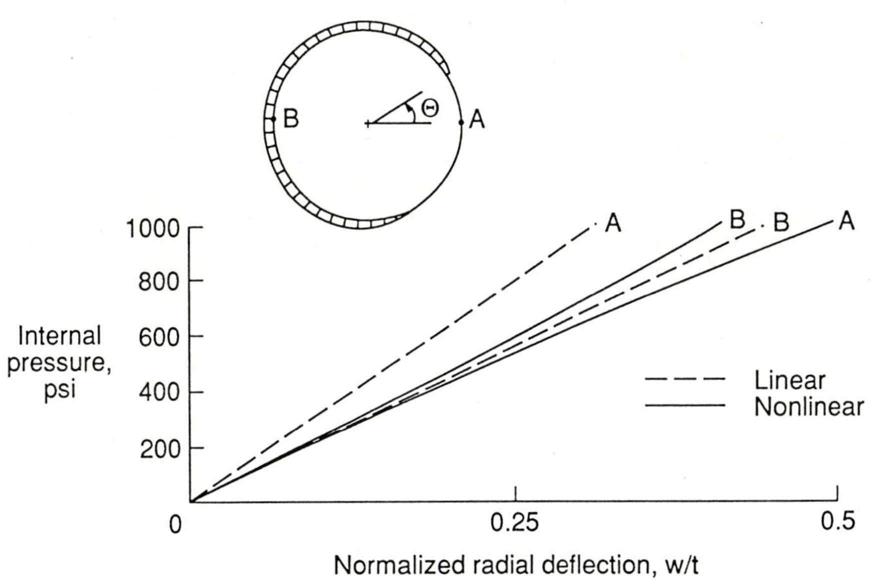

Fig. 9   Nonlinear shell response of the SRB/ETA ring interface region.

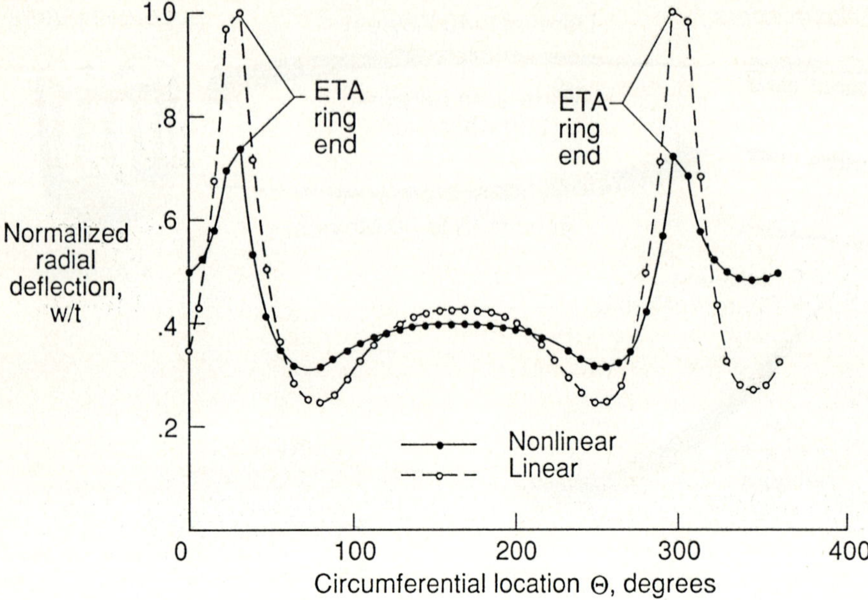

Fig. 10   Radial deflections of the SRM stub ring.

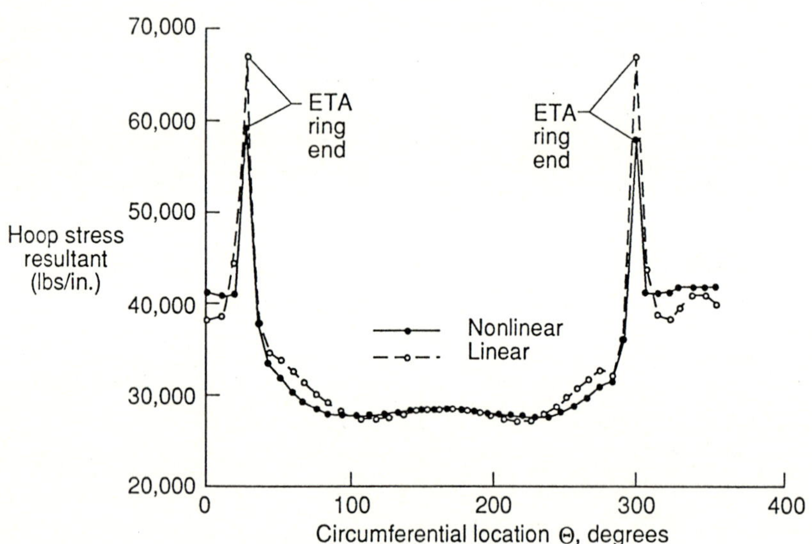

Fig. 11   Hoop stress resultant distributions of the SRM stub ring.

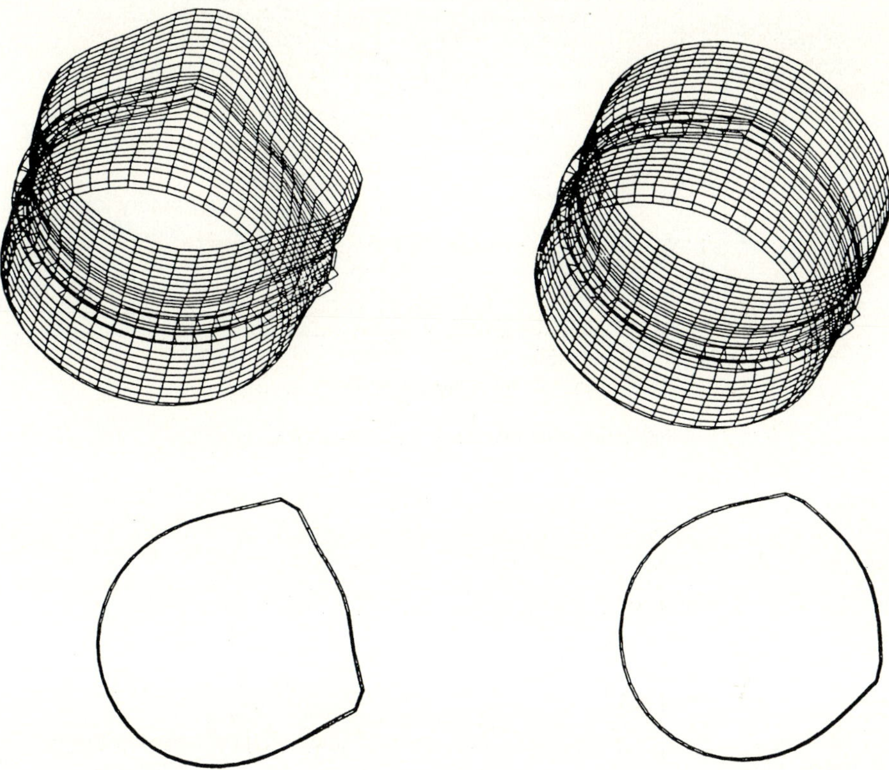

(a) Linear solution.

(b) Nonlinear solution.

Fig. 12 Deformed geometries of the SRB/ETA ring interface
region finite element model.

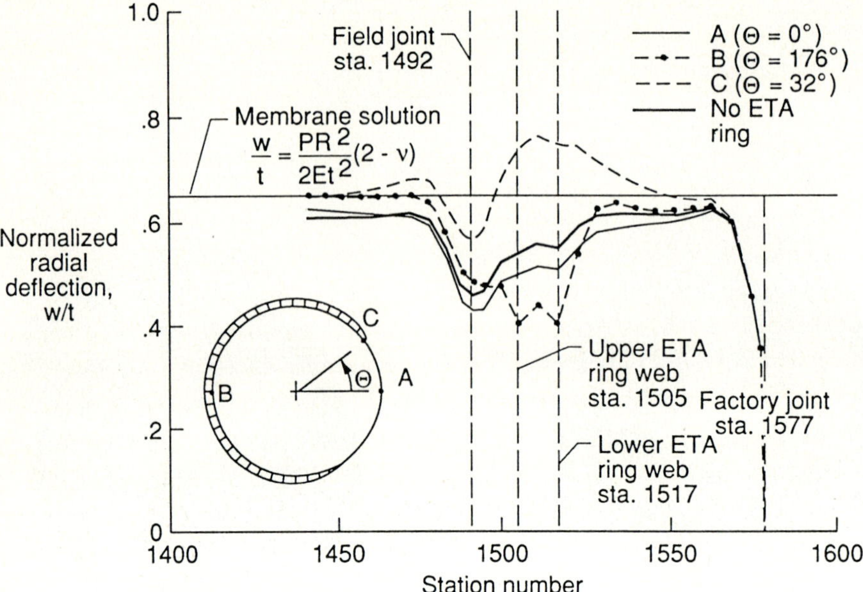

Fig. 13   Axial distribution of the nonlinear radial deflections.

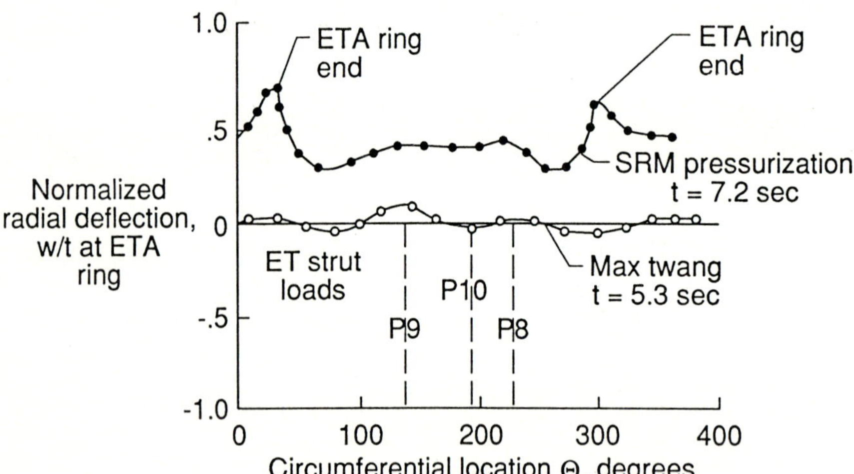

Fig. 14   Nonlinear radial deflections of the SRM stub ring
subjected to selected STS 51-L pre-liftoff loads.

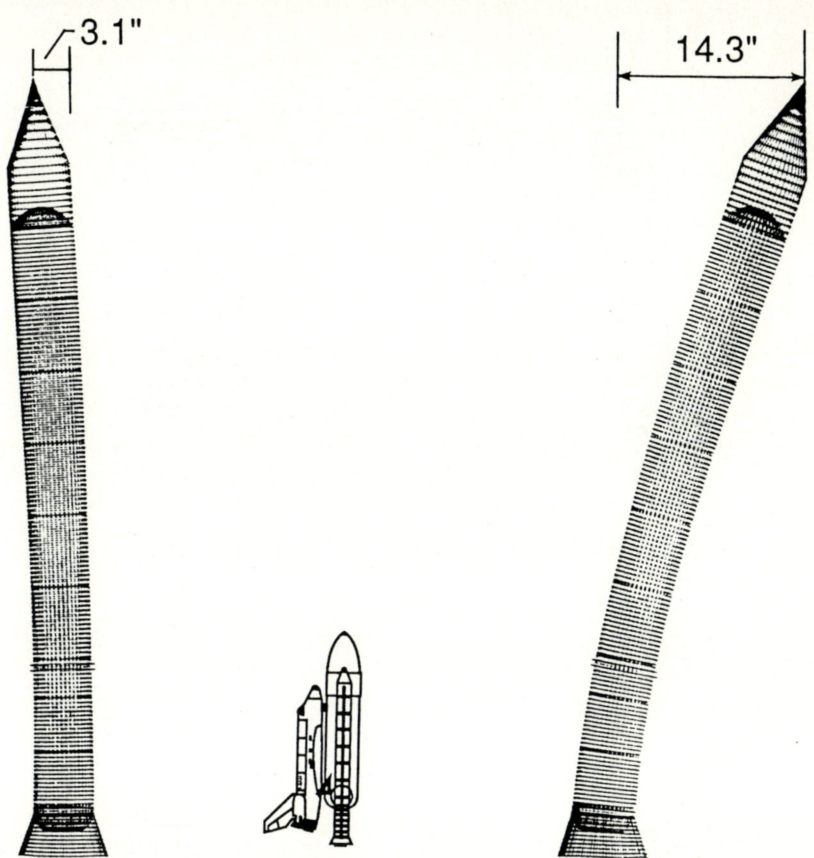

(a) Time t=0 (prior to SSME ignition).

(b) Time t=5.3 sec. ("max twang").

Fig. 15   Summary of the deformed geometries of the SRB
for the four time-consistent load cases considered.

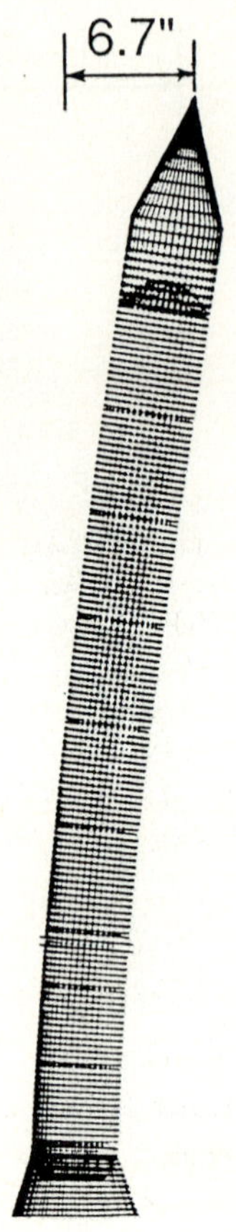

(c) Time t=6.6 sec. (SRM ignition).　　　　(d) Time t=7.2 sec. (liftoff).

Fig. 15 Concluded.

Nonlinear Analysis of Aircraft Tires via Semianalytic Finite Elements

AHMED K. NOOR, KYUN O. KIM and JOHN A. TANNER
George Washington University
NASA Langley Research Center, Hampton, VA

Summary

A computational procedure is presented for the geometrically nonlinear analysis of aircraft tires. The tire was modeled by using a two-dimensional laminated anisotropic shell theory with the effects of variation in material and geometric parameters included. The four key elements of the procedure are: 1) semianalytic finite elements in which the shell variables are represented by Fourier series in the circumferential direction and piecewise polynomials in the meridional direction; 2) a mixed formulation with the fundamental unknowns consisting of strain parameters, stress-resultant parameters, and generalized displacements; 3) multilevel operator splitting to effect successive simplifications, and to uncouple the equations associated with different Fourier harmonics; and 4) multilevel iterative procedures and reduction techniques to generate the response of the shell.

Introduction

Because of the axial symmetry of undeformed tires, it is desirable to exploit, in their modeling and analysis, the substantial capability that currently exists for the numerical analysis of shells of revolution. The most commonly used approach for the analysis of shells of revolution is based on the representation of the shell variables and loads by a Fourier series in the circumferential coordinate θ, combined with the use of a numerical discretization technique (such as finite elements, finite differences or numerical integration) in the meridional direction (see, for example, Goldberg [1], Cohen [2], Bushnell [3], Noor and Stephens [4], Sen and Gould [5], Grigorenko [6], and Padovan and Lestingi [7]). Such an approach has the major advantages of accuracy and stability (no locking or spurious modes), over two-dimensional shell elements. Moreover, for linear problems of shells with uniform circumferential properties, the Fourier series representation permits separation of variables and the equations uncouple in harmonics. However, when applied to the analysis of tires it has the following drawbacks:

1) For geometrically nonlinear problems, the unknowns associated with different harmonics are coupled (see Schaeffer and Ball [8], and Wunderlich, et al [9]).

2) Even for linear problems, because of the anisotropy of the cord-tire composites, the

328

symmetric and antisymmetric responses (with respect to $\theta=0$), associated with each harmonic are coupled.

3) For the case of localized loading (e.g., contact pressure on the tire), a large number of harmonics is needed to accurately predict the response.

The aforementioned drawbacks can make the computational cost of the geometrically non-linear analysis of tires quite expensive. Research on tire modeling and analysis at NASA Langley Research Center has focused on developing accurate and cost-effective strategies for predicting tire response. Included in the research is the development of analysis procedures for substantial reduction of the computational expense resulting from: a) the harmonic and anisotropic couplings (items 1 and 2 above); and b) the generation of the response associated with large number of harmonics (item 3 above). The present paper summarizes the status of the development activities and present for the first time, response data for an actual aircraft tire obtained by the foregoing strategy.

Mathematical Formulation

In the present study a space shuttle nose-gear tire was modeled using a moderate-rotation Sanders-Budiansky shell theory with the effects of transverse shear deformation and laminated anisotropic material response included (Sanders [10] and Budiansky [11]). A total Lagrangian formulation was used and the fundamental unknowns consist of the five general-ized displacements, the eight stress resultants, and the corresponding eight strain components of the middle surface. The sign convention for the different tire stress resultants and general-ized displacements is shown in Fig. 1. The concepts presented in the succeeding sections can be extended to higher-order shear deformation theories, as well as to three-dimensional continuum theory.

<u>Spatial discretization of the tire.</u> Each of the generalized displacements, the stress resultants, and the strain components is expanded in a Fourier series of the circumferential coordinate θ. The discretization in the meridional direction is performed by using a three-field mixed finite element model. The following expressions are used for approximating the external loading, generalized displacements, stress resultants, and strain components within each element:

$$p(s,\theta) = N^i \left(\sum_{n=0}^{\infty} p_n^i \cos n\theta + \sum_{n=1}^{\infty} \bar{p}_n^i \sin n\theta \right) \tag{1}$$

$$E(s,\theta) = \bar{N}^l \left(\sum_{n=0}^{\infty} E_n^l \cos n\theta + \sum_{n=1}^{\infty} \bar{E}_n^l \sin n\theta \right) \tag{2}$$

$$H(s,\theta) = \overline{N}^{\iota}\left(\sum_{n=0}^{\infty} H_n^{\iota}\cos n\theta + \sum_{n=1}^{\infty} \overline{H}_n^{\iota}\sin n\theta\right) \tag{3}$$

$$X(s,\theta) = N^i\left(\sum_{n=0}^{\infty} X_n^i\cos n\theta + \sum_{n=1}^{\infty} \overline{X}_n^{\iota}\sin n\theta\right) \tag{4}$$

where N^i are the shape functions used in approximating the generalized displacements and external loading in the meridional direction; $\overline{N}^{\iota}$ are the shape functions used in approximating the strain components and stress resultants; $p_n^i, \overline{p}_n^i$ and $X_n^i, \overline{X}_n^i$ refer to the load and generalized displacement coefficients associated with the Fourier harmonic n; $E_n^{\iota}, \overline{E}_n^{\iota}$ and $H_n^{\iota}, \overline{H}_n^{\iota}$ refer to the strain and stress-resultant parameters associated with the harmonic n. Note that the degree of the polynomial shape functions $\overline{N}^{\iota}$ is lower than that of N^i. Moreover, the continuity of the strain components and stress resultants is not imposed at the interelement boundaries and, therefore, the strain and stress-resultant parameters can be eliminated on the element level.

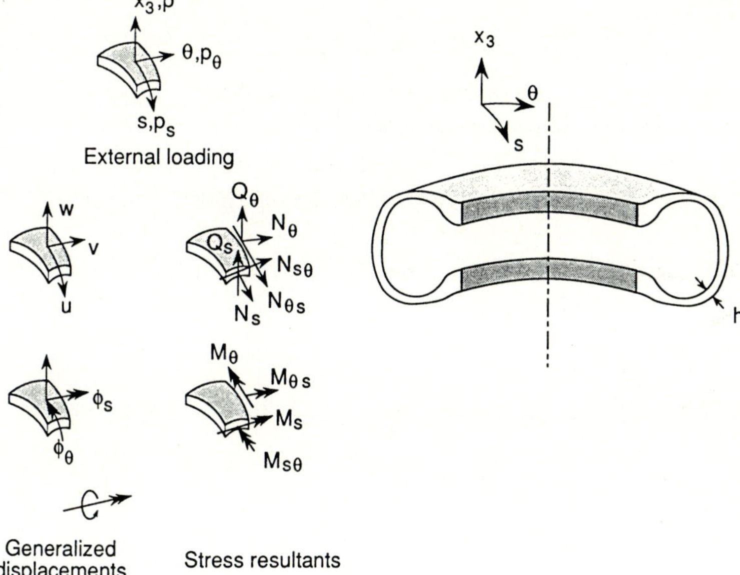

Figure 1 - Two-dimensional shell model of the tire and sign convention for the external loading, generalized displacements and stress resultants

In Eqs. 1-4, the range of the superscript i is 1 to m, the number of displacement nodes in the element; the range of superscript ι is 1 to s, the number of parameters used in approximating each of the strain components and stress resultants. The shell variables without a bar are the coefficients of the cosine series, the barred shell variables are the coefficients of the sine series and a repeated superscript denotes summation over its entire range. Henceforth, the vectors of the 10 generalized displacement parameters; of the 16 stress-resultant parameters; and of the 16 strain parameters associated with the harmonic n are denoted by $\{X\}_n, \{H\}_n, \{E\}_n$, respectively. These vectors can be decomposed into symmetric and antisymmetric sets (with respect to $\theta=0$) as shown in Table 1.

Table 1 - Symmetric and antisymmetric tire parameters
with respect to $\theta=0$

	Symmetric Set	Antisymmetric Set
	Strain components	
$\{E\}_n$	$\varepsilon_{s,n},\, \varepsilon_{\theta,n},\, 2\bar{\varepsilon}_{s\theta,n},\, \kappa_{s,n},$ $\kappa_{\theta,n},\, 2\bar{\kappa}_{s\theta,n},\, 2\varepsilon_{s3,n},\, 2\bar{\varepsilon}_{\theta3,n}$	$\bar{\varepsilon}_{s,n},\, \bar{\varepsilon}_{\theta,n},\, 2\varepsilon_{s\theta,n},\, \bar{\kappa}_{s,n},$ $\bar{\kappa}_{\theta,n},\, 2\kappa_{s\theta,n},\, 2\bar{\varepsilon}_{s3,n},\, 2\varepsilon_{\theta3,n}$
	Stress resultants	
$\{H\}_n$	$N_{s,n},\, N_{\theta,n},\, \bar{N}_{s\theta,n},\, M_{s,n},$ $M_{\theta,n},\, \bar{M}_{s\theta,n},\, Q_{s,n},\, \bar{Q}_{\theta,n}$	$\bar{N}_{s,n},\, \bar{N}_{\theta,n},\, N_{s\theta,n},\, \bar{M}_{s,n},$ $\bar{M}_{\theta,n},\, M_{s\theta,n},\, \bar{Q}_{s,n},\, Q_{\theta,n}$
	Generalized displacements	
$\{X\}_n$	$u_n,\, \bar{v}_n,\, w_n,\, \phi_{s,n},\, \bar{\phi}_{\theta,n},$	$\bar{u}_n,\, v_n,\, \bar{w}_n,\, \bar{\phi}_{s,n},\, \phi_{\theta,n},$

<u>Governing Equations</u>. The governing discrete equations of the tire are obtained by applying the three-field, Hu-Washizu mixed variational principle. If the number of terms (harmonics) retained in the Fourier series is $N+1$, the governing equations can be written in the following compact form:

$$\begin{Bmatrix} f^{(0)} \\ f^{(1)} \\ \vdots \\ f^{(N)} \end{Bmatrix} = \begin{bmatrix} K^{(0)} & & & \\ & K^{(1)} & & \\ & & \ddots & \\ & & & K^{(N)} \end{bmatrix} \begin{Bmatrix} Z_0 \\ Z_1 \\ \vdots \\ Z_N \end{Bmatrix} + \begin{Bmatrix} G^{(0)}(Z_0, Z_1, ..., Z_N) \\ G^{(1)}(Z_0, Z_1, ..., Z_N) \\ \vdots \\ G^{(N)}(Z_0, Z_1, ..., Z_N) \end{Bmatrix} - \begin{Bmatrix} P^{(0)} \\ P^{(1)} \\ \vdots \\ P^{(N)} \end{Bmatrix} = 0 \quad (5)$$

where $\{Z\}_n$ $(n = 0, 1, ..., N)$ is the vector of unknowns associated with the nth harmonic, which includes strain parameters $\{E\}_n$, stress-resultant parameters $\{H\}_n$, and generalized displacements $\{X\}_n$; $[K]^{(n)}$ are linear matrices, $\{G\}^{(n)}$ are vectors of nonlinear terms, and $\{P\}^{(n)}$ are consistent load vectors. The following observations can be made about the governing equations (Eqs. 5):

1. The first matrix on the left side of Eqs. 5 is block diagonal, which is a direct consequence of the orthogonality of the trigonometric functions. The orthogonality of trigonometric functions leads to uncoupling of the equations associated with the different Fourier harmonics for the linear case. For the nonlinear case, the vectors $\{G\}^{(n)}$ couple the unknowns associated with *all* the harmonics (see, for example, Wunderlich, et al [9]).

2. The contributions of the different Fourier harmonics and the anisotropic (nonorthotropic) material coefficients to the governing equations can be identified as follows:

a. *Fourier harmonics* - The block-diagonal matrices $[K]^{(n)}$ ($n \geq 1$) in Eqs. 5 are linear in the Fourier harmonic n. Therefore, $[K]^{(n)}$ can be expressed as the sum of two matrices as follows:

$$[K]^{(n)} = [\tilde{K}] + n\left[\hat{K}\right] \tag{6}$$

where both $[\tilde{K}]$ and $\left[\hat{K}\right]$ are independent of n. The nonlinear vectors $\{G\}^{(n)}$ are quadratic in n.

b. *Anisotropy (Nonorthotropy)* - A unique feature of the mixed formulation used herein is that the anisotropic (nonorthotropic) material coefficients are included only in the linear matrices $[K]^{(n)}$. For the linear case, these anisotropic coefficients result in the coupling between the symmetric and antisymmetric shell parameters (see Noor and Peters [12], and Table 2).

Table 2 - Different types of coupling in the analysis of
tires using semianalytic finite elements

Response	Material	Governing finite element equations
Linear	Isotropic or orthotropic	Uncoupled in harmonics Symmetric and antisymmetric variables uncoupled
	Anisotropic	Uncoupled in harmonics Symmetric and antisymmetric variables uncoupled
Nonlinear	Anisotropic	Coupled in harmonics Symmetric and antisymmetric variables coupled

3. If the vectors $\{Z\}_n$ are partitioned into subvectors of parameters of strains, stress resultants and generalized displacements, that is,

$$\{Z\}_n = \begin{Bmatrix} E_n \\ H_n \\ X_n \end{Bmatrix} \tag{7}$$

then the matrix $[K]^{(n)}$ can be written in the following form:

$$[K]^{(n)} = \begin{bmatrix} K_o + K_a & -R & \cdot \\ -R^t & \cdot & S_o + nS \\ \cdot & S_o^t + n\,S^t & \cdot \end{bmatrix} \tag{8}$$

where the submatrices $[K_o]$ and $[K_a]$ contain the contributions of the orthotropic and anisotropic (nonorthotropic) material coefficients. The explicit forms of the matrices $[K_o]$, $[K_a]$, $[S_o]$, $[S]$, and $[R]$ are given in Noor, et al [13] and Noor and Peters [14].

4. The nonlinear vectors $\{G\}^{(n)}$ contain bilinear terms in $\{H\}_n$ and $\{X\}_n$, as well as quadratic terms in $\{X\}_n$.

Generation of the Nonlinear Response of the Tire

For a given external loading, the governing nonlinear equations (Eqs. 5) are solved by using the Newton-Raphson iterative technique. The recursion formulas for the rth iterational cycle are:

$$\left(\begin{bmatrix} K^{(0)} & & & \\ & K^{(1)} & & \\ & & \ddots & \\ & & & K^{(N)} \end{bmatrix} + \begin{bmatrix} \overline{K}^{(00)} & \overline{K}^{(01)} & \dots & \overline{K}^{(0N)} \\ & 0 & \dots & \overline{K}^{(1N)} \\ & & \ddots & \vdots \\ & & & 0 \end{bmatrix} \right)^{(r)} \begin{Bmatrix} \Delta Z_0 \\ \Delta Z_1 \\ \vdots \\ \Delta Z_N \end{Bmatrix}^{(r)} = - \begin{Bmatrix} f^{(0)} \\ f^{(1)} \\ \vdots \\ f^{(N)} \end{Bmatrix}^{(r)} \tag{9}$$

and

$$\begin{Bmatrix} Z_0 \\ Z_1 \\ \vdots \\ Z_N \end{Bmatrix}^{(r+1)} = \begin{Bmatrix} Z_0 \\ Z_1 \\ \vdots \\ Z_N \end{Bmatrix}^{(r)} + \begin{Bmatrix} \Delta Z_0 \\ \Delta Z_1 \\ \vdots \\ \Delta Z_N \end{Bmatrix}^{(r)} \tag{10}$$

where

$$[\overline{K}]^{(IJ)} = \frac{\partial}{\partial Z_J} \{G\}^{(I)} \quad (I,J = 1 \text{ to } N) \tag{11}$$

For each Newton-Raphson iteration (represented by Eqs. 9 and 10), another iteration loop is performed using the preconditioned conjugate gradient (PCG) technique to account for the coupling between the different harmonics (i.e., the submatrices $[\overline{K}]^{(IJ)}$). In the inner iteration loop the following uncoupled equations are solved:

$$\left([K]^{(0)} + [\overline{K}]^{(00)\,(r)}\right)\{\Delta Z\}_0^{(r)} = -\{f\}^{(0)\,(r)} - \check{\lambda}\left([\overline{K}]^{(01)}\{\Delta Z\}_1\right.$$

$$\left. + [\overline{K}]^{(02)}\{\Delta Z_2\} + ... + [\overline{K}]^{(0N)}\{\Delta Z_N\}\right)^{(r)}$$

$$[K]^{(1)}\{\Delta Z\}_1^{(r)} = -\{f\}^{(1)\,(r)} - \check{\lambda}\left([\overline{K}]^{(10)}\{\Delta Z\}_0 + [\overline{K}]^{(12)}\{\Delta Z\}_2\right.$$

$$\left. + ... + [\overline{K}]^{(1N)}\{\Delta Z\}_N\right)^{(r)} \tag{12}$$

$$\vdots$$

$$[K]^{(N)}\{\Delta Z\}_N^{(r)} = -\{f\}^{(N)\,(r)} - \check{\lambda}\left([\overline{K}]^{(N0)}\{\Delta Z\}_0 + [\overline{K}]^{(N1)}\{\Delta Z\}_1 + ...\right)^{(r)}$$

where $\check{\lambda}$ is a tracing parameter which identifies the coupling between the different Fourier harmonics. When $\check{\lambda}=1$, Eqs. 12 are equivalent to Eqs. 9, and when $\check{\lambda}=0$ the equations uncouple in harmonics. Note that because of the special structure of the Jacobian matrix in Eqs. 9, only the left-side associated with the zeroth harmonic needs to be updated in each iteration. An efficient technique is described in the next subsection for solving Eqs. 12.

Efficient Generation of the Response Associated with Different Harmonics

An efficient procedure is presented herein for generating the tire responses associated with different harmonics (solution of Eqs. 12). The basic idea of this procedure is to approximate the tire response associated with the range of Fourier harmonics, $1 \leq n \leq N$, by a linear combination of a few global approximation vectors that are generated at a particular value of the Fourier harmonic within that range. The full equations of the finite element model are solved for only a single Fourier harmonic, and the responses corresponding to the other Fourier harmonics are generated using a reduced system of equations with considerably fewer degrees of freedom. The proposed procedure can be conveniently divided into two phases: 1) restructuring Eqs. 12, for $1 \leq n \leq N$, to delineate the dependence on the Fourier harmonic n, and 2) generation of global approximation vectors (or modes) to approximate the response associated with a range of values of the Fourier harmonic, and determination of the amplitudes of the modes. Application of the procedure to stress and vibration problems of anisotropic shells of revolution is described in Noor and Peters [15], and Noor and Tanner [16]. Its application to the solution of Eqs. 12 is outlined subsequently.

    <u>Restructuring of the governing equations.</u> If Eqs. 6 and 8 are used, the governing equations for the harmonic n ($1 \leq n \leq N$) can be embedded in a single-parameter family of equations and written in the following compact form:

$$\left([\tilde{K}] + n[\hat{K}]\right)\{\Delta Z\}_n = \{P(n)\} + \check{\lambda}\{\hat{P}(n)\} \tag{13}$$

The two vectors $\{P(n)\}$ and $\{\hat{P}(n)\}$ are quadratic in n.

<u>Basis reduction and reduced system of equations.</u> The basis reduction is achieved by approximating the vectors $\{\Delta Z\}_n$, for a certain range of Fourier harmonics, $1 \leq n \leq N$, by a linear combination of a few global approximation vectors which are generated at a particular value of the Fourier harmonic within that range. The approximation is expressed by the following transformation:

$$\{\Delta Z\}_n = [\Gamma]\{\psi\}_n \tag{14}$$

where $[\Gamma]$ is a transformation matrix whose columns are the preselected approximation vectors, and $\{\psi\}_n$ is a vector of unknown parameters representing the amplitudes of the global approximation vectors for the harmonics n. The number of components of $\{\psi\}_n$ is much less than the number of components of $\{\Delta Z\}_n$.

A Bubnov-Galerkin technique is now used to replace the original equations (Eqs. 13) by the following reduced equations in $\{\psi\}_n$:

$$([\tilde{k}] + n[\hat{k}])\{\psi\}_n = \{q\} + \check{\lambda}\{\hat{q}\} \tag{15}$$

where

$$[\tilde{k}] = [\Gamma]'[\tilde{K}][\Gamma] \tag{16}$$

$$[\hat{k}] = [\Gamma]'[\hat{K}][\Gamma] \tag{17}$$

$$\{q\} = [\Gamma]'\{P(n)\} \tag{18}$$

$$\{\hat{q}\} = [\Gamma]'\{\hat{P}(n)\} \tag{19}$$

<u>Selection and generation of global approximation vectors.</u> The global approximation vectors are selected to be the response associated with a single Fourier harmonic n_o and its various-order derivatives with respect to n. Henceforth, the derivatives of the response with respect to n are referred to as *path derivatives*. The matrix $[\Gamma]$ in Eqs. 14 is therefore given by:

$$[\Gamma] = \left[\{\Delta Z\} \quad \frac{\partial}{\partial n}\{\Delta Z\} \quad \frac{\partial^2}{\partial n^2}\{\Delta Z\} \;... \right]_{n_o} \tag{20}$$

The path derivatives are obtained by successive differentiation of the governing equations (Eqs. 13). The recursion relations for the first three global approximation vectors can be written in the following form:

$$([\tilde{K}] + n_o[\hat{K}])\{\Delta Z\}_{n_o} = \frac{\partial}{\partial n}\{P\} + \check{\lambda}\frac{\partial}{\partial n}\{\hat{P}\} \tag{21}$$

$$([\tilde{K}] + n_o[\hat{K}])\frac{\partial}{\partial n}\{\Delta Z\}_{n_o} = \frac{\partial^2}{\partial n^2}\{P\} + \check{\lambda}\frac{\partial^2}{\partial n^2}\{\hat{P}\} - [\hat{K}]\{\Delta Z\}_{n_o} \tag{22}$$

$$\left([\tilde{K}] + n_o [\hat{K}] \right) \frac{\partial^2}{\partial n^2} \{\Delta Z\}_{n_o} = \frac{\partial^3}{\partial n^3} \{P\} + \check{\lambda} \, \frac{\partial^3}{\partial n^3} \, \{\hat{P}\} - 2[\hat{K}] \frac{\partial}{\partial n} \{\Delta Z\}_{n_o} \qquad (23)$$

Note that the left-side matrix in Eqs. 21 to 23 is the same, and therefore, it needs to be decomposed only once in the process of generating all the global approximation vectors.

Comments on proposed procedure. The following comments are made concerning the foregoing procedure for generating the responses associated with different harmonics:

1. The particular choice of the global approximation vectors used herein provides a direct quantitative measure of the sensitivity of the different response quantities of the tire to the circumferential wave number (the Fourier harmonic) n.

2. For problems requiring large numbers of Fourier harmonics (e.g., 100 or more), the range of n is divided into intervals of fewer (e.g., 7) harmonics each; the global approximation vectors and reduced equations are generated at an intermediate value of n within each interval, and the responses associated with the values of n within that interval are generated by the foregoing procedure. Note that higher accuracy of the reduced solutions can be obtained by marching backward as well as forward in the n-space with the reduced equations.

3. The foregoing procedure can be directly applied to the solution of the governing nonlinear equations, Eqs. 5. This is accomplished by using a reduction method with the control parameter selected to be load, displacement or arc-length in the solution space, and the global approximation vectors selected to be the various-order derivatives of the response quantities with respect to the control parameter, see Noor, et al [13]. The global approximation vectors are obtained by successive differentiation of the governing equations, Eqs. 5, with respect to the control parameter. The left-side matrix of those equations has the same form as that of Eqs. 9. If the global approximation vectors are evaluated at zero value for the control parameter, the matrices $[\overline{K}]^{(IJ)}$ on the left side of Eqs. 9 vanish and the equations uncouple in harmonics. The application of the foregoing procedure considerably reduces the computational effort in generating the global approximation vectors, and greatly enhances the effectiveness of the reduction method.

4. The computational effort can be further reduced by using the procedure outlined in Noor, et al [13] to uncouple the equations associated with the symmetric and antisymmetric shell parameters (with respect to $\theta=0$). The procedure is based on transferring the anisotropic (nonorthotropic) terms (submatrices $[K_a]$ in Eqs. 8) to the right sides of Eqs. 12, and adding another level of PCG iterations to account for them.

Numerical Studies

Numerical studies were performed to assess the accuracy of the two-dimensional shell model of the shuttle nose-gear tire, and the effectiveness of the computational procedure described in the preceding section for generating the response associated with different harmonics. Herein, the application of the model and the computational procedure to the space shuttle nose-gear tire are presented. The geometric and material characteristics of the tire are given in Fig. 2.

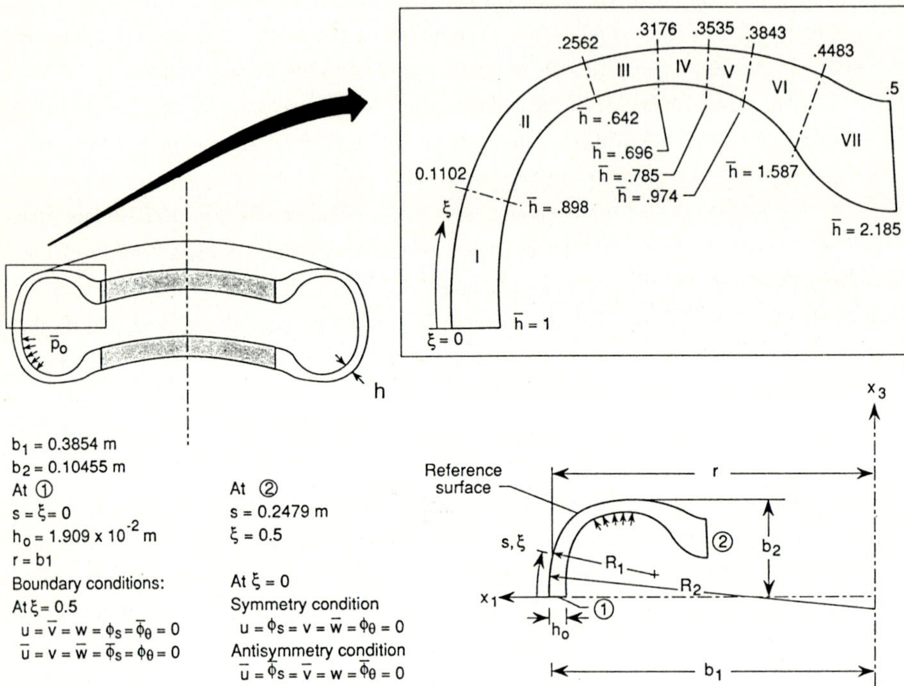

Figure 2 - Characteristics of the space shuttle orbiter nose-gear tire used in the present study. $\bar{h} = h/h_o$

Measurements were made to determine the shape of the tire cross section, and the thickness variation. The cord-rubber composite was treated as laminated anisotropic material. The material properties of the different layers were obtained using the mechanics of materials approach, which has been widely applied to rigid composites (see Jones [17] and Walter [18]). Because of symmetry only half the tire cross section was modeled. It was divided into seven segments (see Fig. 2) with different number of layers, different material properties (corresponding to different cord content in the composite), and different fiber orientation. Spline interpolation was used to smooth out the measured data, and to obtain the geometric and material characteristics of the two-dimensional shell model. The outer surface of the tire

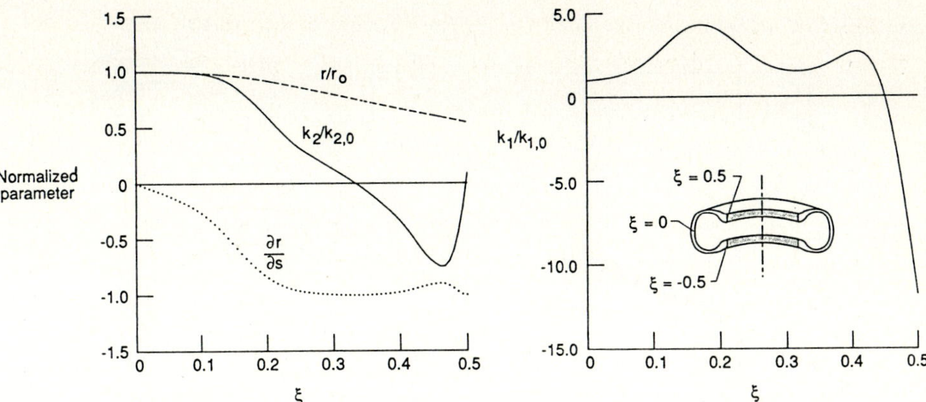

Figure 3 - Meridional variation of the geometric parameters of the two-dimensional shell model of the space shuttle orbiter nose-gear tire shown in Fig. 2. r_o=0.3854 m; $k_{2,0}$=2.5946 m^{-1}, $k_{1,0}$=4.2937 m^{-1}. Reference surface chosen to be the outer surface.

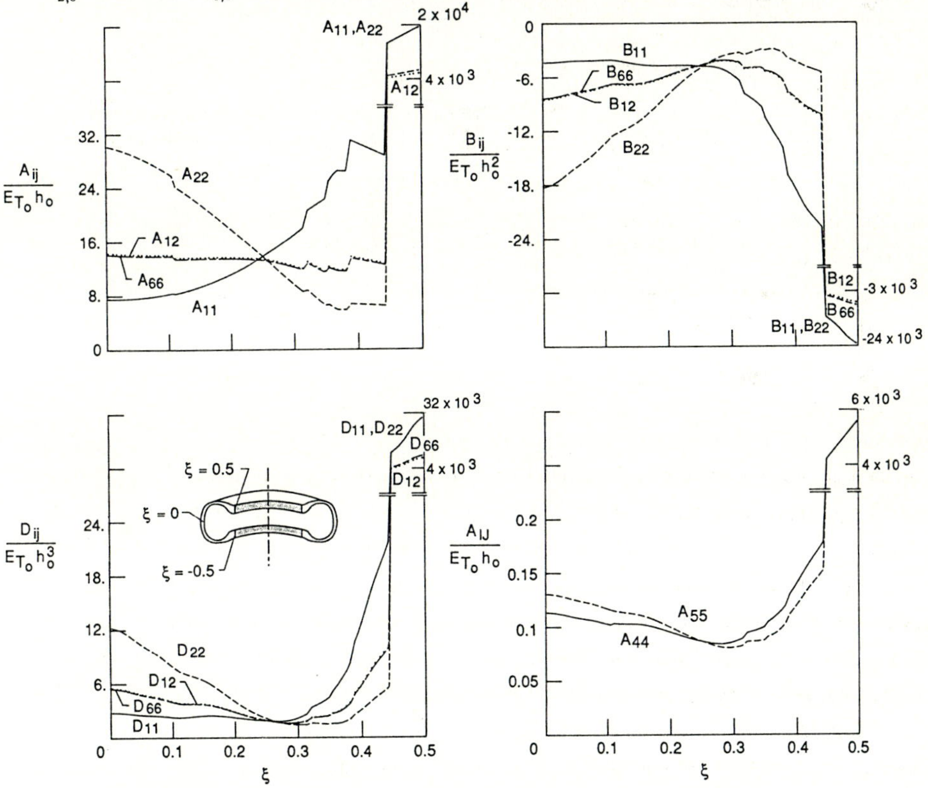

a) stiffness coefficients associated with uncoupled (orthotropic) response.

Figure 4 - Meridional variation of the stiffness coefficients of the two-dimsional shell model of the space shuttle orbiter nose-gear tire shown in Fig. 2. E_{T_o}=8×10^6 Pa

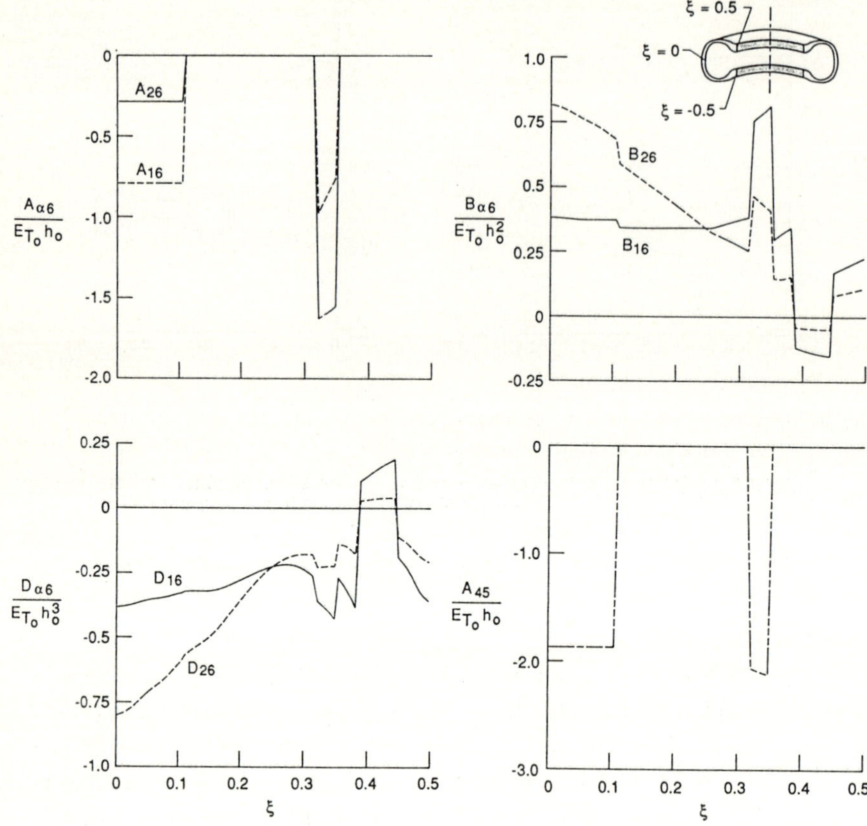

b) coupling (nonorthotropic) stiffness coefficients.

Figure 4 - (concluded)

was chosen to be the reference surface for the two-dimensional shell model. The meridional variations of both the geometric characteristics of the reference (outer) surface and the stiffness coefficients of the shell model are shown in Figs. 3 and 4.

The numerical studies were performed using three-field mixed finite element models for the discretization of the tire in the meridional direction. Linear interpolation functions were used for approximating each of the stress resultants and strain components, and quadratic Lagrangian interpolation functions are used for approximating each of the generalized displacements. The integrals in the governing equations were evaluated using a two-point Gauss-Legendre numerical quadrature formula. Because of the symmetry of the shell meridian and loading, only one half of the tire meridian was analyzed. The finite element models used are shown in Fig. 5.

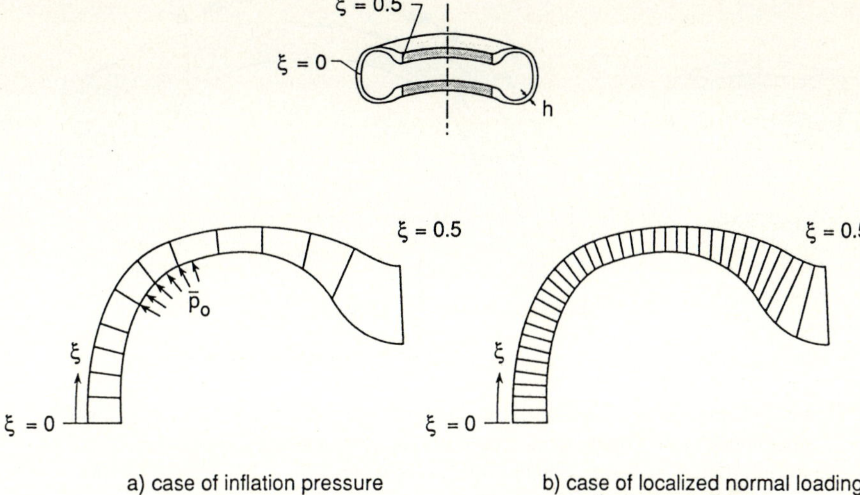

a) case of inflation pressure   b) case of localized normal loading
                                              on the outer surfaces

Figure 5 - Finite element models used in the present study.

To assess the accuracy of the shell model of the tire, the deformations produced by uniform inflation pressure of $\bar{p}_o = 2.2063 \times 10^6$ *Pa*, acting normal to the inner surface, were calculated using the geometrically nonlinear shell theory. Twelve finite elements were used in modeling half the cross section (a total of 384 strain parameters, 384 stress-resultant parameters, and 243 nonzero generalized displacements - see Fig. 5a). Comparison was made with the experimental data obtained on the shuttle nose-gear tire (see Fig. 2). The results are summarized in Figs. 6 to 8. Close agreement between the predicted deformations and experimental results is demonstrated in Fig. 6. Figures 7 and 8 show the meridional variations of the generalized displacements, stress resultants and strain energy densities. As can be seen from Fig. 8, for the case of inflation pressure, the transverse shear strain energy density is considerably smaller than the extensional/bending energy density.

To assess the effectiveness of the computational procedure, linear solutions were obtained for a localized normal loading on the outer surface simulating contact pressure. The normal loading in pascals is given by the following equations, which model experimental data obtained at NASA Langley on the shuttle tire:

$$p = \begin{cases} -\dfrac{p_o \beta}{\pi} - \displaystyle\sum_{n=1}^{10} p_n \cos n\theta \, , & -0.2 < \xi < 0.2 \\[2em] 0 \, , & |\xi| > 0.2 \end{cases} \tag{24}$$

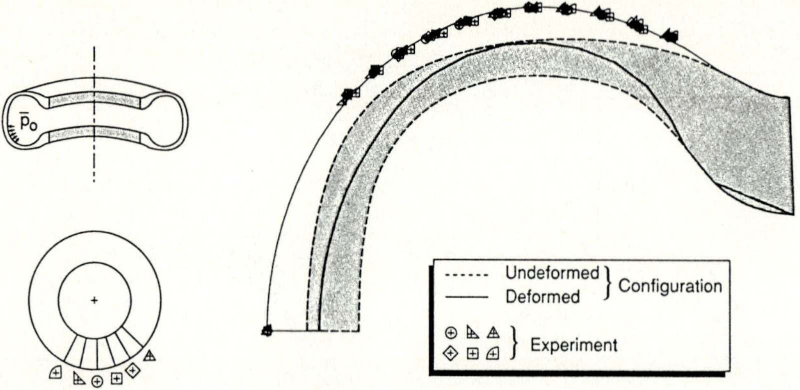

Figure 6 - Comparison of deformed tire configuration predicted by two-dimensional shell model with experiments. Space shuttle nose-gear tire shown in Fig. 2. Uniform inflation pressure acting normal to the inner surface, $\bar{p}_o = 2.2063 \times 10^6$ Pa.

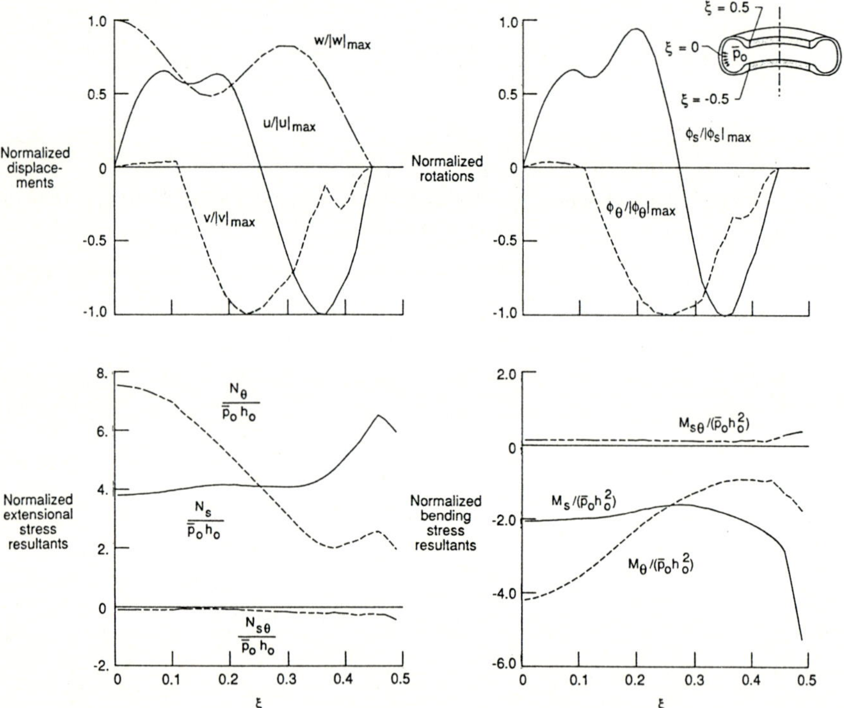

Figure 7 - Meridional variation of the generalized displacements and stress resultants produced by inflation pressure. Two-dimensional shell model of the space shuttle nose-gear tire shown in Fig. 2. $|u|_{max} = 3.657 \times 10^{-3}$ m, $|v|_{max} = 6.007 \times 10^{-4}$ m, $|w|_{max} = 1.445 \times 10^{-2}$ m, $|\phi_s|_{max} = 0.309$, $|\phi_\theta|_{max} = 6.509 \times 10^{-2}$.

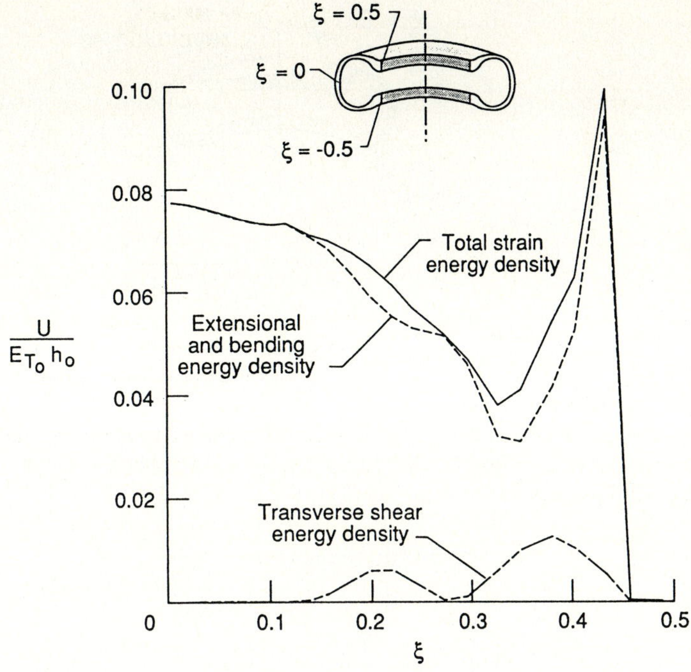

Figure 8 - Meridional variation of the strain energy densities produced by inflation pressure. Two-dimensional shell model of the space shuttle nose-gear tire shown in Fig. 2.

$$\text{where } p_n = \frac{2p_o}{n\pi} \sin n\beta \tag{25}$$

and p_o and β are functions of ξ as shown in Fig. 9.

Because of the symmetry of the shell meridian and loading, only one-half of the meridian is analyzed using 37 elements (a total of 1184 stress-resultant parameters, 1184 strain parameters, and 743 nonzero displacement degrees of freedom - see Fig. 5b). The boundary conditions at the centerline are taken to be the symmetric or antisymmetric conditions. Typical results are presented in Figs. 10 and 11 and in Tables 3 and 4.

The foregoing procedure was applied to this problem, and 10 global approximation vectors were evaluated at $n_o = 5$ and used to generate the tire response in the range $n = 1$ to 10. Accuracy of the generalized displacements obtained by the foregoing strategy with 8, 10 and 15 global approximation vectors is indicated in Figs. 10 and 11. Each generalized displace-

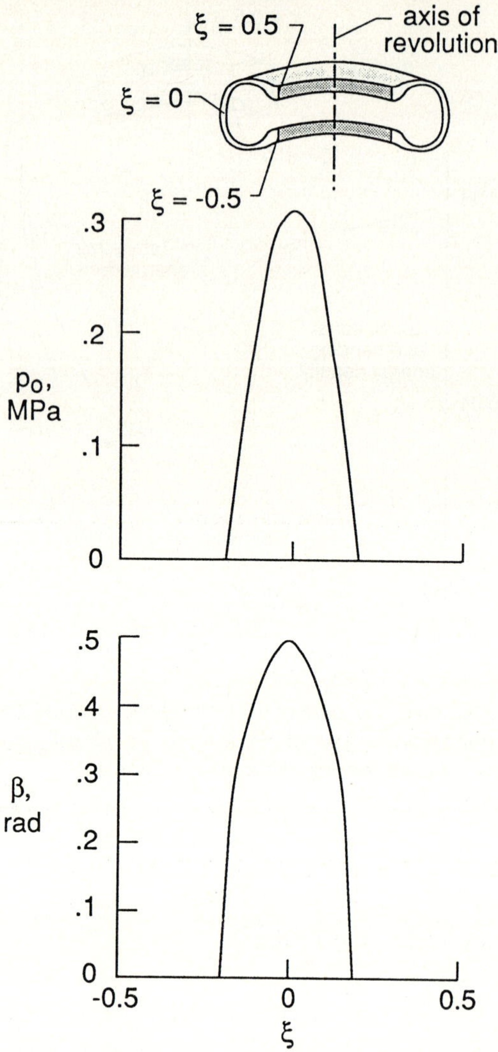

Figure 9 - Variation of contact pressure and angle in the meridional direction.
$p_{o_{max}} = 3.07 \times 10^5$ Pa. $\beta_{max} = 0.49375$ rad.

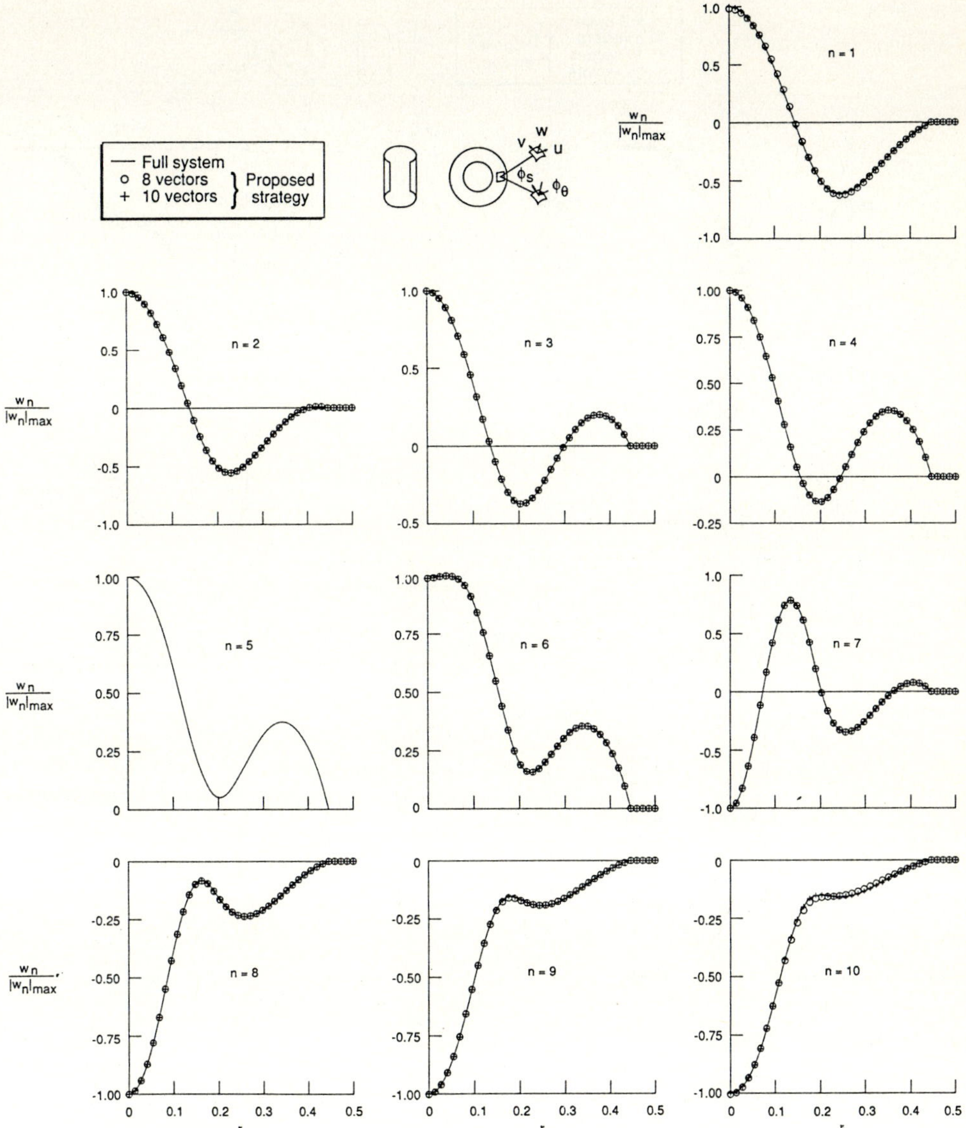

Figure 10 - Accuracy of normal displacements w_n obtained by the proposed procedure. Two-dimensional shell model of the space shuttle orbiter nose-gear tire shown in Fig. 2.

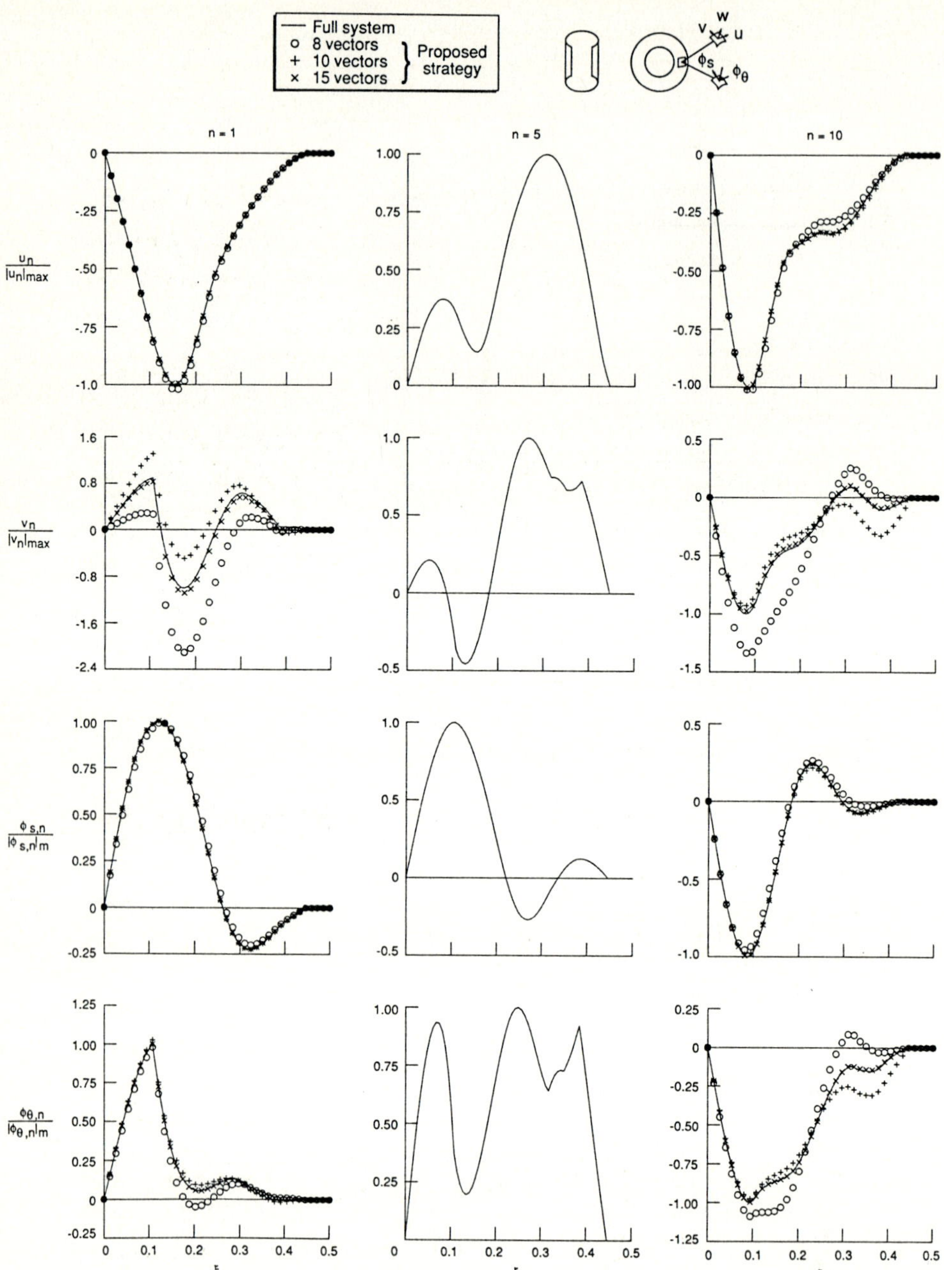

Figure 11 - Accuracy of in-plane displacements and rotation components obtained by the proposed procedure. Two-dimensional shell model of the space shuttle orbiter nose-gear tire shown in Fig. 2.

ment in Figs. 10 and 11 is normalized by dividing by its maximum absolute value given in Tables 3 and 4. Generalized displacements predicted by the foregoing procedure with 15 vectors are almost indistinguishable from the direct finite element solution.

Table 3 - Maximum absolute values of the normal displacement components w_n [two-dimensional shell model for the space shuttle nose-gear tire shown in Fig. 2; $p_{o_{max}}$ =3.07 × 10^5 Pa, h_o =0.01908 m, E_{T_o} =8 × 10^6 Pa]

Fourier Harmonic n	$w_n\, E_{T_o}/\!\left(p_{o_{max}} h_o\right)$	Fourier Harmonic n	$w_n\, E_{T_o}/\!\left(p_{o_{max}} h_o\right)$
1	18.08	6	1.095
2	11.41	7	0.3122
3	7.110	8	1.134
4	4.841	9	1.424
5	2.906	10	1.313

Table 4 - Maximum absolute values of the in-plane displacements and rotation components [two-dimensional shell model for the space shuttle nose-gear tire shown in Fig. 2; $p_{o_{max}}$ =3.07 × 10^5 Pa, h_o =0.01908 m, E_{T_o} =8 × 10^6 Pa]

Displacements and Rotations	Fourier Harmonic		
	n=1	n=5	n=10
$u_n\, E_{T_o}/\!\left(p_{o_{max}} h_o\right)$	7.584	0.3266	0.1229
$v_n\, E_{T_o}/\!\left(p_{o_{max}} h_o\right)$	0.04162	0.01139	3.773×10^{-3}
$\phi_{s,n}\, E_{T_o}/p_{o_{max}}$	3.602	0.4840	0.1304
$\phi_{\theta,n}\, E_{T_o}/p_{o_{max}}$	0.1805	0.01271	6.982×10^{-3}

Summary and Conclusions

A computational procedure is presented for the geometrically nonlinear analysis of aircraft tires. The space shuttle nose-gear tire was modeled using a two-dimensional laminated anisotropic shell theory with the effects of variation in material and geometric parameters included.

The governing discrete equations of the tire are obtained by applying the three-field, Hu-Washizu mixed variational principle. The multilevel operator splitting is used to 1) uncouple the equations associated with different harmonics, 2) identify the effects of different Fourier harmonics, and 3) delineate the effect of anisotropic (nonorthotropic) material properties. The nonlinear governing finite element equations of the tire are solved using the Newton-Raphson iterative procedure. An efficient procedure is presented for the solution of the resulting algebraic equations at each iteration, associated with different Fourier harmonics. The effectiveness of this procedure is demonstrated by means of a numerical example of the linear response of the space shuttle orbiter nose-gear tire. The tire is subjected to localized normal loading on the outer surface (simulating the contact pressure).

Results of the present study suggest the following conclusions relative to the two-dimensional shell model used in simulating the response of the tire, and the proposed computational procedure for generating the tire response associated with different Fourier harmonics.

1. A two-dimensional shell model with variable geometric and stiffness characteristics accurately predicts the deformation of the tire when subjected to inflation pressure.

2. Use of path derivatives (derivatives of the response with respect to the Fourier harmonic) as global approximation vectors leads to accurate solutions with a small number of vectors. Therefore, the time required to solve the reduced equations was relatively small and the total time required to generate the response for a range of 10 Fourier harmonics was little more than that required for a *single* Fourier harmonic.

3. Global approximation vectors provide a direct measure of the sensitivity of the different response quantities to the circumferential wave (harmonic) number. Sensitivity of the global response can also be assessed with these vectors.

4. The reduction method used in the proposed computational procedure exploits the best elements of the finite element method and of the Bubnov-Galerkin technique, as follows:

a) The finite element method was used as a general approach for generating global approximation vectors. The full finite element equations are solved only for a single Fourier harmonic.

b) The Bubnov-Galerkin technique was used as an efficient procedure for minimizing and distributing the error throughout the structure.

5. The reduction method extends the range of applicability of the Taylor series expansion by relaxing the requirement of using small changes in the circumferential wave number.

Acknowledgment

The work of the first and second authors was supported by a NASA Grant No. NAG1-852.

References

1. Goldberg, J. E.: Computer analysis of shells, in Proc. Symposium on the Theory of Shells to Honor L. H. Donnell. Univ. of Houston, Houston, TX 1967, 3-22.

2. Cohen, G. A.: FASOR - A second generation shell of revolution code. Comp. Struct. 10 (1979) 301-309.

3. Bushnell, D.: Analysis of buckling and vibration of ring-stiffened segmented shells of revolution. Int. J. Solids Struct. 6 (1970) 157-181.

4. Noor, A. K.; Stephens, W. B.: Comparison of finite difference schemes for analysis of shells of revolution. NASA TN-D-7337 (1973).

5. Sen, S. K.; Gould, P. L.: Free vibration of shells of revolution using FEM. J. Eng. Mech. Div. ASCE 100 (1974) 283-303.

6. Grigorenko, Y. M.: Laminated isotropic and anisotropic shells of revolution with variable stiffness. Izdatel'stvo Naukova Dumka, Kiev, USSR (1973) (in Russian).

7. Padovan, J.; Lestingi, J. F.: Complex numerical integration procedure for static loading of anisotropic shells of revolution. Comp. Struct. 4 (1974) 1159-1172.

8. Schaeffer, H. G.; Ball, R. E.: Nonlinear deflections of asymmetrically loaded shells of revolution. AIAA Paper No. 68-292 (1968).

9. Wunderlich, W.; Cramer, H.; Obrecht, H.: Application of ring elements in the nonlinear analysis of shells of revolution under nonaxisymmetric loading. Comput. Methods Appl. Mech. and Eng. 51(1-3) (1985) 259-275.

10. Sanders, J. L.: Nonlinear theories for thin shells. Quart. Appl. Math. 21 (1963) 21-36.

11. Budiansky, B.: Notes on nonlinear shell theory. J. Appl. Mech. 35 (1968) 393-401.

12. Noor, A. K.; Peters, J. M.: Analysis of laminated anisotropic shells of revolution. J. Eng. Mech. Div. ASCE 113 (1987) 49-65.

13. Noor, A. K.; Andersen, C. M.; Tanner, J. A.: Exploiting symmetries in the modeling and analysis of tires. NASA TP-2649 (1987).

14. Noor, A. K.; Peters, J. M.: Vibration analysis of laminated anisotropic shells of revolution. Comput. Methods Appl. Mech. & Eng. 61(3) (1987) 277-301.

15. Noor, A. K.; Peters, J. M.: Stress and vibration analyses of anisotropic shells of revolution. Int. J. Num. Methods Eng. 26 (1988) 1145-1167.

16. Noor, A. K.; Tanner, J. A.: Advances in contact algorithms and their application to tires. NASA TP-2781 (1988).

17. Jones, R. M.: Mechanics of composite materials. New York: Hemisphere Pub. Co. 1975.

18. Walter, J. D.: Cord reinforced rubber, ch. 3 in Mechanics of pneumatic tires (S. K. Clark, ed.), Dept. of Transportation, National Highway Traffic Safety Administration, Washington, D.C. 1981.

Finite Element Analysis of Rubber Membranes

G. BOLZON, B. A. SCHREFLER and R. VITALIANI

Istituto di Scienza e Tecnica delle Costruzioni
Università di Padova, Italy

Introduction

Rubbers and rubber composites are commonly used materials in the machine and tyre industry, as well as for applications in marine and river environments. The related structures consist, for example of shaft couplings and belts, insulators, nozzle attachements, pressure vessels, wires and domes.

Various finite element procedures have been proposed to investigate the behaviour of these structures which may undergo large displacements and finite deformations. An appropriate kinematic description is required and follower forces are also to be taken into account, as well as appropriate material models both for the rubber, whose typical features are finite straining and incompressibility, and for the reinforcements.

3-D [1, 2, 3], plane strain [4, 5], axisymmetric [6, 7, 8], shell [9] and membrane models [10, 11, 12, 13] have been developed, taking also into account layered or discrete reinforcement.

The membrane model, which is presented here, can work very effectively for some flexible structures [14], whose flexural stiffnesses can be neglected.

In the following, the general framework for the analysis of highly deformable bodies is briefly recalled, and the incremental equilibrium equations for hyperelastic incompressible bodies subjected to follower forces are written within the continuum approach.

The theory is then applied to the finite element analysis of Mooney-Rivlin rubber membranes reinforced with extensible cords.

Two examples dealing with the form-finding of rubber and rubber composites are shown, also considering the influence of no-compression in the assumed material models.

<u>Kinematics</u>

Let us consider a body subjected to large displacements and finite deformations due to the action of some applied forces [15, 16, 17, 18].

Let C_o be the reference configuration for the body, in its initially undeformed state, and let C be its deformed configuration (when some load is applied).

With reference to a fixed rectangular cartesian coordinate system, whose base unit vectors are $\underset{\sim}{e}_i$, the position of each point (particle) of the body in the undeformed configuration C_o can be identified by its coordinates x_i, or by the vector:

$$\underset{\sim}{x} = x_i \ \underset{\sim}{e}_i \tag{1}$$

The same point (particle) in the deformed configuration $\underset{\sim}{C}$ is identified by the position vector $\underset{\sim}{z}$, or by the displacement vector $\underset{\sim}{u}$, so that:

$$\underset{\sim}{z} = z_i \, (\underset{\sim}{x}) \ \underset{\sim}{e}_i = \underset{\sim}{x} + \underset{\sim}{u} \tag{2}$$

$$\underset{\sim}{u} = u_i \, (\underset{\sim}{x}) \ \underset{\sim}{e}_i \tag{3}$$

Let the mapping of $\underset{\sim}{x}$ into $\underset{\sim}{z}$ be a single value function, so that this transformation can be represented by the non-singular operator $\underset{\sim}{J}$, which is called the deformation gradient tensor:

$$d\underset{\sim}{z} = \underset{\sim}{J} \ d\underset{\sim}{x} , \tag{4}$$

$$J_{ij} = \frac{\partial z_i}{\partial x_j} = z_{i,j} \tag{5}$$

and also:

$$J_{ij} = \delta^{ij} + \frac{\partial u_i}{\partial x_j} = \delta^{ij} + u_{i,j} \tag{6}$$

where δ^{ij} is the Kronecker delta.

As $\underset{\sim}{J}$ is non-singular, it can be decomposed into a pure stretch and a rigid rotation, through the polar decomposition theorem:

$$J = R\,U = V\,R \tag{7}$$

where U and V are symmetric, positive definite right and left stretch tensors, and R is a rigid rotation tensor, and hence orthogonal.

<u>Strains</u>

As a measure of the straining of each point of the body, in passing from the configuration C_o to C, we assume the Green-Lagrange strain tensor:

$$\varepsilon = \frac{1}{2}\,(G - I) \tag{8}$$

where G is the right Cauchy-Green tensor

$$G = J^T\,J = U^2 \tag{9}$$

and I is the unit tensor.

Hence, the components of ε are:

$$\varepsilon_{ij} = \frac{1}{2}\,(z_{k,i}\,z_{k,j} - \delta_{ij}) = \frac{1}{2}\,(u_{i,j} + u_{j,i} + u_{k,i}\,u_{k,j})\ . \tag{10}$$

The principal invariants of the tensor G, which will be useful in the further developments, are

$$I_1 = \operatorname{tr}(G) \tag{11}$$

$$I_2 = \det(G)\,\operatorname{tr}(G^{-1}) = \operatorname{tr}(\operatorname{co}(G)) \tag{12}$$

$$I_3 = \det(G) \tag{13}$$

or:

$$I_1 = G_{ii} = 3 + 2\,\varepsilon_{ii} \tag{14}$$

$$I_2 = \frac{1}{2}\,(\delta^{ir}\,\delta^{is}\,G_{ri}\,G_{sj} - \delta^{ir}\,\delta^{is}\,G_{ij}\,G_{rs}) = 3 + 4\,\varepsilon_{ii} + 2\,(\varepsilon_{ii}\,\varepsilon_{jj} - \varepsilon_{ij}\,\varepsilon_{ji}) \tag{15}$$

$$I_3 = 1 + 2\,\varepsilon_{ii} + 2\,(\varepsilon_{ii}\,\varepsilon_{jj} - \varepsilon_{ij}\,\varepsilon_{ji}) + \frac{4}{3}\,\alpha^{ijk}\,\alpha^{rst}\,\varepsilon_{ir}\,\varepsilon_{js}\,\varepsilon_{kt} \tag{16}$$

where α^{ijk}, α^{rst} are the permutation symbols.

It is worth noting that a variation of the body volume, v, and of its density, ρ, as the mass is conserved, can be expressed by:

$$\frac{dv}{dv_o} = \frac{\rho_o}{\rho} = \sqrt{I_3}\;. \tag{17}$$

Hence, for isochoric deformations, $I_3 = 1$ since the volume is preserved.

Any change from the configuration C can be identified by the increment Δu of the displacement vector u. The corresponding variation of the strain tensor, $\Delta\underset{\sim}{\varepsilon}$, can be decomposed in the terms $\Delta\underset{\sim}{e}$, which is linear in $\Delta u_{i,j}$, and Δn, so that:

$$\Delta\varepsilon_{ij} = \Delta e_{ij} + \Delta n_{ij} \tag{18}$$

$$\Delta e_{ij} = \frac{1}{2}\,(\Delta u_{i,j} + \Delta u_{j,i} + u_{k,i}\,\Delta u_{k,j} + \Delta u_{k,i}\,u_{k,j}) \tag{19}$$

$$\Delta n_{ij} = \frac{1}{2}\,\Delta u_{k,i}\,\Delta u_{k,j}\;. \tag{20}$$

Finally, the variation of the invariants I_k with respect to the strain components ε_{ij} can be expressed by:

$$\frac{\partial I_1}{\partial \varepsilon_{ij}} = 2\,\delta^{ji} \tag{21}$$

$$\frac{\partial I_2}{\partial \varepsilon_{ij}} = 4\,[\delta^{ji}\,(1 + \varepsilon_{kk}) - \delta^{ir}\,\delta^{js}\,\varepsilon_{rs}] \tag{22}$$

$$\frac{\partial I_3}{\partial \varepsilon_{ij}} = 2\,\delta^{ji}\,(1 + 2\,\varepsilon_{kk}) - 4\,\delta^{ir}\,\delta^{js}\,\varepsilon_{rs} + 4\,\alpha^{imn}\,\alpha^{jst}\,\varepsilon_{ms}\,\varepsilon_{nt} \tag{23}$$

or

$$\frac{\partial I_1}{\partial \underset{\sim}{\varepsilon}} = 2\,\underset{\sim}{I} \tag{24}$$

$$\frac{\partial I_2}{\partial \underset{\sim}{\varepsilon}} = 2 \, (I_1 \, \underset{\sim}{I} - \underset{\sim}{G}) \tag{25}$$

$$\frac{\partial I_3}{\partial \underset{\sim}{\varepsilon}} = 2 \, I_3 \, \underset{\sim}{G}^{-1} \; . \tag{26}$$

Stresses

Let df_i be the differential force acting on the differential area (da) in the deformed body, and let n_i be the direction cosines of the unit outward normal to (da). As it is well known, the true stresses or Cauchy stresses, σ_{ij}, are defined by the equilibrium condition:

$$df_i = (da) \, n_j \, \sigma_{ij} \; . \tag{27}$$

The tensor $\underset{\sim}{\sigma}$ is symmetric when, as usual, no body couples are present.

Conjugate to the Green-Lagrange strain components are instead the 2nd Piola-Kirchhoff stresses, S_{ij}.

These stresses are the result of the application of a force vector

$$\underset{\sim}{df_o} = \underset{\sim}{J}^{-1} \, \underset{\sim}{df} \tag{28}$$

on the oriented area $(da_o) \, \underset{\sim}{n}^o$, which is the image in the undeformed configuration of the oriented area (da) $\underset{\sim}{n}$.

This transformation on the forces is required to maintain the symmetry of the stress tensor.

Then:

$$df_i = (da_o) \, n_j^o \, S_{jk} \, z_{i,k} \; . \tag{29}$$

The relation between σ_{ij} and S_{ij} is hence:

$$\sigma_{ij} = \frac{1}{\det \, (\underset{\sim}{J})} \, z_{i,m} \, S_{m,n} \, z_{j,n} \tag{30}$$

or

$$\underset{\sim}{\sigma} = \frac{1}{\det \, (\underset{\sim}{J})} \, \underset{\sim}{J} \, \underset{\sim}{S} \, \underset{\sim}{J}^T \; . \tag{31}$$

Hyperelastic Incompressible Materials

For any material, the rate of increase of internal energy, called also the stress-working rate [18, 19], per unit volume of a deformed body, in the configuration C, is defined by:

$$\dot{W} = \frac{1}{2}\, \sigma_{ij}\, (\dot{E}_{ij} + \dot{E}_{ji}) = \sigma_{ij}\, \dot{E}_{ij} \tag{32}$$

where $\dot{E} = \dot{J}\, J^{-1}$, and σ is the true (Cauchy) symmetric stress tensor. The symmetric part of $\dot{E}$ represents the strain rate referred to the configuration C.

For highly deformable materials, it is more convenient to define, instead of $\dot{W}$, the stress-working rate $\dot{W}_o$ measured per unit volume of the body in its known reference configuration C_o. Since mass is conserved:

$$\dot{W}_o = \frac{dv}{dv_o}\, \dot{W} = \frac{\rho_o}{\rho}\, \dot{W} = \sqrt{I_3}\, \dot{W} \ , \tag{33}$$

$$\dot{W}_o = \sqrt{I_3}\, \sigma_{ij}\, \dot{E}_{ij} \ . \tag{34}$$

From the definition of the strain and stress measures, ε and S, and their relations with σ and $\dot{E}$, follows also that

$$\dot{W}_o = S_{ij}\, \dot{\varepsilon}_{ij} \ , \tag{35}$$

so that S and $\dot{\varepsilon}$ are energetically conjugate.

For perfectly elastic materials, the deformations are reversible and are independent of the previous strain-history. Hence, it is possible to define a differentiable strain energy function W_o (per unit undeformed volume, dv_o) which depends only on the current state of deformation of the body. This function represents a potential for the stresses which can be obtained directly by derivation from W_o [15, 17]:

$$S_{ij} = \frac{\partial W_o}{\partial \varepsilon_{ij}} \ . \tag{36}$$

The stresses thereby form a conservative system and are obviously independent of the deformation path and of the strain rate.

When the material is homogeneous and isotropic, the elastic potential is a function of the three principal strain invariants only, or, more conveniently, of the principal invariants of the Cauchy-Green tensor $\underset{\sim}{G}$:

$$W_o = W_o (I_1, I_2, I_3) \ . \tag{37}$$

With the further assumption of incompressibility of the material, W_o depends on $\underset{\sim}{\varepsilon}$ just through the invariants I_1 and I_2, since $I_3 = 1$. However, in this case the components of the strain tensor are no more independent, since they are related by the constraint $I_3 = \det (\underset{\sim}{G}) = 1$.

Moreover, only the deviatoric components of the stress tensor can be obtained from the strain energy of the body, since any hydrostatic state of stress can be added to the real one without changing the volume of the body and therefore its strain energy. Hence, the true hydrostatic pressure can be obtained only from the boundary conditions on the forces (tractions) acting on the body, not from the motion.

Incremental Equilibrium Equations for Hyperelastic Incompressible Bodies Subjected to Follower Forces

Let $\underset{\sim}{b}$, $\underset{\sim}{t}$ and $\underset{\sim}{f}$ be the volume, surface and point forces acting on the body, referred to its deformed configuration C. Let this body be isotropic, homogeneous, hyperelastic and incompressible. In the following, attention is focused on the motion of the body, relative to its initial undeformed configuration C_o, under the action of the applied loads.

The incompressibility condition requires the consideration of two independent fields of unknowns: the hydrostatic pressures, and the displacements. These latter are moreover constrained by the condition

$$\det (\underset{\sim}{G}) = 1 \ . \tag{38}$$

The system of the internal forces (stresses) is conservative because of the material model assumed, and their potential per unit undeformed volume is expressed by the strain energy function W_o as

$$W_o = W_o (I_1, I_2) \ . \tag{39}$$

It is convenient to modify this function by introducing explicitly the incompressibility condition through a Lagrange multiplier p and the constraint function $f(I_3)$, which will be zero when $I_3 = 1$ [4, 6].
The modified potential, $\tilde{W}_o$, can hence be written as

$$\tilde{W}_o = W_o (I_1, I_2) + pf(I_3) \ . \tag{40}$$

The Lagrange multiplier p can be identified, by giving an energetic meaning to $\tilde{W}_o$, with the hydrostatic pressure acting on the material, which gives no contribution to the strain energy for isochoric deformations.
Otherwise, by simply assuming

$$f (I_3) = I_3 - 1 \ , \tag{41}$$

the last term in $\tilde{W}_o$ represents the internal work done by the hydrostatic stress component for any change in the volume.
The equilibrium is now imposed by means of the virtual work principle, since the applied forces, which are generally configuration dependent when the displacements are large, don't arise from a potential.

Let hence $\delta\underset{\sim}{\varepsilon}$ be the virtual increase in the strain tensor related to a virtual displacement $\delta\underset{\sim}{u}$ from the equilibrium configuration C.

The corresponding virtual internal work can be expressed as a variation of the potential function $\tilde{W}_o$.

$$\delta L_i = \int_{V_o} \delta \tilde{W}_o \ dv = \int_{V_o} \frac{\partial \tilde{W}_o (\underset{\sim}{u}, p)}{\partial \varepsilon_{ij}} \delta\varepsilon_{ij} \ dv = \int_{V_o} S_{ij} \ \delta\varepsilon_{ij} \ dv \tag{42}$$

where

$$S_{ij} (\underset{\sim}{u}, p) = \frac{\partial W_o}{\partial \varepsilon_{ij}} + p \frac{\partial f(I_3)}{\partial \varepsilon_{ij}} \ . \tag{43}$$

The virtual work of the external forces, for the same displacement δu, can be expressed with reference to the undeformed configuration C_o by

$$\delta L_e = \int_{V_o} b_i^o \, \delta u_i \, dv + \int_{A_o} t_i^o \, \delta u_i \, da + f_i \, \delta u_i \ , \tag{44}$$

where b^o and t^o are the transformations in C_o of the real applied forces, b and t, so that

$$\int_{V_o} b_i^o \, \delta u_i \, dv = \int_V b_i \, \delta u_i \, dv \tag{45}$$

and

$$\int_{A_o} t_i^o \, \delta u_i \, da = \int_A t_i \, \delta u_i \, da \ . \tag{46}$$

Generally, b^o and t^o, as well as b, t and f, depend upon the configuration C through the displacements u of their application points.

To effectively solve the problem, it is convenient to linearize the equilibrium equations obtained from the equality

$$\delta L_i = \delta L_e \ . \tag{47}$$

This can be done by an incremental procedure [6, 27].

Let $\bar{u}$ be a known displacement field, which identifies the configuration $\bar{C}$, close to C, where the external applied forces, $\bar{b}, \bar{t}$ and $\bar{f}$ are in equilibrium with the stresses $\bar{S}$ and $\bar{p}$, and $\bar{\varepsilon}$ are the corresponding strains. It is possible to consider the decomposition

$$S_{ij} = \bar{S}_{ij} + \Delta S_{ij} \tag{48}$$

where

$$\Delta S_{ij} = \frac{\partial S_{ij}}{\partial \varepsilon_{rs}} \, \Delta\varepsilon_{rs} + \frac{\partial S_{ij}}{\partial p} \, \Delta p = C_{ijrs} \, \Delta\varepsilon_{rs} + \frac{\partial f(I_3)}{\partial \varepsilon_{ij}} \, \Delta p \tag{49}$$

and

$$C_{ijrs} = \frac{\partial^2 W_o}{\partial \varepsilon_{ij} \partial \varepsilon_{rs}} + p \frac{\partial^2 f(I_3)}{\partial \varepsilon_{ij} \partial \varepsilon_{rs}} \quad . \tag{50}$$

$\underset{\sim}{C}$ is hence the 4th order tensor of the constitutive properties of the material.

In the same way:

$$\varepsilon_{ij} = \bar{\varepsilon}_{ij} + \Delta \varepsilon_{ij} \tag{51}$$

$$\delta \varepsilon_{ij} = \delta \Delta \varepsilon_{ij} = \delta \Delta e_{ij} + \delta \Delta n_{ij} \quad . \tag{52}$$

Hence:

$$\delta L_i = \int_{V_o} S_{ij} \, \delta \varepsilon_{ij} \, dv = \int_{V_o} C_{ijrs} \, \Delta \varepsilon_{rs} \, \delta \Delta \varepsilon_{ij} \, dv + \int_{V_o} \bar{S}_{ij} \, \delta \Delta \varepsilon_{ij} \, dv +$$

$$+ \int_{V_o} \Delta p \, \frac{\partial f(I_3)}{\partial \varepsilon_{ij}} \, \delta \Delta \varepsilon_{ij} \, dv \approx \int_{V_o} C_{ijrs} \, \Delta e_{rs} \, \delta \Delta e_{ij} \, dv + \tag{53}$$

$$+ \int_{V_o} \bar{S}_{ij} \, \delta \Delta n_{ij} \, dv + \int_{V_o} \bar{S}_{ij} \, \delta \Delta e_{ij} \, dv + \int_{V_o} \Delta p \, \frac{\partial f(I_3)}{\partial \varepsilon_{ij}} \, \delta \Delta e_{ij} \, dv \quad .$$

In the latter equality, the first two integrals contain the terms which are quadratic in the displacement gradients, $\Delta u_{i,j}$, while the remaining two contain only the linear ones.

For the external forces [20, 21]:

$$b_i^o = \bar{b}_i^o + \frac{\partial b_i^o}{\partial \mu} \Delta \mu + \frac{\partial b_i^o}{\partial u_j} \Delta u_j = \bar{b}_i^o + \frac{\Delta \mu}{\bar{\mu}} \bar{b}_i^o + b_{i,J}^o \Delta u_J \tag{54}$$

where $\Delta \mu$ represents the increase of the load multiplier.

Working in the same way on t^o and $\underset{\sim}{f}$:

$$\delta L_e = \int_{V_o} b_{i,J}^o \, \Delta u_J \, \delta \Delta u_i \, dv + \int_{A_o} t_{i,J}^o \, \Delta u_J \, \delta \Delta u_i \, da + f_{i,J} \, \Delta u_J \, \delta \Delta u_i +$$

$$\tag{55}$$

$$+ (1 + \frac{\Delta \mu}{\bar{\mu}}) \left[\int_{V_o} b_i^o \, \delta \Delta u_i \, dv + \int_{A_o} t_i^o \, \delta \Delta u_i \, da + f_i \, \delta \Delta u_i \right] \quad .$$

Once again, the first integrals contain terms which are quadratic, this time in the displacement increments Δu_i. These terms are typical for follower forces. Indeed, when load is independent of the deformation, all the $b^o_{i,J}$, $t^o_{i,J}$ and $f_{i,J}$ are zero.

Finally, giving a variation δp to p, one can get the incompressibility condition in the form:

$$\int_{V_o} f(I_3)\, \delta p = 0 \ . \tag{56}$$

This equation is needed to complete the set of equations which allows for the complete solution of the problem. In incremental form:

$$\int_{V_o} \left[\frac{\partial f(I_3)}{\partial \varepsilon_{ij}}\, \Delta\varepsilon_{ij} + \bar{f}(I_3) \right] \delta\Delta p = 0 \ . \tag{57}$$

The Mooney-Rivlin Relationship for Rubber

Natural and vulcanized rubbers are among the most common highly elastic materials. Rubber is an isotropic, effectively incompressible and substantially homogeneous material [17].

Because of the numerous applications in industry, the behaviour of this material is from long time conveniently investigated by experiments, also in the range of finite deformations.

The stress-strain relationship for a typical rubber, although elastic, is non linear. Hooke's law is adequate to describe the behaviour of this material only for simple shear [22, 23, 24].

From these considerations, Mooney [25] suggested a two-constant expression for the strain energy function per unit undeformed volume, W_o, of this material:

$$W_o = C_{10}\,(I_1 - 3) + C_{01}\,(I_2 - 3); \qquad I_3 = 1 \ . \tag{58}$$

The same equation can be obtained by the first order expansion of the polynomial later suggested by Rivlin [26] for describing the large elastic deformations of hyperelastic materials:

$$W_o = \sum_{i,\, j,\, k\, =\, 0}^{\infty} C_{ijk}\, (I_1 - 3)^i\, (I_2 - 3)^j\, (I_3 - 1)^k \tag{59}$$

when the incompressibility condition is introduced.

The constants which appear in these equations are such to give zero strain energy in the undeformed configuration. The parameters C_{10} and C_{01} in (58) characterize each material and are experimentally determined. These parameters can be related to the usual shear modulus G by

$$G = 2\,(C_{10} + C_{01})\,, \tag{60}$$

and to the initial elastic modulus E, as the ratio between the stress and the corresponding strain for small (infinitesimal) monoaxial deformation, by:

$$E = 6\,(C_{10} + C_{01})\,. \tag{61}$$

Equation (58) is called the Mooney-Rivlin strain energy function. It has been found to agree with the predictions of the statistic molecular theory of rubber elasticity, and to give a very good fit to experimental data up to stretches of about 250-300%. Beyond these values it is more suitable to assume for W_o higher order expansions. For further applications it is however useful to refer to the modified strain energy function $\tilde{W}_o$:

$$\tilde{W}_o = W_o\,(I_1,\, I_2) + p\,f\,(I_3)\,. \tag{62}$$

Let $f(I_3)$ be

$$f(I_3) = \frac{I_3 - 1}{2}\,. \tag{63}$$

Hence:

$$\tilde{W}_o = C_{10}\,(I_1 - 3) + C_{01}\,(I_2 - 3) + \frac{p}{2}\,(I_3 - 1) \tag{64}$$

and:

$$\underset{\sim}{S} = \frac{\partial \tilde{W}_o}{\partial \underset{\sim}{\varepsilon}} = 2\,C_{10}\,\underset{\sim}{I} + 2\,C_{01}\,(I_1\,\underset{\sim}{I} - \underset{\sim}{G}) + p\,I_3\,\underset{\sim}{G}^{-1}\,. \tag{65}$$

Introducing explicitly the incompressibility condition

$$\underset{\sim}{S} = 2\,C_{10}\,\underset{\sim}{I} + 2\,C_{01}\,(I_1\,\underset{\sim}{I} - \underset{\sim}{G}) + p\,\underset{\sim}{G}^{-1}\,. \tag{66}$$

Equation (66) is a general one, and is independent of the choice of $f(I_3)$. In fact:

$$\frac{\partial f(I_3)}{\partial \underset{\sim}{\varepsilon}} = \frac{\partial f(I_3)}{\partial I_3} \frac{\partial I_3}{\partial \underset{\sim}{\varepsilon}} = k \frac{\partial I_3}{\partial \underset{\sim}{\varepsilon}} = 2\,k\,I_3\,\underset{\sim}{G}^{-1} = 2\,k\,\underset{\sim}{G}^{-1} \tag{67}$$

where k represents the value attained by $\partial f(I_3)/\partial I_3$ when $I_3 = 1$. In (63) k was assumed for convenience equal to 1/2.

By a further derivation, the 4th order constitutive tensor $\underset{\sim}{C}$ can be obtained from $\underset{\sim}{S}$. Its terms are:

$$C_{ijrs} = \frac{\partial S_{ij}}{\partial \varepsilon_{rs}} = \frac{\partial^2 \tilde{W}_o}{\partial \varepsilon_{ij}\,\partial \varepsilon_{rs}} \, . \tag{68}$$

These terms are functions of the elastic constants C_{10} and C_{01} as well as of p and of the components of the tensor $\underset{\sim}{G}$. The general formulation is reported in [3]. The direct dependence of the tensors $\underset{\sim}{S}$ and $\underset{\sim}{C}$ on p can be avoided when some special boundary conditions on stresses exist. This is the case in membranes and shells.

For example, let M be the curved surface which represents the mid-surface of a membrane however oriented in space. Let $\underset{\sim}{i}_k$ be the unit base vectors of an orthogonal cartesian reference system, in each point of M. This frame is variable in space, and the plane tangent to M in each point is determined by $\underset{\sim}{i}_1$ and $\underset{\sim}{i}_2$, and its normal is coincident with $\underset{\sim}{i}_3$.

Referring the stress and the strain tensors, $\underset{\sim}{S}$ and $\underset{\sim}{\varepsilon}$, to this reference system, the components S_{33}, S_{13} and S_{32} are zero in each point of M, as well as ε_{13} and ε_{32}, and hence G_{13} and G_{32}.

From the condition:

$$S_{33} = 2\,C_{10} + 2C_{01}\,(G_{11} + G_{22}) + p\,(G_{11}\,G_{22} - G_{12}\,G_{21}) = 0 \tag{69}$$

it is possible to obtain p as a function of the state of deformation in the membrane plane.

The corresponding component of the deformation, ε_{33}, instead is not zero, and represents the variation of the thickness of the membrane. This strain must be explicitly determined, since G_{33} appears in S_{11} and S_{22}. G_{33} can be directly obtained from the incompressibility condition

$$\det (\underset{\sim}{G}) = 1 \tag{70}$$

which yields:

$$G_{33} = \frac{1}{(G_{22}\, G_{11} - G_{12}\, G_{21})} \, . \tag{71}$$

In the shell hypothesis, equations (69) and (70) remain the same. Only, G_{33} assumes a rather more complex form since G_{13} and G_{32} are no more zero and contribute to $\det (\underset{\sim}{G})$ [9].

Finite Element Formulation

After the discretization of the body by means of finite elements, the field variables, $\underset{\sim}{u}$ and p, are expressed by interpolation of the corresponding nodal values, using different shape functions for $\underset{\sim}{u}$ and p.

The value of p is assumed uniform within each element, while

$$\underset{\sim}{u} = \underset{\sim}{N}\, \underset{\sim}{q} \tag{72}$$

and likewise

$$\Delta\underset{\sim}{u} = \underset{\sim}{N}\, \Delta\underset{\sim}{q} \tag{73}$$

where $\underset{\sim}{N}$ is the matrix of the shape functions and $\underset{\sim}{q}$ is the vector of the nodal displacements.

For further development, it is more convenient to organize the stress and the strain components in vector form, taking advantage of the symmetry of the respective tensors.

Let $\underset{\sim}{\hat{S}}$ be the vector of the 2nd Piola-Kirchhoff stress components:

$$\hat{\underline{S}}^T = \{S_{11}\ S_{22}\ S_{33}\ S_{12}\ S_{23}\ S_{31}\} \tag{74}$$

and let $\hat{\underline{\varepsilon}}$ be the strain vector:

$$\hat{\underline{\varepsilon}}^T = \{\varepsilon_{11}\ \varepsilon_{22}\ \varepsilon_{33}\ 2\varepsilon_{12}\ 2\varepsilon_{23}\ 2\varepsilon_{31}\}. \tag{75}$$

Likewise, let $\hat{\underline{e}}$, $\hat{\underline{n}}$ and $\hat{\underline{G}}$ be the vectors corresponding to the tensors $\underline{e}$, $\underline{n}$ and $\underline{G}$.

By introducing (73) in (19) and (20), for each element can be written [27]:

$$\Delta\hat{\underline{e}} = \underline{B}_L\ \Delta\underline{q} \tag{76}$$

$$\Delta\hat{\underline{n}} = \Delta\underline{q}^T\ \underline{B}_{NL}^T\ \underline{B}_{NL}\ \Delta\underline{q} \tag{77}$$

where $\underline{B}_L$ and $\underline{B}_{NL}$ are the strain-displacement transformation matrices, whose terms contain the shape functions and their derivatives.

Furthermore, let $\underline{B}^\circ$, $\underline{T}^\circ$ and $\underline{F}$ be the matrices whose terms are the derivatives of the element loads $\underline{b}^\circ$, $\underline{t}^\circ$ and $\underline{f}$ with respect to the displacements $\underline{u}$.

Hence, the integrals appearing in (53) and in (55) can be rewritten as:

$$\int_{V_o} \Delta e_{rs}\ C_{ijrs}\ \Delta e_{ij}\ dv = \Delta\underline{q}^T\ \underline{K}_L\ \Delta\underline{q} \tag{78}$$

$$\int_{V_o} \bar{S}_{ij}\ \Delta n_{ij}\ dv = \Delta\underline{q}^T\ \underline{K}_{NL}\ \Delta\underline{q} \tag{79}$$

$$\int_{V_o} \bar{S}_{ij}\ \Delta e_{ij}\ dv = \Delta\underline{q}^T\ \underline{h} \tag{80}$$

$$\int_{V_o} \Delta p\ \frac{\partial f(I_3)}{\partial \varepsilon_{ij}}\ \Delta\varepsilon_{ij}\ dv = \Delta\underline{q}^T \underline{K}_I\ \Delta p \tag{81}$$

$$\int_{V_o} b_{i,J}^\circ\ \Delta u_J\ \Delta u_i\ dv + \int_{A_o} t_{i,J}^\circ\ \Delta u_J\ \Delta u_i\ da + f_{i,j}\ \Delta u_J\ \Delta u_i = \Delta\underline{q}^T\ \underline{K}_R\ \Delta\underline{q} \tag{82}$$

$$(1+\frac{\Delta\mu}{\bar{\mu}})\left\{\int_{V_o} \bar{b}_i^o \, \Delta u_i \, dv + \int_{A_o} \bar{t}_i^o \, \Delta u_i \, da + \bar{f}_i \, \Delta u_i\right\} = \underset{\sim}{r} \tag{83}$$

$$\int_{V_o} f(I_3) \, dv = i \tag{84}$$

where

$$\underset{\sim}{K}_L = \int_{V_o} \underset{\sim}{B}_L^T \, \underset{\sim}{\hat{C}} \, \underset{\sim}{B}_L \, dv \tag{85}$$

$$\underset{\sim}{K}_{NL} = \int_{V_o} \underset{\sim}{B}_{NL}^T \, \underset{\sim}{\hat{\bar{S}}} \, \underset{\sim}{B}_{NL} \, dv \tag{86}$$

$$\underset{\sim}{K}_R = \int_{V_o} \underset{\sim}{N}^T \underset{\sim}{B}^o \, \underset{\sim}{N} \, dv + \int_{A_o} \underset{\sim}{N}^T \, \underset{\sim}{T}^o \, \underset{\sim}{N} \, da + \underset{\sim}{N}^T \underset{\sim}{F} \, \underset{\sim}{N} \tag{87}$$

$$\underset{\sim}{K}_I = \int_{V_o} 2 \, \underset{\sim}{B}_L^T \, \underset{\sim}{G}^{-1} dv \tag{88}$$

$$\underset{\sim}{h} = \int_{V_o} \underset{\sim}{B}_L^T \, \underset{\sim}{\hat{\bar{S}}} \, dv \tag{89}$$

$$\underset{\sim}{r} = (1+\frac{\Delta\mu}{\bar{\mu}})\left\{\int_{V_o} \underset{\sim}{N}^T \, \bar{b}^o \, dv + \int_{A_o} \underset{\sim}{N}^T \, \bar{t}^o \, da + \underset{\sim}{N}^T \, \underset{\sim}{\bar{f}}\right\}. \tag{90}$$

At element level, equations (47) and (56) become:

$$[\underset{\sim}{K}_L + \underset{\sim}{K}_{NL} + \underset{\sim}{K}_R] \, \Delta\underset{\sim}{q} + \underset{\sim}{K}_I \, \Delta p = \underset{\sim}{r} - \underset{\sim}{h} \tag{91}$$

$$\underset{\sim}{K}_I^T \, \Delta\underset{\sim}{q} = i. \tag{92}$$

Let K_T be the tangent stiffness matrix:

$$\underset{\sim}{K}_T = \underset{\sim}{K}_L + \underset{\sim}{K}_{NL} + \underset{\sim}{K}_R. \tag{93}$$

The presence of K_R, which can be unsymmetric in many cases of practical interest, can make K_T itself unsymmetric [21]. Within an iterative solution scheme, as will be

364

used here, the unsymmetric component of K_R, or the whole K_R, is often neglected. The respective contributions are included in the r.h.s. of (91), by updating the vector of the equivalent nodal forces r at each iteration [28].

The whole system can be rewritten in a more concise form as

$$\begin{bmatrix} K_T & K_I \\ K_I^T & \emptyset \end{bmatrix} \begin{Bmatrix} \Delta q \\ \Delta p \end{Bmatrix} = \begin{Bmatrix} r - h \\ i \end{Bmatrix} . \tag{94}$$

Often it is more convenient to reduce the unknowns to the displacements only. Hence the parameter Δp must be eliminated by condensation. This can be carried on by a penalty procedure [6].

By introducing a penalty parameter α in (92):

$$K_I^T \, \Delta q \, - \, \alpha \, \Delta p = i \, , \tag{95}$$

Δp can be obtained as

$$\Delta p = \frac{1}{\alpha} \left(K_I^T \, \Delta q \, - \, i \right) . \tag{96}$$

The penalty parameter α is reduced close to zero as the iterative procedure converges.

The system (91) is then reduced to

$$\tilde{K}_T \, \Delta q \, = \, r - \tilde{h} \tag{97}$$

where

$$\tilde{K}_T \, = \, K_T + \frac{1}{\alpha} \, K_I \, K_I^T \tag{98}$$

and

$$\tilde{h} = h - \frac{1}{\alpha} \, K_I \, . \tag{99}$$

This complication may be avoided when p can be exactly evaluated from the strain components, as in (69).

<u>Finite Elements Employed</u>

The finite elements employed in the analyses are isoparametric mono- and bi-dimensional elements in R^3 space with linear and/or parabolic interpolation functions for the geometry and the displacements.

These elements are obtained from the degeneration of 3-D elements. The initial position and the displacements of a generic point are expressed as a function of the same parameters at particular points of the element axis or of the element mid-plane, in compliance with the kinematical assumptions for the displacements. Moreover, the same equations valid for the continuum approach are used to state the equilibrium condition [27, 29, 30].

The integration of the terms arising from these equations are performed explicitly over the thickness or on the cross sections respectively of the mono- and bi-dimensional elements and numerically on the mid-plane or along the element axis.

The further assumption of uniform deformation on the cross-wise planes is made both for cables and for membranes, so that all the bending effects are neglected. The geometry of each element is defined by an isoparametric transformation into the real space of the parent elements which are defined in the (ξ, η, ζ) space.   The base unit vectors of this latter space are $\underline{l}$, $\underline{m}$ and $\underline{n}$ (fig.1).

In this transformation the initially rectangular intrinsic convected frame (ξ, η, ζ) is mapped into a curvilinear coordinate system which in general is no more rectangular.

However, the axis ζ is assumed to remain orthogonal to the plane identified by

$\underline{l}$ and $\underline{m}$ which is the tangent plane for 2-D elements.

Moreover, the axis η of the cable elements is assumed orthogonal in each point to ξ

and to the global reference axis x_3, whose direction is $\underline{e}_3$. In this manner the convected frame remains orthogonal, even though same fictitious torsion in the elements can be introduced. The employed constitutive relationship eliminates this effect.

Under these hypotheses, the angles ϕ_3 and ϕ_4 are sufficient for defining the isoparametric transformation of the cable element, (fig. 2), where:

$$\text{sen } \phi_3(\xi) = \frac{1}{A} \frac{\partial x_3}{\partial \xi} \qquad A = \left[\left(\frac{\partial x_1}{\partial \xi} \right)^2 + \left(\frac{\partial x_2}{\partial \xi} \right)^2 + \left(\frac{\partial x_3}{\partial \xi} \right)^2 \right]^{1/2} \qquad (100)$$

$$\cos \phi_3(\xi) = \left[1 - \text{sen}^2 \phi_3(\xi) \right]^{1/2} \qquad (101)$$

$$\text{sen } \phi_4(\xi) = \frac{1}{B} \frac{\partial x_2}{\partial \xi} \qquad B = \left[\left(\frac{\partial x_1}{\partial \xi} \right)^2 + \left(\frac{\partial x_2}{\partial \xi} \right)^2 \right]^{1/2} \qquad (102)$$

$$\cos \phi_4(\xi) = \left[1 - \text{sen}^2 \phi_4(\xi) \right]^{1/2} . \qquad (103)$$

Hence, along the element axis:

$$\underline{m}^T = \{ - \text{sen } \phi_4(\xi), \cos \phi_4(\xi), 0 \} \qquad (104)$$

$$\underline{n}^T = \{ - \text{sen } \phi_3(\xi) \cos \phi_4(\xi), - \text{sen } \phi_3(\xi) \, \text{sen } \phi_4(\xi), \cos \phi_3(\xi) \} \qquad (105)$$

The coordinates are then interpolated from the nodal values $\underline{X}^k$ of each finite element by means of the shape functions N_k as:

$$x_i = \sum_k N_k(\xi) \, X_i^k + \frac{a}{2} \eta \, m_i(\xi) + \frac{b}{2} \zeta \, n_i(\xi) \qquad (106)$$

The coordinates X_i^k are defined in the global reference system.

Four angles are instead needed for identifying the local intrinsic convected frame for two dimensional elements. These are the angles ϕ_1 and ϕ_5 which define the position of the axis ξ, and the angles ϕ_2 and ϕ_6 which define the position of the axis η.

In each point

$$\text{sen } \phi_1(\xi, \eta) = \frac{1}{A} \frac{\partial x_3}{\partial \xi} \ , \qquad \text{sen } \phi_5(\xi, \eta) = \frac{1}{B} \frac{\partial x_2}{\partial \xi} \qquad (107, 108)$$

$$\text{sen } \phi_2(\xi, \eta) = \frac{1}{C} \frac{\partial x_3}{\partial \eta} \ , \qquad \text{sen } \phi_6(\xi, \eta) = \frac{1}{D} \frac{\partial x_2}{\partial \eta} \qquad (109, 110)$$

where

$$C = \left[\left(\frac{\partial x_1}{\partial \eta}\right)^2 + \left(\frac{\partial x_2}{\partial \eta}\right)^2 + \left(\frac{\partial x_3}{\partial \eta}\right)^2\right]^{1/2} \qquad D = \left[\left(\frac{\partial x_1}{\partial \eta}\right)^2 + \left(\frac{\partial x_2}{\partial \eta}\right)^2\right]^{1/2}.$$

The direction cosines of $\underline{n}$ can then be obtained from the trigonometric functions of these angles and of the auxiliary angles $\phi_1^{\cdot}$ and $\phi_2^{\cdot}$, which can be related to (107) - (110) by simple geometrical considerations:

$$\underline{n}^T = \{-\cos\phi_2^{\cdot}\ \text{sen}\ \phi_1\ \cos\phi_5 + \cos\phi_1^{\cdot}\ \text{sen}\ \phi_2^{\cdot}\ \text{sen}\ \phi_5,$$

$$-\cos\phi_2^{\cdot}\ \text{sen}\ \phi_1\ \text{sen}\ \phi_5 + \cos\phi_1^{\cdot}\ \text{sen}\ \phi_2^{\cdot}\ \cos\phi_5,\ \cos\phi_2^{\cdot}\ \cos\phi_1\} \qquad (111)$$

So, as in (106):

$$x_i = \sum_k N_k(\xi,\eta)\ X_i^k + \frac{t}{2}\ \zeta\ n_i(\xi,\eta). \qquad (112)$$

The jacobian matrix of the transformation from (ξ, η, ζ) to (x_1, x_2, x_3) for cable elements can hence be expressed by

$$\underline{J}_c = \begin{bmatrix} \sum_k \dfrac{\partial N_k}{\partial \xi}\ X_1^k + \dfrac{a}{2}\ \eta\ P_1(\xi) + \dfrac{b}{2}\ \zeta\ P_2(\xi) & \dfrac{a}{2}\ m_1 & \dfrac{b}{2}\ n_1 \\[2ex] \sum_k \dfrac{\partial N_k}{\partial \xi}\ X_2^k + \dfrac{a}{2}\ \eta\ P_3(\xi) + \dfrac{b}{2}\ \zeta\ P_4(\xi) & \dfrac{a}{2}\ m_2 & \dfrac{b}{2}\ n_2 \\[2ex] \sum_k \dfrac{\partial N_k}{\partial \xi}\ X_3^k + \dfrac{b}{2}\ \zeta\ P_5(\xi) & 0 & \dfrac{b}{2}\ n_3 \end{bmatrix} \qquad (113)$$

where $P_1(\xi, \eta), \ldots P_5(\xi, \eta)$ are the functions, obtained by the derivation of $\underline{m}$ (104) and $\underline{n}$ (105) with respect to ξ.

For two-dimensional elements, the jacobian coordinate matrix can be expressed by:

$$
\underset{\sim}{J}_c = \begin{bmatrix}
\sum_k \dfrac{\partial N_k}{\partial \xi} X_1^k + \dfrac{1}{2}\zeta P_1(\xi,\eta) & \sum_k \dfrac{\partial N_k}{\partial \eta} X_1^k + \dfrac{1}{2}\zeta P_2(\xi,\eta) & \dfrac{t}{2} n_1 \\[3mm]
\sum_k \dfrac{\partial N_k}{\partial \xi} X_2^k + \dfrac{t}{2}\zeta P_3(\xi,\eta) & \sum_k \dfrac{\partial N_k}{\partial \eta} X_2^k + \dfrac{t}{2}\zeta P_4(\xi,\eta) & \dfrac{t}{2} n_2 \\[3mm]
\sum_k \dfrac{\partial N_k}{\partial \xi} X_3^k + \dfrac{1}{2}\zeta P_5(\xi,\eta) & \sum_k \dfrac{\partial N_k}{\partial \eta} X_3^k + \dfrac{1}{2}\zeta P_6(\xi,\eta) & \dfrac{t}{2} n_3
\end{bmatrix} \tag{114}
$$

where $P_1(\xi,\eta), \ldots P_6(\xi,\eta)$ are the functions obtained from $\underset{\sim}{n}$ (111) by derivation with respect to ξ and to η.

The geometry of the body in the deformed configuration can be interpolated in the same way, by replacing the initial nodal coordinates X^k with the current coordinates $\underset{\sim}{Z}^k$.

The displacements $\underset{\sim}{u}$ are also represented by the interpolation of the nodal displacements, which are the basic unknowns in the finite element formulation.

Since for cables and membranes, uniform deformation is assumed in the cross-wise plane or over the thickness, the displacements are there uniform too, and the kinematical unknowns are the nodal displacements $\underset{\sim}{U}^k$ of the element mid-plane or of the element axis only. These are interpolated by the same shape functions N^k as in (106) and (112).

The deformation can be obtained by expressing each component of the corresponding tensor in the form:

$$
\varepsilon_{ij} = \frac{1}{2}\left(z_{k,i}\, z_{k,j} - \delta_{ij}\right). \tag{115}
$$

For the derivation of the current coordinates with respect to the initial ones, the coordinate jacobian matrix $\underset{\sim}{J}_c$ is used:

$$
\frac{\partial N^k(\xi,\eta,\zeta)}{\partial(x_1,x_2,x_3)} = \underset{\sim}{J}_c^{-1}\,\frac{\partial N^k(\xi,\eta,\zeta)}{\partial(\xi,\eta,\zeta)} \tag{116}
$$

Finally, the stresses can be simply evaluated from their corresponding deformations referred to the intrinsic coordinate system by

$$\hat{\underline{S}}' = \hat{\underline{C}}' \, \hat{\underline{\varepsilon}}' \; . \tag{117}$$

In fact, the stress-strain relationships can be easily defined in the mid-plane or along the axis of each element, but for the assembling of the element matrices these relationships must be transformed in the global reference system:

$$\hat{\underline{S}} = \hat{\underline{C}} \, \hat{\underline{\varepsilon}} \; . \tag{118}$$

This transformation can be carried out by means of the matrix $\underline{T}_\varepsilon$ [28, 31], linking $\hat{\underline{\varepsilon}}'$ and $\hat{\underline{\varepsilon}}$,

$$\hat{\underline{\varepsilon}}' = \underline{T}_\varepsilon \, \hat{\underline{\varepsilon}} \tag{119}$$

and hence

$$\hat{\underline{C}} = \underline{T}_\varepsilon^T \, \hat{\underline{C}}' \, \underline{T}_\varepsilon \; . \tag{120}$$

The terms which appear in $\underline{T}_\varepsilon$ are functions of the direction cosines of the axes of the intrinsic convected frame with respect to the global reference system.

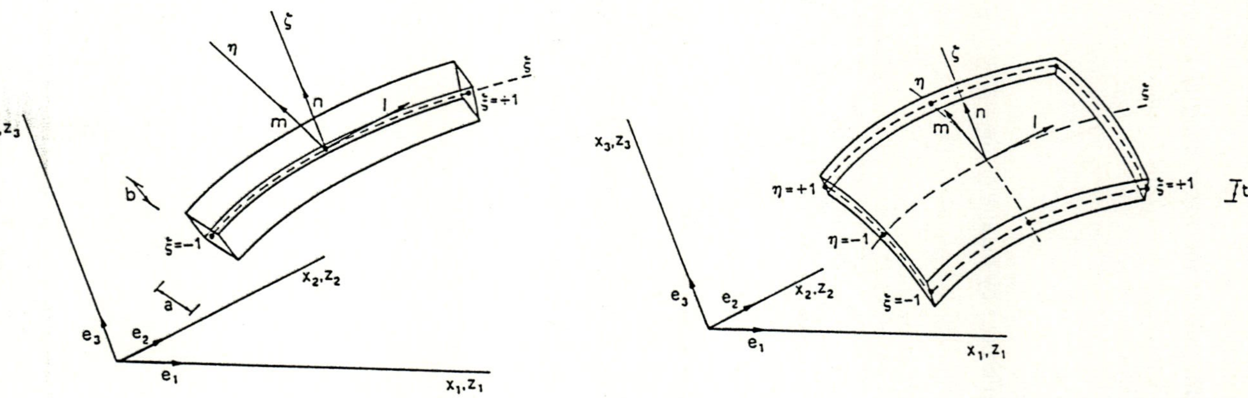

Fig. 1. Intrinsic and global reference systems for cable and membrane elements.

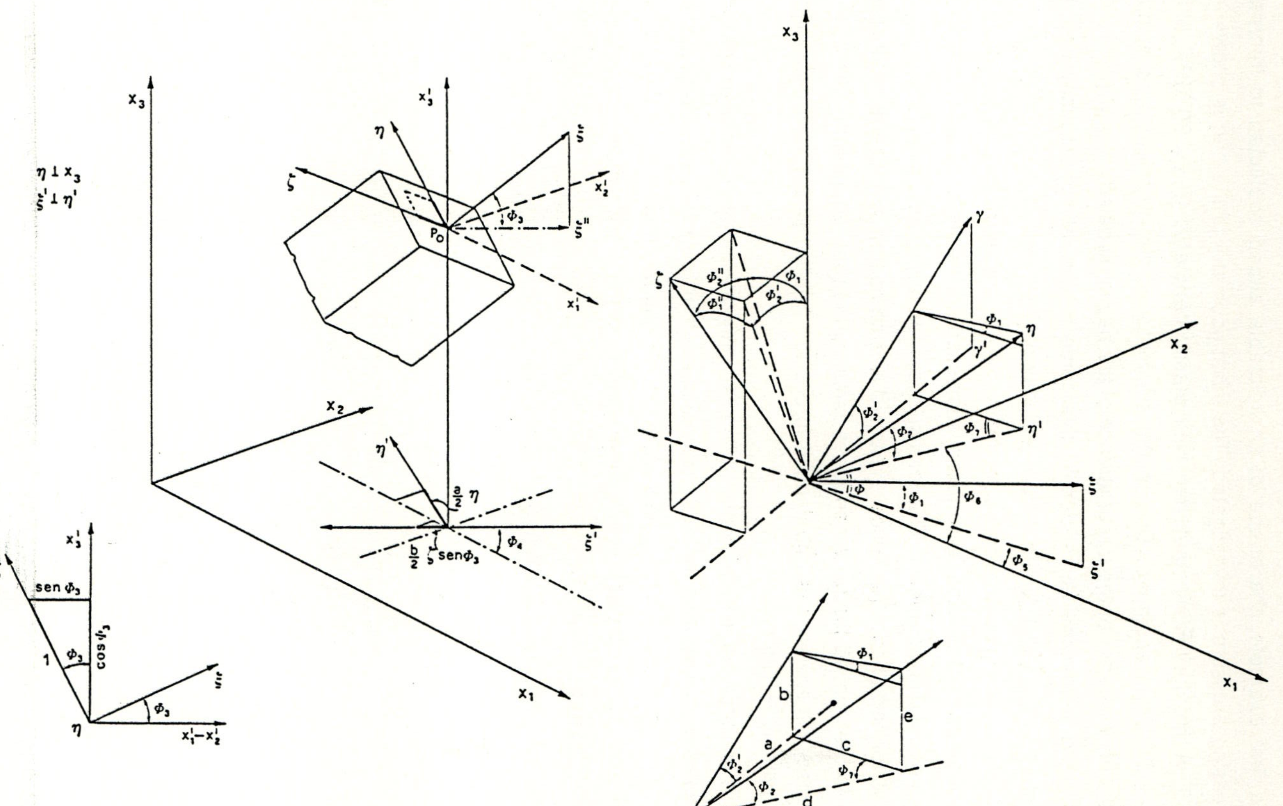

Fig. 2. Definition of the angles between the global and the intrinsic reference systems.

Examples

1. As a first example, the shape of an inflatable airbag subjected to internal increasing pressure is investigated. The initial geometry of the bag is shown in fig. 3. A Mooney-Rivlin material, with elastic constants $C_{10} = 90$ MPa and $C_{01} = 8$ MPa is assumed. Due to the symmetry of the geometry and of the load, only a quarter of the upper surface of the bag is modeled.

   A starting procedure is needed to overcome the initial hypostaticity of the structure. This is due to the lack of stiffness of the membrane model out of its lying plane. The global stiffness matrix in this flat underformed configuration is hence singular. An initial deformation is imposed on the structure by prescribing some nodal displacements. The forces (nodal reactions) necessary to maintain this deformed shape are hence evaluated and compared with the nodal forces due to the applied loads. When these forces are equal, equilibrium has been reached. Otherwise, the unbalanced forces are applied and a new deformed geometry is calculated.

   In this way, the equilibrium configuration of highly deformable structures can be effectively calculated. When the displacements are very large, the computational cost of such a procedure can also be much lower than those of a conventional incremental procedure starting from the undeformed initial shape of the structure.

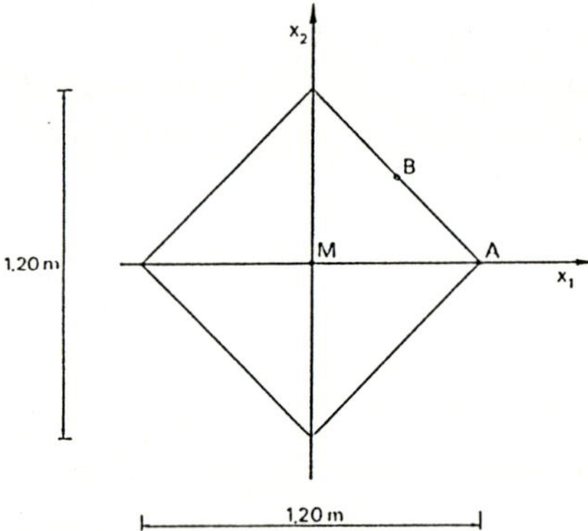

Fig. 3. Inflatable airbag. Undeformed configuration.

The deformed geometries in equilibrium with the internal applied pressures of 0.005 MPa and 0.05 MPa for the examined structure are shown in fig. 4.

The same problem was previously analyzed by Bauer [32] and by Contri and Schrefler [28]. They assumed a linear elastic material model for the bag, with elastic modulus E = 588 MPa and Poisson's ratio ν = 0.4. The initial elastic modulus of the rubber membrane considered here is close to this value, and the results agree well enough with those in [28] and [32], considering the different assumptions.

The present analyses show folds developing in the zones where one of the principal stresses is tensile and the other compressive or zero. The wrinkling hypothesis assumed for the analyses is the same as in [32]. In [28], a no-compression material model is also assumed, so that in the wrinkled zones one principal stress is tensile and the other zero. The resulting wrinkles are closer to those of a physical model. To deal with this no-compression material model, the equilibrium configuration under the applied loads is first determined for the structure supposed to resist both compressions and tensions. The principal directions of the corresponding stress state is then evaluated in each Gauss point. When some component results in compression, this is set to zero but the directions of the principal stresses are maintained. The nodal forces corresponding to this new stress state and their difference with the initial ones are calculated in each element. The contributions from the adjoining elements are added in each node and iterations are performed till the unbalanced forces vanish. This procedure does not update the strains when some stress component is set to zero. This does not allow the application of such an algorithm to hyperelastic materials, whose constitutive tensor is affected by the strains. Hence Bauer's modified wrinkling condition is applied here for rubber material.

2. An initially flat rubber membrane is considered, reinforced with extensible cords. The Mooney-Rivlin constants for rubber are respectively 56 and 16 kPa. The cords are assumed to be made of linear elastic material with elastic modulus equal to 42.000 kPa. The ratio between the initial elastic moduli of the rubber matrix and of the reinforcing cords is equal to 1000. The cords do not resist compression. The same algorithm as above for no-compression materials is used for these cords.

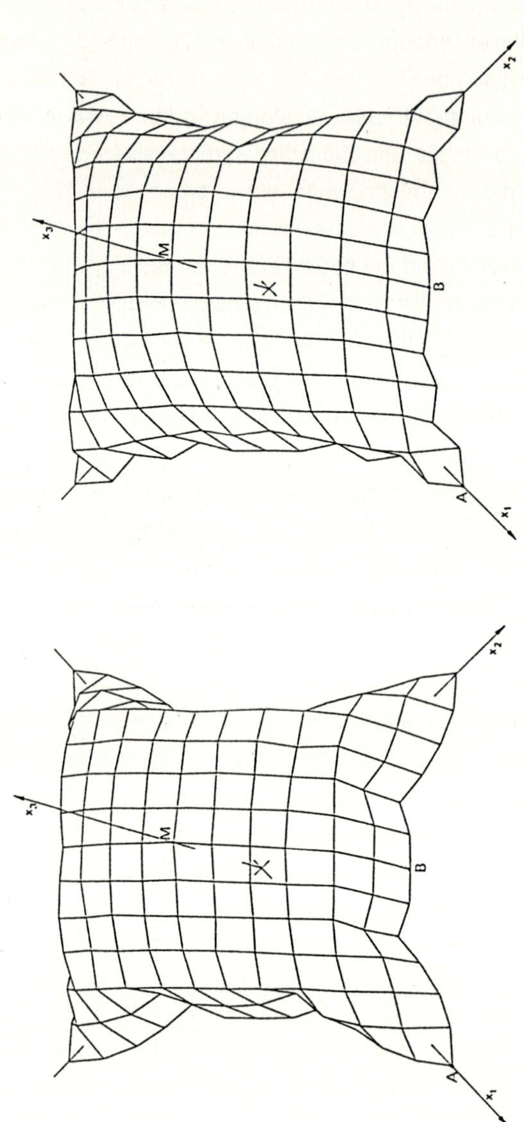

Fig. 4. Inflatable airbag. Deformed configurations for internal pressure of 0.005 MPa (left) and 0.05 MPa (right).

The acting forces are uniform pressures up to 5 kPa, acting orthogonally to the initial lying plane of the membrane. The composite ply is firmly restrained on the corners, and supported by flexible cords along the edges. These cords have the same cross section and elastic modulus of the reinforcement. Since such restraints are rather flexible, great geometry changes can develop, as shown in fig. 5. Furthermore the deformation is increased in some zones by the loss of stiffness of the inactive cords.

Some other examples, dealing with hydrostatic distributions of pressures on weirs and dams and other inflatable structures, can be found in [12] and [13].

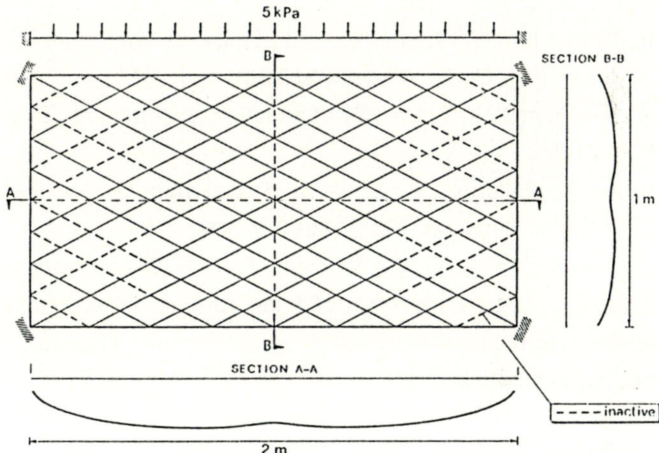

Fig. 5. Cord reinforced rubber membrane fixed at the corners and its deformed shape under pressure. The same scale for displacements and geometry is used.

References

[1]   Watanabe, Y.; Kaldjian, M. J.: Modeling and analysis of bias-ply motorcycle tires. Comp. Struct. 17, 5-6 (1983) 653-658.

[2]   Peeken, H.; Döpper, R; Orschall, B.: A 3-D rubber material model verified in a user-supplied subroutine. Comp. Struct. 26, 112 (1987) 181-189.

[3]   Bolzon, G.; Vitaliani, R.: Derivation of hyperelastic incompressible material constitutive tensor within a total Lagrangian framework. Comp. Struct. (1989) to appear.

[4] Scharnhorst, T.; Pian, T. H. H.: Finite element analysis of rubber-like materials by a mixed model. Int. J. Num. Meth. Engng 12 (1978) 665-676.

[5] Batra, R. C.: Finite plane strain deformations of rubber like materials. Int. J. Num. Meth. Engng 15 (1980) 145-160.

[6] Jankovich, E.; Leblanc, F.; Durand, M.; Bercovier, M.: A finite element method for the analysis of rubber parts, experimental and analytical assessment. Comp. Struct. 14, 5-6 (1981) 385-391.

[7] Häggblad B.; Sundberg, J. A.: Large strain solutions of rubber components. Comp. Struct. 17, 5-6 (1983) 835-843.

[8] Swanson, S. R.; Christensen, L. W.; Ensign, M.: Large deformation finite element calculations for slightly compressive hyperelastic materials. Comp. Struct. 21, 112 (1985) 81-88.

[9] Cesar de Sa, J. M. A.; Owen, D. R. J.: Finite element analysis of reinforced rubber shells. Eng. Comp. 4 (1987) 318-331.

[10] Tang, S. C.: Large strain analysis of an inflating membrane. Comp. Struct. 15, 1 (1982) 71-78.

[11] Tabbador, F.; Stafford, J. R.: Some aspects of rubber composite finite element analysis. Comp. Struct. 21, 1/2 (1985) 327-339.

[12] Bolzon, G.; Schrefler, B. A.; Vitaliani, R.: A 3-D geometrically non linear analysis of inflated cord-reinforced membranes of rubber-like materials. Computational Mechanics '88 (S.N. Atluri and G. Yagawa eds) Springer-Verlag, Berlin (1988) 27ii: 1-4.

[13] Bolzon, G.; Schrefler, B. A.; Vitaliani, R.: Finite element analysis of flexible dams made of rubber-textile composites. Computer Modelling in Ocean Engineering (B. A. Schrefler and O. C. Zienkiewicz eds) Balkema, Rotterdam (1988) 591-596.

[14] Burckhart, B.; Hafner, E.; Hennicke, J.: Flexible Schlauchkonstruktionen - Membranen in Wasser. Preprints of the 2nd Int. Symp. on Widespan Surface Structures, Stuttgart (1977) SFB64: 2.4.1.-2.4.7.

[15] Oden, J. T.: Finite elements of nonlinear continua. Mc Graw-Hill, London (1972).

[16] Eringen, A. C.: Nonlinear theory of continuous media. Mc Graw-Hill, London, (1962).

[17] Green, A. E.; Adkins, J. E.: Large elastic deformations. Clarendon Press, Oxford (1960).

[18] Atluri, S. N.: Alternate stress and conjugate strain measures and mixed variational formulations involving rigid rotations for computational analyses of finitely deformed solids with application to plates and shells - I. Theory. Comp. Struct. 18, 1 (1984) 93-116.

[19] Truesdell, C.; Noll, W.: The nonlinear field theories of mechanics. Encyclopedia of Physics (S Flügge edt.) III/3. Springer-Verlag, Berlin (1965).

[20] Argyris, J.; Symeonidis, S.: Nonlinear finite element analysis of elastic systems under nonconservative loading - natural formulation. Part I. Quasistatic problems. Comp. Meth. Appl. Mech. Engng 26 (1981) 75-123.

[21] Schweizerhof, K.; Ramm, E.: Displacement dependent pressure loads in nonlinear finite element analysis. Comp. Struct. 18 (1984) 1099-1114.

[22] Gent, A. A.: Rubber and rubber elasticity: a review. J. Polym. Sci., Symposium n. 48 (1974) 1-17.

[23] Treloar, L. R. G.: The mechanics of rubber elasticity. J. Polym. Sci., Symposium n. 48 (1974) 107-123.

[24] James, A. G.; Green, A.; Simpson, G. M.: Strain energy functions of rubber - I. Characterization of gum vulcanizates. J. Appl. Polym. Sci. 19 (1975) 2033-2058.

[25] Mooney, M.: A theory of large elastic deformation. J. Appl. Phys. 11 (1940) 582-592.

[26] Rivlin, R. S.: Large elastic deformations of isotropic materials - I. Fundamental concepts. Phil. Trans. Roy. Soc., London, A240 (1948) 459-490.

[27] Bathe, K. J.: Finite element procedures in engineering analysis. Prentice-Hall, Englewood Cliffs (1982).

[28] Contri, P.; Schrefler, B. A.: A geometrically nonlinear finite element analysis of wrinkled membrane surfaces by a no-compression material model. Com. Appl. Num. Meth. 4 (1988) 5-15.

[29] Wood, R.; Zienkiewicz, O.: Geometrically non linear finite element analysis of beams, frames, arches and axisymmetric shells. Comp. Struct. 7 (1977) 725-735.

[30] Schrefler, B. A.; Wood, R. S.; Odorizzi, S.: A total Lagrangian geometrically non linear analysis of combined beam and cable structures. Comp. Struct. 17 (1983) 115-127.

[31] Brebbia, C. A.; Connor, J. J.: Fundamentals of finite element techniques. Butterworths, London (1973).

[32] Bauer, N.: Zur Darstellung von Falten in Membranen mit Hilfe der Methode der finiten Elementen. Stuttgart (1975) SFB64. 33.

Long-Time Deformations and Creepbuckling
of Prestressed Concrete Shells

H. Walter, G. Hofstetter and H.A. Mang
Institute for Strength of Materials, Technical University of Vienna, Austria

Summary

The paper contains a report on the theoretical fundamentals of the extension of a computer program for geometrically and physically nonlinear finite element analysis of reinforced concrete shells to prestressed concrete shells, taking into account long-time deformations of the concrete and of the prestressing steel. The numerical investigation serves the purpose of demonstrating the capacity of this program extension.

1. Introduction

The design of thin concrete shells is a challenge for the architect, because the material and its structural properties offer a great deal of freedom in adapting the shape to the architect's aesthetic and functional conception. The assessment of the structural response of such shells under service conditions and of their load carrying capacity under extreme conditions is far from being a routine work for the civil engineer either.

As a tool for this kind of analyses, the finite element program FESIA (Finite Element Shell Instability Analysis) was developed at the Institute for Strength of Materials at the Technical University of Vienna, Austria. In the course of this development the first step towards realistic modeling of reinforced concrete shells was made by Floegl [1]. He implemented a computer code capable of treating both, the geometrically nonlinear behavior of shells, and the nonlinear material behavior of reinforced concrete. This computer code was extended to enable consideration of time-dependent constitutive properties of concrete and to model the prestress.

Chapter 2 of this paper contains a summary of the employed material models for the prestressing steel, the reinforcing steel and the concrete, taking into account short-time as well as long-time behavior. Chapter 3 summarizes the most important aspects of the finite element model with an emphasis on the modeling of the tendons and the long-time behavior of concrete, respectively. Chapter 4 consists of a numerical study which serves to judge the applicability of the computer program. Two shell structures have been chosen for the analysis, the first being a shallow spherical cap subjected to hydrostatic pressure and the second being the model of a prestressed reactor containment. In both cases the results of the analysis will be compared with experimental data.

2. Material models

2.1 Prestressing steel

The stress-strain relationship of prestressing steel is continuously curved. Therefore, a bilinear idealization (which is frequently used for reinforcing steel) is inadequate. Kang [2] and Van Greunen [3] use a multilinear approximation for the stress-strain relationship. The material properties known from the Austrian or German specifications of a certain type of prestressing steel - Young's modulus $E_z^{(0)}$, the stress at 0.01% plastic strain, $\beta_{0.01}$, the yield stress at 0.2% plastic strain, $\beta_{0.2}$, the tensile strength β_z and the corresponding strain values - are sufficient to set up a trilinear stress-strain curve.

In order to improve the correlation with the actual stress-strain relationship (without requiring additional data besides those known from the specifications) the following analytical representation has been chosen for FESIA:

Up to $\sigma_z = \bar{\sigma}_z = 0,8\,\beta_{0.01}$ (see fig.1) a linear stress-strain relation is specified:

$$\sigma_z = E_z^{(0)}\epsilon_z, \qquad \sigma_z \leq \bar{\sigma}_z = 0.8\,\beta_{0.01}. \tag{1}$$

The value of $\bar{\sigma}_z$ permits an optimal match with the physical properties of the steel. For the nonlinear part of the stress-strain diagram the function

$$\sigma_z(\epsilon_z) = \sum_{i=1}^{8} a_i\,\epsilon_z^{-(i-1)}, \qquad \sigma_z \geq 0.8\,\beta_{0.01}, \tag{2}$$

is taken. This function has to match the known values $(\bar{\epsilon}_z;\ \bar{\sigma}_z)$, $(\bar{\epsilon}_z;\ d\sigma_z/d\epsilon_z = E_z^{(0)})$, $(\bar{\epsilon}_z;\ d^2\sigma_z/d\epsilon_z^2 = 0)$, $(\epsilon_{0.01};\ \beta_{0.01})$, $(\epsilon_{0.2},\ \beta_{0.2})$, $(\delta;\ \beta_z)$, $(\delta;\ d\sigma_z/d\epsilon_z = 0)$, $(\delta;\ d^2\sigma_z/d\epsilon_z^2 = 0)$. Unloading and reloading are modeled as linear functions with the slope $E_z^{(0)}$.

At stress levels higher than 55 % of the yield stress at 0.1% plastic strain, long-time deformations are observed in the prestressing steel. In the context with the interaction of concrete and prestressing steel, the relaxation of prestress at constant strain is particularly important. Experimental investigations resulted in relaxation functions specified in codes of

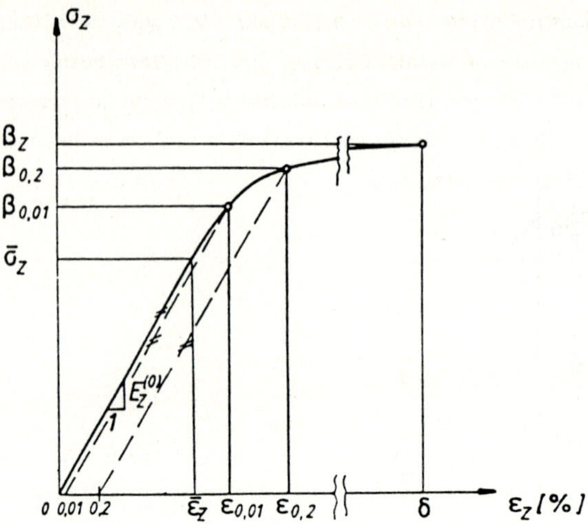

Fig.1. Stress-strain diagram for prestressing steel

practice, either via graphs (e.g. CEB/FIP) or via analytical functions. The implementation in FESIA uses an expression devised by Magura, Sozen and Siess [4] :

$$\frac{\sigma_z}{\sigma_z^I} = 1 - \frac{\log t}{10} \left(\frac{\sigma_z^I}{\beta_{0,1}} - 0,55 \right), \quad \text{if} \quad \frac{\sigma_z^I}{\beta_{0,1}} \geq 0,55 \ . \tag{3}$$

In eq.(3) σ_z^I symbolizes the initial stress of the prestressing steel, σ_z the stress at time instant t (in hours) and $\beta_{0.1}$ the yield stress at 0.1 % plastic strain.

The following scheme is applied to calculate the changes of prestress within the framework of a solution algorithm for incrementally-iterative finite element analyses. These changes are computed at the beginning of the first iteration step, corresponding to a new time step. This is done before the prestress changes due to creep and shrinkage are evaluated. Fig.2 serves as an illustration of the scheme:

— The prestress $\sigma_{z\,(1)} = \sigma_{z\,(1)}^I$, applied at time t_1, decreases to $\bar{\sigma}_{z\,(2)}$ within the time interval $[t_1, t_2]$

— Prestress changes due to other reasons cause a drop of the prestress from $\bar{\sigma}_{z\,(2)}$ to $\sigma_{z\,(2)}$. (If external loads are applied at the time instant t_2, the prestressing force may increase.)

— Eq.(3) cannot be applied directly to calculate the prestress changes in the time interval $[t_{i-1}, t_i]$, $i > 2$, because in eq.(3) the changes of prestress are always deduced from the initial values σ_z^I (at instant t_1). Therefore, the value of a fictitious initial stress $\sigma_{z\,(i)}^I$ has to be computed. This is accomplished by rewriting eq.(3) so as to make σ_z^I a function of $\sigma_{z\,(i-1)}$ and t_{i-1}.

— Using the fictitious initial stress $\sigma^{I}_{z\,(i)}$ instead of σ^{I}_{z} in eq.(3), yields the new stress $\bar{\sigma}_{z\,(i)}$.

— Finally, $\sigma_{z\,(i)}$ is obtained after computation of all other prestress changes.

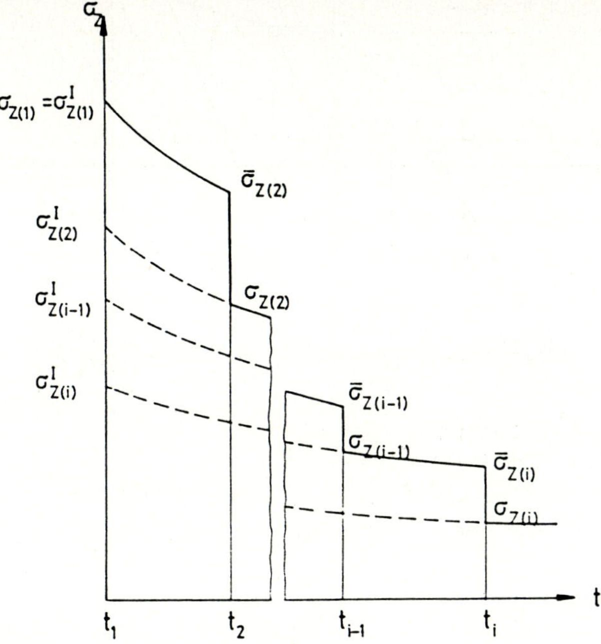

Fig.2. Relaxation of the prestressing steel

2.2 Reinforcing steel

A bilinear stress-strain relation is used to model the material behavior of the reinforcing steel. Elastic unloading and reloading is accounted for, assuming that a Bauschinger-effect exists. The solid line in fig.3 refers to the stress-strain relation for loading. The dashed lines are part of a parallelogram, representing the envelope of possible loading paths. Young's modulus E_s determines the stress-strain relation up to the yield stress β_y and the corresponding strain ϵ_y. Between β_y and the ultimate stress β_u (with the corresponding strain ϵ_u) hardening is assumed to occur. The tangent of the slope of the stress-strain relation is E_H. Unloading is modeled by means of a linear stress-strain relation, where E_s is the modulus of elasticity. If one of the dashed lines with the slope arctg E_H is reached during unloading, additional hardening will begin. Beyond the ultimate stress β_u and $-\beta_u$, respectively, the stresses and the stiffness drop to zero and remain zero.

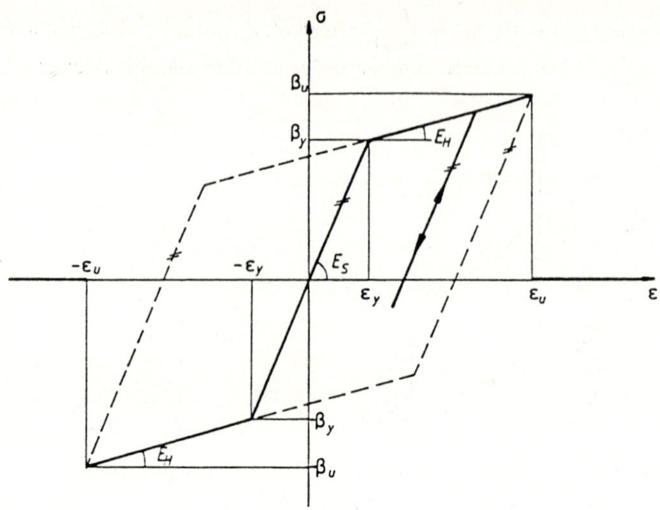

Fig.3. Stress-strain diagram for the reinforcing steel

2.3 Concrete

2.3.1 Short time behavior

The material model for concrete used in the analysis is an extension of the model implemented by Floegl [1]. It is based on experiments conducted by Kupfer [5], restricted to plane
states of stress and to proportional, monotonic loading. The analytical representation of the
stress-strain relation depends on the ratio $\alpha_2 = \sigma_1/\sigma_2$ of the principal stresses and on the
corresponding ultimate stresses σ_{iu} and strains ϵ_{iu}.

In order to be able to distinguish between loading and unloading, the original relation
between total strains and total stresses (secant constitutive law) was replaced by a relation
between strain increments and stress increments (tangent constitutive law). The incremental
stress-strain relation for unloading is characterized as linearly elastic, where the modulus of
elasticity is equal to the initial Young's modulus $E_B^{(0)}$. For the case of cyclic loading, this
would be a very crude approximation; however, for partial unloading due to a stress redistri-
bution caused by time-dependent effects (within the context of long-time behavior of concrete), this specification is sufficiently accurate.

One important aspect of the modification of the original constitutive model is a technique
allowing to distinguish between elasto-plastic loading and elastic unloading. A method analogous to plasticity models with isotropic hardening was employed. Curves which are affine to
the failure envelope are utilized to determine the state of loading (subsequently, these curves
will be called "loading surfaces"). The thick, full line in fig.4 shows the failure envelope in

the principal stress plane (σ_1-σ_2-plane). The thick, full line and the dashed line represent two affine curves. Given a certain ratio $\alpha_2 = \sigma_1/\sigma_2 = $ const. (proportional loading), for each of the principal directions a stress-strain relation (σ_1-ϵ_1 and σ_2-ϵ_2, respectively) is defined (see fig.4). Because of the assumed affinity to the failure envelope, the loading surfaces can be described by one parameter, s_{max}, as

$$s_{max} = \sigma_1/\sigma_{1u} = \sigma_2/\sigma_{2u} = \text{const.} \tag{4}$$

From each step of the analysis to the next step, the parameter s_{max} is stored as a state variable for each integration point. In order to decide whether a strain increment $\Delta\epsilon_{(q)}$ causes loading or unloading, at first, linearly elastic behavior is assumed. (This assumption corresponds to the dashed lines in the stress-strain relations in fig.4. However, fig.4 is only valid for proportional loading.) The resulting stresses define a trial value of $s_{max(q)}$, denoted as $\hat{s}_{max(q)}$, where q is an index of the current calculation step. For the case of $\sigma_{max(q)} \leq \sigma_{max(q-1)}$, the assumption of "linearly elastic behavior" is correct. For the case of $\hat{\sigma}_{max(q)} > \sigma_{max(q-1)}$ — as is shown in fig.4 — elasto-plastic loading occurs and a nonlinear relation has to be applied to determine the stress increment $\Delta\sigma_{(q)}$ and the correct value of $\sigma_{max(q)}$. The algorithm for the computation of stress increments for the case of loading and constitutive models for cracked and crushed concrete are described in detail in [6,7].

2.3.2 Long-time behavior

In order to account for the time-dependent behavior of concrete, it is useful to split up the strains at a certain instant of time t as follows [8]:

$$\epsilon(t) = \epsilon^E(t) + \epsilon^P(t) + \epsilon^C(t) + \epsilon^S(t) + \epsilon^T(t), \tag{5}$$

where $\epsilon^E(t)$ denotes the elastic strains, $\epsilon^P(t)$ the plastic strains, $\epsilon^C(t)$ the creep strains, $\epsilon^S(t)$ the strains caused by shrinkage and $\epsilon^T(t)$ the temperature strains.

Both, the sum $\epsilon^I = \epsilon^E + \epsilon^P$, representing the instantaneous strains, and the creep strains ϵ^C depend on the stress level. Fig.5 shows the stress-dependent strains and their changes with time for the special case of a constant stress σ applied at time instant t_1 and removed at t_2. Usually (for exceptions, see [8]) $\epsilon^0(t) = \epsilon^S(t) + \epsilon^T(t)$ are considered to be stress-independent. Eq.(5) implies mutual independence of the individual contributions to ϵ, which is a simplification of the actual physical situation. This simplification is essential to keep the number of parameters for experimental work within a feasible range. One of the most important observations concerning creep is the validity of the relation

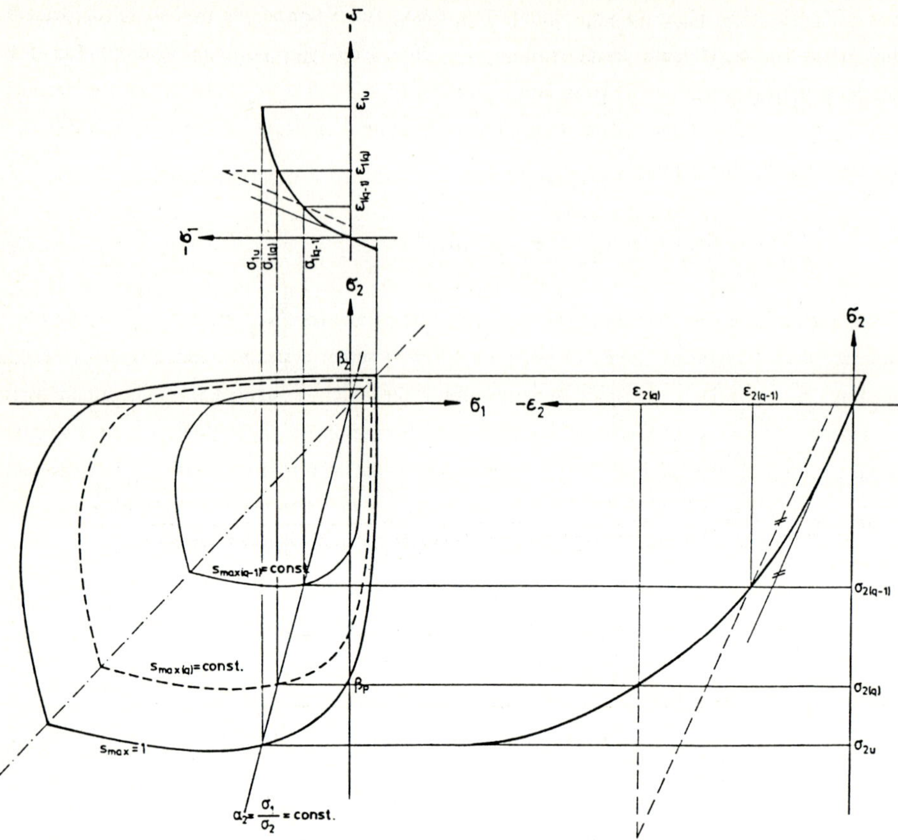

Fig.4. Loading curves, fracture envelope and stress-strain diagrams for a certain ratio of $\alpha_2 = \sigma_1/\sigma_2$

$$\epsilon^\sigma(t) = \epsilon^I(t) + \epsilon^C(t) = \sigma\, J(t,t_0) \tag{6}$$

for a constant stress below 30 % of the uniaxial compression strength f_c. $J(t,t_0)$ is the compliance function; its parameters t and t_0 denote the current instant of time and the time of application of the stress σ. For a stress $\sigma < 0.3\, f_c$, the stress-strain relation of concrete can be considered as being linear; therefore, the compliance function can be split up into

$$J(t,t_0) = 1/E_B(t_0) + C(t,t_0). \tag{7}$$

Accordingly, the stress-dependent strains can be split up into

$$\epsilon^\sigma = \sigma\, J(t,t_0) = \sigma/E_B(t_0) + \sigma\, C(t,t_0) = \epsilon^I(t_0) + \epsilon^C(t,t_0). \tag{8}$$

$E_B(t_0)$ denotes Young's modulus of concrete at time t_0. $C(t,t_0)$ is the creep compliance function; it is equal to the strain at time t caused by a constant unit stress acting in the time interval $[t_0,t]$.

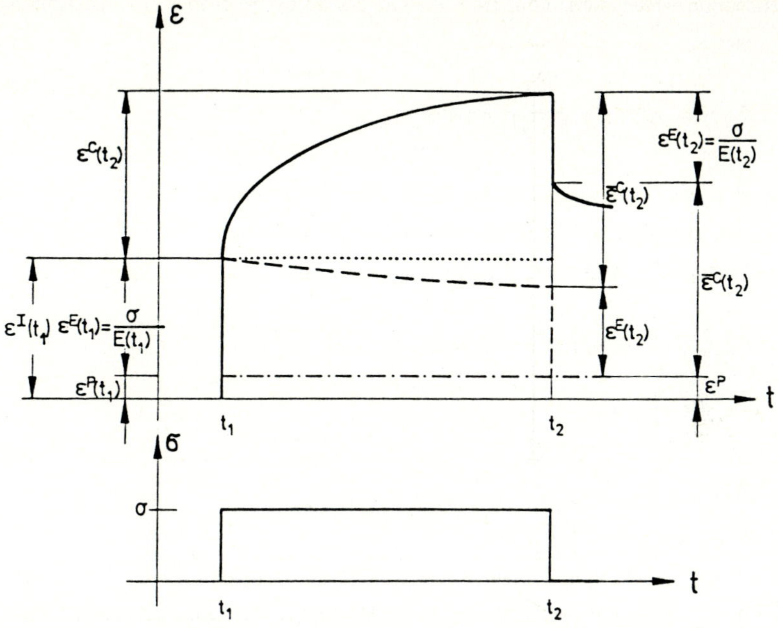

Fig.5. Split-up of the strains

Two generalizations of eq.(8) are introduced in the following. The first refers to stresses varying with time; Boltzmann's superposition law is assumed to be valid:

$$\epsilon(t) = \int_0^t J(t,t_0)\, d\sigma(t_0) + \epsilon^0(t). \tag{9}$$

The second generalization refers to the nonlinear stress-strain relationship for the short-time behavior of concrete:

$$\epsilon^\sigma(t) = \epsilon^E(t_0) + \epsilon^P(\sigma(t_0)) + \epsilon^C(t,t_0) = \sigma/E_B(t_0) + \epsilon^P(\sigma(t_0)) + \sigma\, C(t,t_0). \tag{10}$$

The long-time part of the strains is assumed to be proportional to the stress (see the last term in eq.(10)). The implemented algorithm, however, enables the application of stress-dependent creep compliance functions $C = C(t,t_0,\sigma(t_0))$. This can be particularly useful in the case of stress reversals, where the superposition law is not valid any longer [8,9].

Young's modulus and the strength of concrete are also time-dependent quantities. The increase of the modulus of elasticity due to aging will result in different changes (discontinuities) of strains, if the same change of stress occurs at different instants of time. Fig.5 shows an example of this state of affairs: The change of stress from 0 to σ at time t_1 causes an elastic strain which is different from the one caused by the change of stress from σ to 0 at time t_2, that is,

$$\epsilon^E(t_1) = \sigma/E(t_1) \neq \epsilon^E(t_2) = \sigma/E(t_2). \tag{11}$$

In order to preserve the validity of eq.(10) for each instant of time, maintaining the definition of plastic strains as time-independent strains, the definitions of creep strains and of the creep compliance function must be modified as follows:

$$\sigma = E(t)\,\epsilon^E = E(t)\,(\epsilon^I - \epsilon^P) = E(t)\,(\epsilon - \bar{\epsilon}^C - \epsilon^P), \tag{12}$$

where

$$\bar{\epsilon}^C = \epsilon^C + \sigma(1/E(t_1) - 1/E(t)) \tag{13}$$

and

$$\bar{C}(t,t_0,\sigma(t_0)) = C(t,t_0,\sigma(t_0)) + 1/E(t_0) - 1/E(t). \tag{14}$$

To account for the time-dependence of the concrete strength $f_c = f_c'(t_0)$, the constitutive model for the short-time behavior has to be adapted. The failure envelope is assumed to expand proportionally to the increase of the prism strength of concrete β_P:

$$\frac{\sigma_{iu}(\alpha_2,t_0)}{\sigma_{iu}(\alpha_2,28)} = \frac{\beta_P(t_0)}{\beta_{P28}}, \tag{15}$$

where $\beta_{P28} \equiv \beta_P(t = 28\text{ days})$. A change in size of the failure envelope does not affect the applicability of the incremental stress-strain relation. However, the loading criterion has to be modified: A stress, which is kept constant within the time interval $[t_0,t]$ must not result in a change of the state of loading within that interval. Because of the change of the concrete strength, the value of s_{max} has to be adapted before stress calculations are performed at any new instant of time:

$$\sigma_{max}(t) = \frac{\sigma_i(\alpha_2,t_0)}{\sigma_{iu}(\alpha_2,t)} = \frac{s_{max}(t_0)}{\dfrac{\beta_P(t)}{\beta_P(t_0)}} . \tag{16}$$

The implementation of the program code allows the user to choose either user-defined functions for the creep compliance as well as for the increase of Young's modulus and the compressive strength with time, or respective functions specified by the ACI-Committee 209 [10] or in the CEB/FIP-code [9].

3. Finite element discretization

3.1 Spatial discretization

The spatial discretization of surface structures is accomplished by means of triangular thin-shell finite elements developed by Thomas and Gallagher [11] on the basis of a shear-rigid shell theory. The use of cubical shape functions results in 30 degrees of freedom per element; three Lagrange multipliers (one at each element boundary line) allow satisfaction of the requirement of interelement C_1-continuity.

The middle surface of a shell is described by two parameters α^1 and α^2 as

$$x = x(\alpha^1,\alpha^2) \tag{17.1}$$

The equations

$$x = x(\alpha^1,\alpha^2=\text{const.}), \quad x = x(\alpha^1=\text{const.},\alpha^2) \tag{17.2}$$

describe a mesh of parameter lines. The tangent vectors to the parameter lines are defined as

$$\mathbf{a}_i = \frac{\partial x}{\partial \alpha^i} = x_{,i} \;, \qquad i = 1,2 \; ; \tag{18.1}$$

the direction, normal to the surface, is defined as

$$\mathbf{a}_3 = \frac{\mathbf{a}_1 \times \mathbf{a}_2}{|\mathbf{a}_1 \times \mathbf{a}_2|} \;, \quad |\mathbf{a}_3| = 1, \tag{18.2}$$

(see fig.6), the covariant components of the metric tensor are given as

$$a_{ij} = \mathbf{a}_i.\mathbf{a}_j, \quad i,j = 1,2. \tag{18.3}$$

A shell theory of "small displacements but moderately large rotations", devised by Koiter [12] enables consideration of geometric nonlinearity: The covariant components of the tensor of membrane strains consist of a linear part,

$$e_{ij} = \frac{1}{2}(u_{i;j} + u_{j;i}) - u_3 b_{ij}, \quad i,j = 1,2 \;, \tag{19.1}$$

and a nonlinear contribution

$$\eta_{ij} = \frac{1}{2}\varphi_i\varphi_j + \frac{1}{2}a_{ij}\omega^2, \quad i,j = 1,2 \;. \tag{19.2}$$

In equation (19.1) $u_{i;j}$ denotes the covariant derivative of the component u_i of the displacement vector with respect to α^j. b_{ij} is a covariant component of the curvature tensor. With the

388

help of

$$\frac{da_3}{d\alpha^i} = a_{3,i},$$
(20)

b_{ij} can be written as

$$b_{ij} = a_{3,i} \cdot a_j$$
(21)

In eq.(19.2), φ_1 and φ_2 are the rotations of the normal vector about the covariant base vectors a_1 and a_2; ω denotes the rotation of the tangent plane to the middle surface about the normal vector a_3. For further details of the employed shell theory, Ref.12 should be consulted.

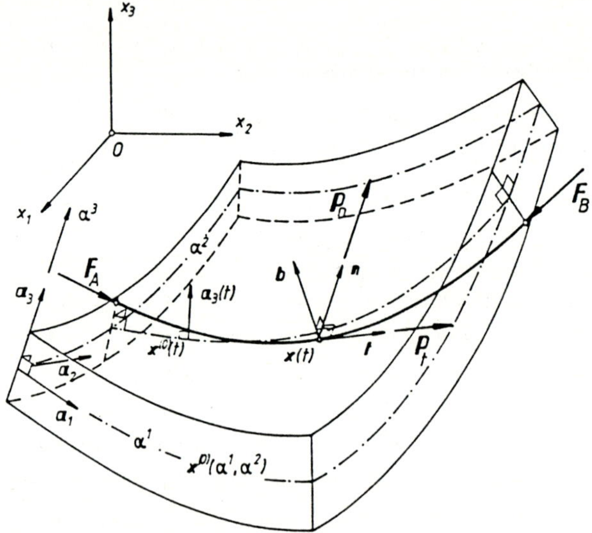

Fig.6. Tendon located in a thin shell

For the purpose of modeling the nonlinear material behavior, the finite elements are subdivided into several thin layers. For each of these layers the assumption of a plane state of stress is made. The thicknesses of the layers and their distances from the middle surface of the shell are chosen such as to provide an optimal scheme for the numerical integration over the thickness of the shell. Seven Gauss-points are used for the numerical integration over the middle surface of a layer of a finite element.

The axes of the tendons are described analytically. A tendon axis will be specified as a surface curve, if the axis is embedded in the middle surface of the shell. It will be specified as a space curve, if the axis of the tendon is eccentric with respect to the middle surface. Fig.6 refers to the general case of a space curve. The analytical formulation of a space curve describing the course of the axis of a tendon is given as

$$x = x(t) = x^{(0)}(t) + \alpha^3(t)a_3(t), \tag{22}$$

where $x^{(0)}(t) = x^{(0)}(\alpha^1(t), \alpha^2(t))$ is the orthogonal projection of the tendon axis onto the middle surface of the shell. Thus, the complete geometric description of the course of a tendon axis must contain an analytical expressions for the surface curve,

$$\alpha^1 = \alpha^1(t), \quad \alpha^2 = \alpha^2(t), \tag{23.1}$$

and the eccentricity

$$\alpha^3 = \alpha^3(t), \tag{23.2}$$

which must be less than one half of the thickness h of the shell.

3.2 Discretization in time

In addition to the spatial discretization, a discretization in time is required for long-time analyses. The time-dependent changes of the material properties of concrete will cause a change of the state of stress and the state of deformation of a structure, even if the external loads are kept constant. Fig.7 shows a typical load history — three load increments are applied at the time instant t_1 and two load increments are added at the time instant $t_4 = t_5$.

Fig.7 also illustrates the corresponding stress history. Between two instants of time at which external loads are applied (and at which, consequently, the stress is discontinuous) the stresses change gradually. The user of the FE-program has to specify instants of time at which the response of the structure shall be determined. Quantities depending on the stresses and on the duration of their action are determined by replacing actual stress histories by piecewise constant stress histories characterized by one stress jump within each time interval. In the time interval $[t_i, t_{i+1}]$, the stress jump is specified at the time instant $t_{i+0.5} = \sqrt{t_i \, t_{i+1}}$ [8]. In fig.7 this idealization is referred to as a dashed line in the stress history. Instants of time at which external loads are applied or removed are treated as time intervals of zero length. The described idealizations allow the use of a global solution algorithm, which originally has been devised for incremental short-time analyses, for time-dependent analyses without the need for major changes: From the viewpoint of the solution algorithm it is insignificant whether a stress- or a strain increment has originated from a change of the external loading or from a change of the material properties with time. Making use of the described

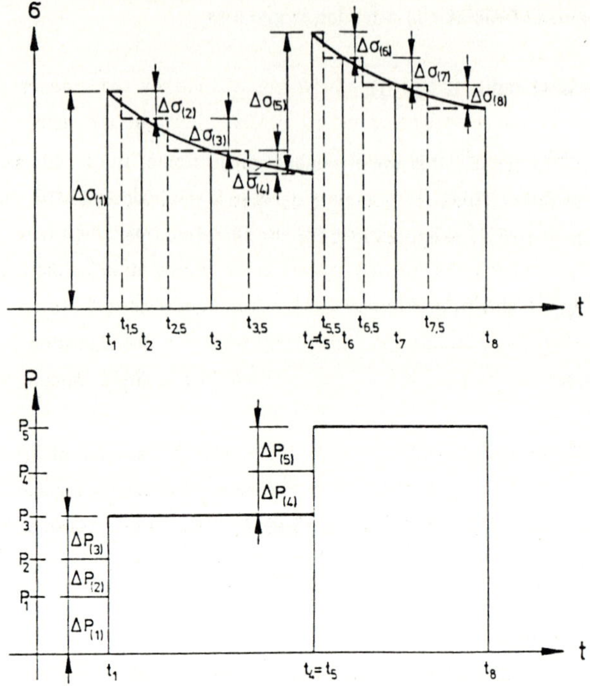

Fig.7. Discretization in time

idealization of continuous stress changes within a time interval $[t_i, t_{i+1}]$ as a stress jump $\Delta\sigma_i$ at $t_{i+0.5}$, yields the following expression for the creep strains:

$$\epsilon^C(t) = \int_0^t C(t,t_0,\sigma(t_0))\ d\sigma(t_0) \approx \sum_{i=1}^n C(t,t_{i-0.5}, (\sigma_{(i)}+\sigma_{(i-1)})/2)\ \Delta\sigma_{(i)}. \tag{24}$$

The computation of $\epsilon^C(t)$ according to eq.(24) requires storage of the total stress history. In order to save disk and memory space, special techniques (e.g. [8],[13]) were developed, which only require storage of a number of "hidden" state variables instead of the total stress history. In the computer program FESIA, however, such techniques are not utilized. The additional flexibility — the algorithm allows the use of arbitrary linear as well as nonlinear compliance functions — and the conceptual clarity justify the additional amount of required computer resources.

3.3 Consideration of prestress

Prestress is taken into account by the forces exerted by the tendons on the respective ducts. Thus, the structure is treated as a free body without the tendons. The mentioned forces result in FEM-specific work-equivalent node forces.

The first step required for the determination of these node forces is computation of the coordinates of all points of intersection of each tendon axis with the boundaries of the elements crossed by this axis. An automatic scheme of computation of these points is a necessary prerequisite for the practical applicability of a pertinent FE-program. The doctoral thesis of Hofstetter [14] contains a detailed description of the algorithm developed for this purpose. Subsequently, the part of a tendon axis within one finite element will be denoted as "tendon segment".

The tendon force does not only depend on the prestress applied at the anchors, on friction between the prestressing tendon and the duct and on overstressing and releasing the tendon. In addition, the tendon force is also influenced by subsequent deformations of the structure. Depending on the method of prestress, the latter category of influence differs considerably.

Fig.8 shows the forces acting on an infinitesimal arc element ds of the axis of a spatially curved tendon, assuming that the tendon force F increases with increasing arc length s.

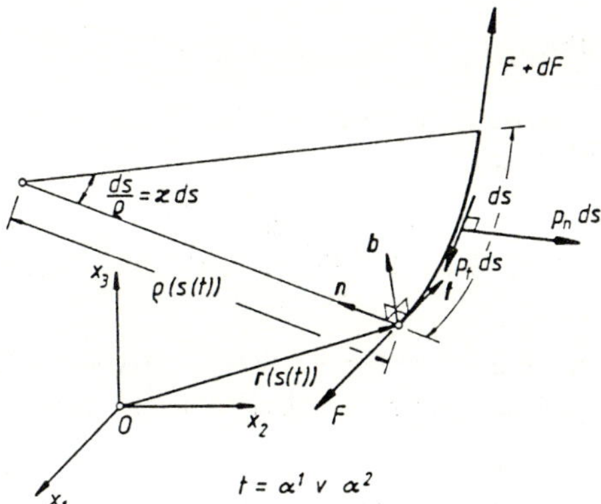

Fig.8. Forces acting on an infinitesimal element of a spatially curved tendon

The equations of equilibrium in the direction of the normal vector **n** and the tangent vector **t** are given as

$$F\varkappa ds - dN = 0, \quad dF - dR = 0. \qquad (25),(26)$$

In the eqs. (25) and (26), F = F(s) denotes the tendon force F, which is a function of s, $\varkappa$ is the curvature of the tendon axis, dN is the reaction force exerted by the duct on the tendon and dR is the friction force acting in the opposite direction of the sliding velocity of the tendon relative to the duct. Expressing dN and dR in terms of p_n and p_t (see fig.8) and introducing Coulomb's friction law, results in

$$dN = p_n ds \qquad dR = p_t ds = \mu dN, \tag{27}$$

where μ is the coefficient of sliding friction. Replacing dN in Eq.(25) by

$$dN = (1/\mu)dR = (1/\mu)dF, \tag{28}$$

yields

$$dF = \mu F \kappa ds. \tag{29}$$

Integration of eq.(29) within the interval $[P(s=s_P), Q(s=s_Q)]$, $s_Q > s_P$, assuming F(s) to be a smooth function in the considered interval, results in

$$F_Q = F_P e^{\mu \int_{s_P}^{s_Q} \kappa ds} = F_P e^{\mu \int_{t_P}^{t_Q} \kappa \dot{s} dt} \tag{30}$$

with $F_P \equiv F(s=s_P)$ and $F_Q \equiv F(s=s_Q)$. The curvature $\kappa = \kappa(t)$ is given as

$$\kappa(t) = |\frac{dt(s)}{ds}| = |\frac{dt(s(t))}{dt} \frac{1}{\dot{s}}|, \qquad \dot{s} = \dot{s}(t) = \frac{ds}{dt} . \tag{31}$$

ds can be derived from eq.(22) as

$$ds = \sqrt{\frac{dx(t)}{dt} \frac{dx(t)}{dt}} \, dt = \sqrt{\frac{dx^{(0)}}{dt} \frac{dx^{(0)}}{dt} + 2\alpha^3 \frac{da_3}{dt} \frac{dx^{(0)}}{dt} + (\frac{d\alpha^3}{dt})^2 + (\alpha^3)^2 \frac{da_3}{dt} \frac{da_3}{dt}} \, dt. \tag{32}$$

For the special case of a panel or a slab, the terms in eq.(32), which are underlined once, must vanish. If the axis of the tendon is embedded in the middle surface of a shell, all underlined terms in eq.(32) will be zero.

In order to account for wobble $\beta^0[°/m]$ of the tendon in the duct in a heuristic manner, the expression for the curvature in eq.(31) is modified as follows:

$$\kappa^*(t) = \kappa(t) + \kappa^0, \quad \kappa^0 = \beta^0 \pi/180 = \text{const.} \tag{33}$$

If the points P and Q coincide with the points of intersection of the tendon axis with the boundaries of a finite element, eq.(30) can be used to compute the tendon force at one point of intersection, provided the tendon force at the other point of intersection is known. The integration in eq.(30) is performed numerically, using a three-point Gauss-Legendre integration scheme. Within an element the tendon force is assumed to vary linearly. This approxi-mation has proved to be sufficiently accurate.

For the case of a tendon, which is prestressed at both ends, the anchor forces $F_A \equiv F(s=s_A)$ and $F_B \equiv F(s=s_B)$ are known. They serve as starting values for the automatic compu-tation of the tendon force at all points of intersection of the tendon axis and the boundaries of the finite elements passed through by the tendon (fig.9). As shown in fig.(9), the approxi-mation of the tendon force within an element as a linear function is assumed to be valid also for the segment a, where the friction force changes its direction.

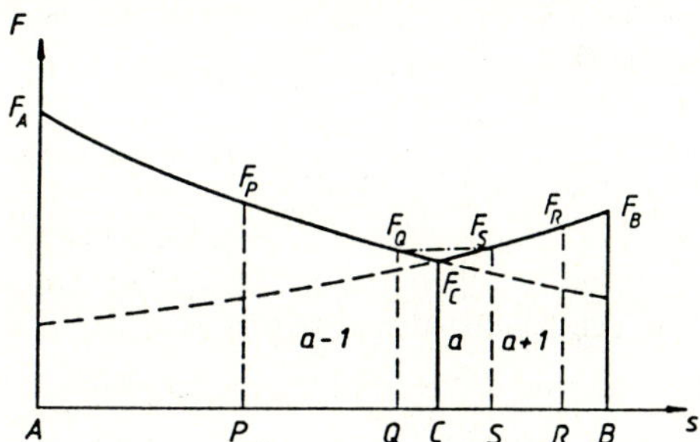

Fig.9. Tendon force as a function of the arc-length for the case of prestressing from both ends of the tendon

The algorithms for determination of prestress losses due to anchor slip and for accounting for stress changes caused by overstressing and releasing are explained in detail in [14].

3.4 Consideration of prestress changes

Prestress changes may have time-independent or time-dependent reasons. Concrete de-formations due to load or temperature changes belong to the first group. Creep and shrinkage of concrete and relaxation of the prestressing steel belong to the second group.

For the case of pretensioning, prestress losses occur already at the transfer of the pres-tress forces from rigid anchor blocks to the concrete. For the case of posttensioning, prestress losses due to friction between the tendon and the duct and to slip of the anchorage occur during the tensioning operation. If this operation is carried out sequentially, additional pres-tress losses will occur. Such losses in tendons which already have been prestressed result from the shortening of the concrete caused by tensioning of yet untensioned tendons. Independent of the method of prestress, load or temperature changes and time-dependent effects result in additional changes of the tendon forces.

Because of the assumption of full bond between the tendon and the duct, the strain ϵ_z at an arbitrary point of a *bonded* tendon is given as

$$\epsilon_z^{(i,j)} = \epsilon_z^{(i,j-1)} + \Delta\epsilon_{BZ}^{(i,j)}. \tag{34}$$

The superscripts i and j denote the i-th load increment and the j-th iteration step of the incremental-iterative solution algorithm. $\Delta\epsilon_{BZ}^{(i,j)}$ represents the change of concrete strain in the direction of the tangent vector $\mathbf{t}$ to the axis of the tendon, resulting from a displacement increment $\Delta u^{(i,j)}$.

By contrast with bonded tendons, the strains of *unbonded* tendons cannot be related directly to the concrete strains. However, the total change of the length of a tendon must be equal to the total change of length of the concrete surrounding the duct. This incremental continuity condition can be formulated mathematically as

$$\int_\ell \Delta\epsilon_{BZ}^{(i,j)} ds = \int_\ell \Delta\epsilon_Z^{(i,j)} ds \tag{35}$$

where ℓ is the length of the unbonded tendon. The left-hand side of eq.(35), referring to the concrete, can be evaluated as soon as the field of displacement increments $\Delta u^{(i,j)}$ of the structure is known. The strain increment of the tendon on the right-hand side of eq.(35) depends on the increment of the tendon force. If friction is taken into account, this increment will be a function of the arc length s. Considering the small magnitude of the coefficient of friction between the tendon and the duct ($\mu \approx 0.06$ for widely used greased, unbonded tendons), changes of friction forces after prestressing can be neglected, resulting in a prestress increment $\Delta F^{(i,j)}$, which is constant over the length of the tendon. Use of eq.(35) to determine $\Delta F^{(i,j)}$, yields:

$$\int_\ell \Delta\epsilon_{BZ}^{(i,j)} ds = \frac{\Delta F^{(i,j)}}{A_z} \int_\ell \frac{1}{E_z(\epsilon_z)} ds, \tag{36}$$

In eq.(36), A_z is the cross-sectional area of the tendon and E_z is Young's modulus of the prestressing steel. Because of the nonlinear stress-strain relationship of the prestressing steel, eq.(36) must be solved iteratively.

Based on the aforementioned assumption of a linear approximation of the tendon force within a finite element, the forces exerted by the tendon on the duct in the direction of the vectors $\mathbf{n}$ and $\mathbf{t}$ are

$$p_n(t) = \kappa(t)F(t), \qquad t = \alpha^1 \vee \alpha^2,$$

and

$$p_t(t) = \left| \frac{F_Q - F_P}{\ell_a} \right| = \text{const.} \tag{37}$$

In eq.(37), F_P and F_Q stand for the tendon forces at the points of intersection of the tendon with the element boundaries; l_a denotes the length of the tendon segment.

4. Numerical study

4.1 Vandepitte shell

For the purpose of testing the extension of the mentioned computer code for consideration of time-dependent constitutive properties, experiments concerning creep buckling of shallow spherical caps cast of plain concrete were recomputed. These experiments were carried out by Vandepitte and Weymeis [15] at the University of Gent, in 1971. Four series of tests referring to buckling under hydrostatic pressure were run. Fig.10 shows an axial section of such a cap. Material properties are given in [15]. The movably supported edge beam made of steel may be prestressed so as to enforce, approximately, a membrane state of stress within the shell membrane.

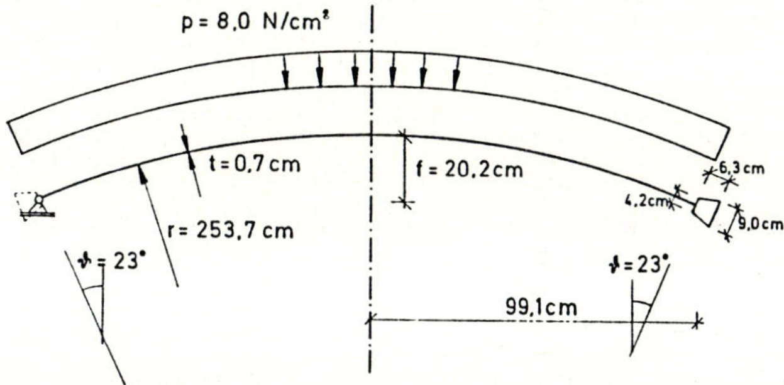

Fig.10. Axial section of a spherical cap

At first, recomputations of results concerning short-time instability of caps without prestressing of the edge beam were conducted in order to determine a suitable finite element discretization. One half of the cap as well as a sector with an aperture angle of 30° were analyzed. The dashed part of fig.11 shows the finite element mesh of this sector. In the figure caption, the abbreviation R.H. stands for environmental relative humidity; t_0 denotes the age of concrete at the onset of shrinkage. In the vicinity of the edge beam a rather fine discretization is necessary to account for the bending moments in this region. The analytically obtained buckling pressure is approximately 15% greater than the experimental result. It is believed that the imperfection-sensitivity of spherical shells is the reason for the discrepancy between experimental and numerical results. The latter are insensitive with respect to the aperture angle of the analyzed shell sector. By contrast with the axisymmetric

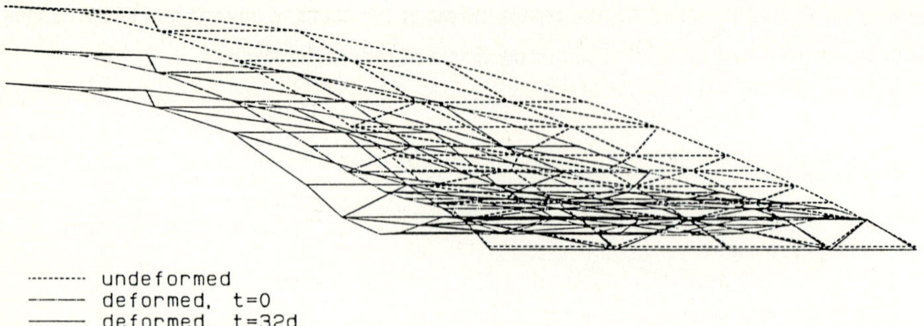

Fig.11. Finite element mesh and deformed configurations (20-fold superelevated) of a sector of the spherical cap; CEB - model code, 40 % R.H., t_0 = 32d, a) ---------- undeformed, b) ——·——·-- deformed, t = 0, c) ———— deformed, t = 32d

buckling modes, the experimental failure modes are unsymmetric.

Since test data concerning the long-time behavior of specimen of the concrete used for casting the model shells (microconcrete) were not available, standardized creep and shrinkage functions (ACI [10] and CEB [9] functions) were taken for the long-time analyses. Therefore, and also because of lack of information about the environmental conditions at the testing site, a series of long-time analyses with different assumptions concerning the environmental parameters was carried out. Fig.12 contains a plot of the displacement of the pole of the spherical cap from the time of applying the load (t = 0) until four weeks afterwards (t = 28d), for different parameter combinations. Two thirds of the tested shells failed before t = 28d. According to the analyses, however, failure would not occur before t = 2 years. Under the assumption of 70% R.H. failure would not occur at all.

In the opinion of the authors, the imperfection sensitivity of the investigated shell is the main reason for overestimating its load carrying capacity by the analyses. Finally, a comment on the sensitivity of the results for the time at which creep buckling of the spherical caps occurs (t_c) is appropriate. An increase of the load by only 10% yields a reduction of t_c by $\approx$ 75%!

4.2 Model of a reactor containment

The analysis of a prestressed reactor secondary containment was considered as another test of the capacity of the developed FE-program. In 1980, McGregor, Rizkalla and Simmonds [16] published the results of experiments conducted on a model of such a containment of the scale 1:14. Fig.13 shows two sections through the experimental setup. The structure consists of a cylindrical part which is prestressed both in the horizontal and vertical direction and a dome with a net of prestressing tendons. A prestressed concrete ring connects the two

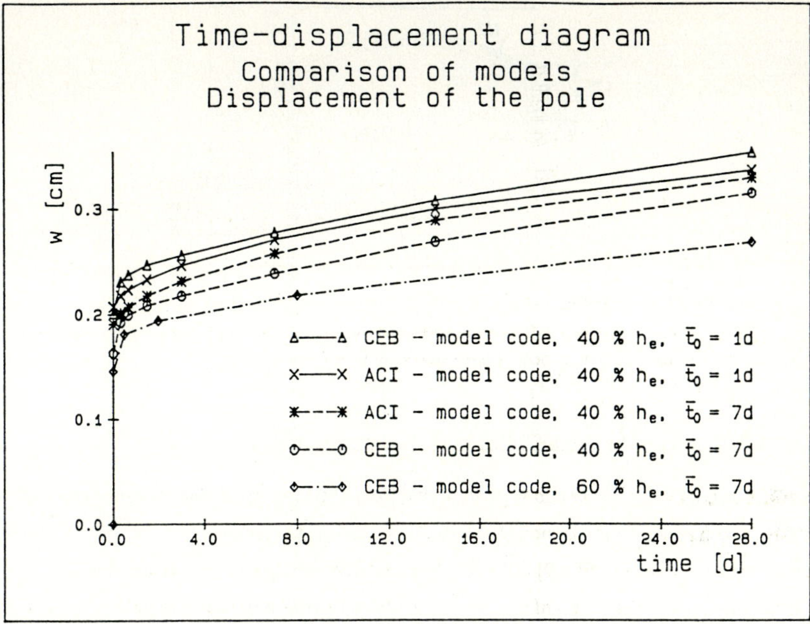

Fig.12. Time-displacement diagrams for the pole of the spherical cap for different assumptions for the analyses

parts. For the purpose of applying the anchorage of the horizontal tendons in the cylindrical part of the containment, the thickness of the wall of the cylinder was increased at four different locations.

Taking the symmetry of the geometry and the loading (dead load and internal hydraulic pressure) into account, two FE-discretizations for an octant of the structure were used for the analyses. Fig.14 shows both discretizations as well as the course of the tendons. (The graphics software utilized to draw this figure connects nodal points and points of intersection by straight lines; in reality, however, the finite elements and the tendon axes are curved.) The material properties used for the analysis and details about the course of the tendons and the reinforcement can be found in [6].

The experimental measurements were not focused for time-dependent effects. However, because of the time span of more than three months between the casting of the dome and the completion of the experiment, time-dependent effects had to be considered in the analysis. As an idealization for the analysis, the entire dead load was assumed to act from the time of removal of the casing of the dome (age of the concrete of the dome t = 7d). Moreover, the tensioning operation, extending over four days, was combined to one single load step (t = 22d). Finally, only the two most significant parts of the load history were taken into account in the analyses. As far as the ignored parts of the load history are concerned, the internal pressure never reached more than 25% of the ultimate load.

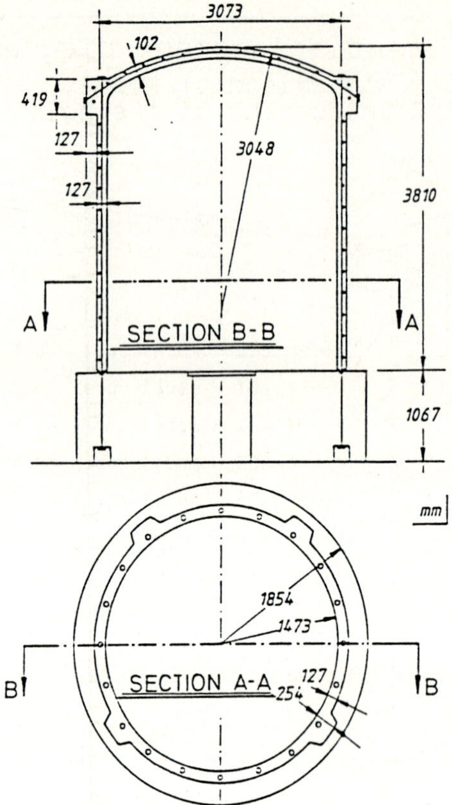

Fig.13. Model of a prestressed reactor secondary containment

The first loading, considered in the analysis, refers to the experiment F, conducted at t = 80d [16]. This increase of the loading was stopped at an internal pressure of 0.55 MPa. For the analysis it was assumed that this pressure was acting for 6h before being removed. The increase of the load up to failure, referring to the experiment G, was conducted at t = 101d; the ultimate pressure was 1.1 MPa.

Fig.15 contains the load-deflection paths for two points of the structure. The complete loading cycle of the experiment F is shown; the illustration of the load-deflection path according to the experiment G is restricted to a load level less than 0.6 MPa. Comparing the curves based on analytical and experimental results, a satisfactory agreement of the two is noted. The unloading branch of the analytical results corresponding to the experiment F agrees only qualitatively with the unloading branch of the exprimental results. The computed failure load is approximately 20% less than the experimental value; the failure mode — tensile failure of a tendon — is predicted correctly.

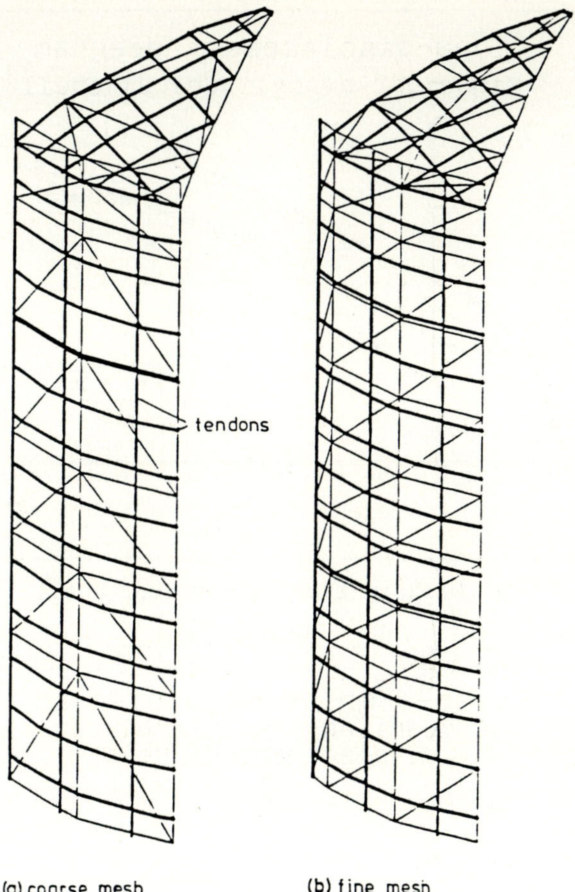

Fig.14. Finite element discretizations and course of the tendons: a) coarse mesh; b) fine mesh

Fig.16 displays crack patterns according to the analysis (within the framework of the concept of "smeared cracks") on the outer surface of the shell (because of the parametric representation of the sector of the dome, its actual shape is distorted to a rectangle). Cracks caused by shrinkage in heavily reinforced regions under dead load (DL) are clearly visible. Some of these cracks close after prestressing (PS).

The state of displacement just after the application of prestress and that immediately before the start of the experiment F, after 58 days of creep, are shown in fig.17. For the purpose of comparison, a plot of the undeformed configuration has been added.

Fig.18 contains a comparison of two time-displacement paths. One of them refers to an analysis without consideration of the experiment F, whereas the other one is characterized by the consideration of this test. Obviously, the slightly different time increments do not influence the response of the structure (the time discretization is sufficiently fine). The expected

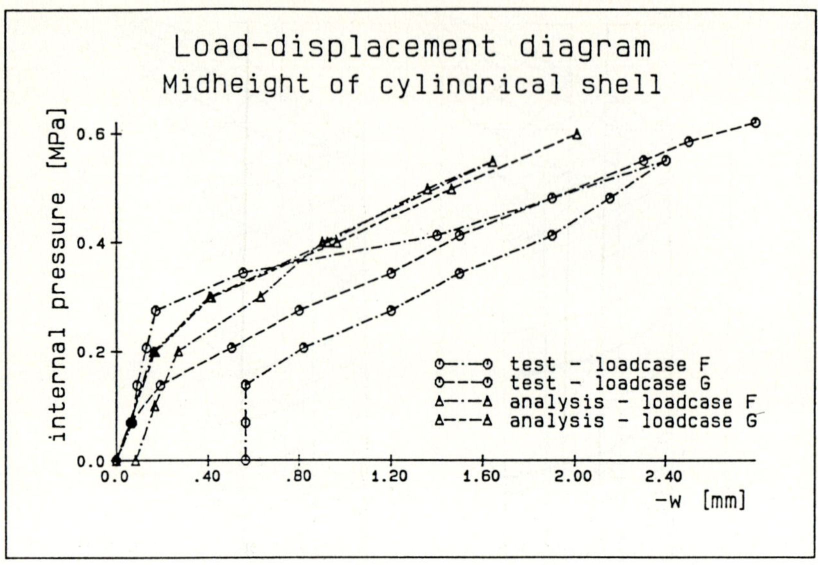

(a) Midheight of cylindrical shell

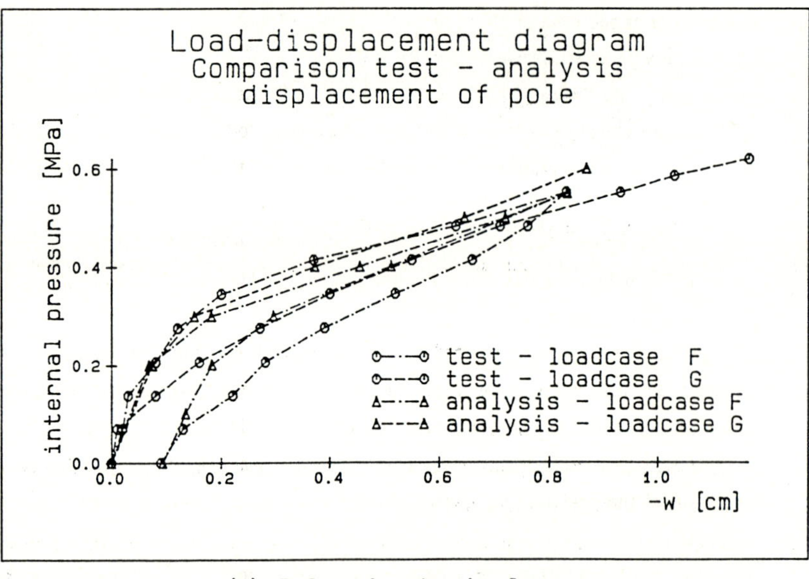

(b) Pole of spherical cap

Fig.15. Load-displacement paths for two points of the structure: a) midheight of the cylindrical part of the shell; b) pole of the spherical cap

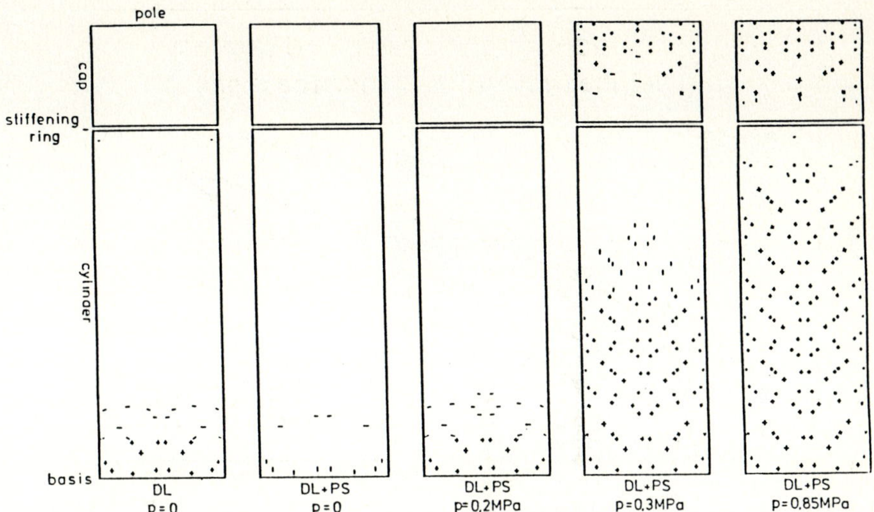

Fig.16. Crack patterns (within the framework of the concept of "smeared cracks") on the outer surface of the shell

weakening of the structure can only be observed in the medium region of the cylindrical part of the shell but not at the pole of the dome.

Additional parameter studies have shown that

— there are only minor differences of the results obtained by the two FE-discretizations; consideration of the aforementioned local increase of wall thickness of the cylindrical shell results in additional stiffness of the structure for the case of the fine mesh

— disregard of aging of concrete practically does not influence the load-displacement paths (obviously the concrete has nearly almost reached its ultimate strength when prestressing is applied)

— changes of the assumptions concerning the relative environmental humidity show a considerable influence on the time-displacement paths; the ultimate load, however, is practically unaffected by this parameter variation. This is due to the fact that the magnitude of the ultimate load is governed by the load-carrying capacity of the reinforcing and the prestressing steel at a time when the concrete is already considerably damaged.

— an increase of the prestressing forces results in larger deformations due to prestress and in smaller deformations under internal pressure at low load levels. However, the relaxation of the prestressing steel prevents an increase of the ultimate load.

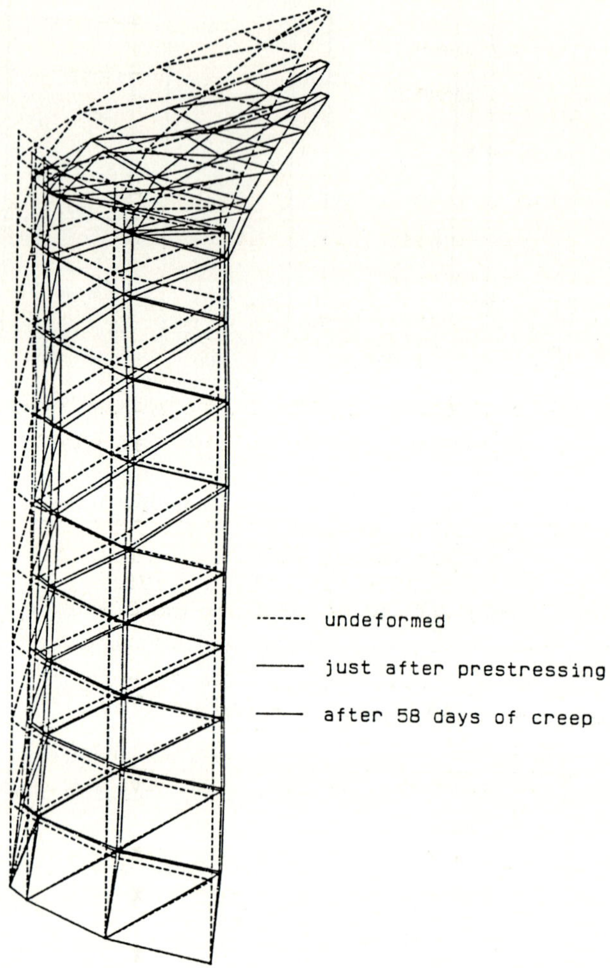

Fig.17. State of displacement: a) just after the application of prestress; b) after 58 days of creep (300-fold superelevated)

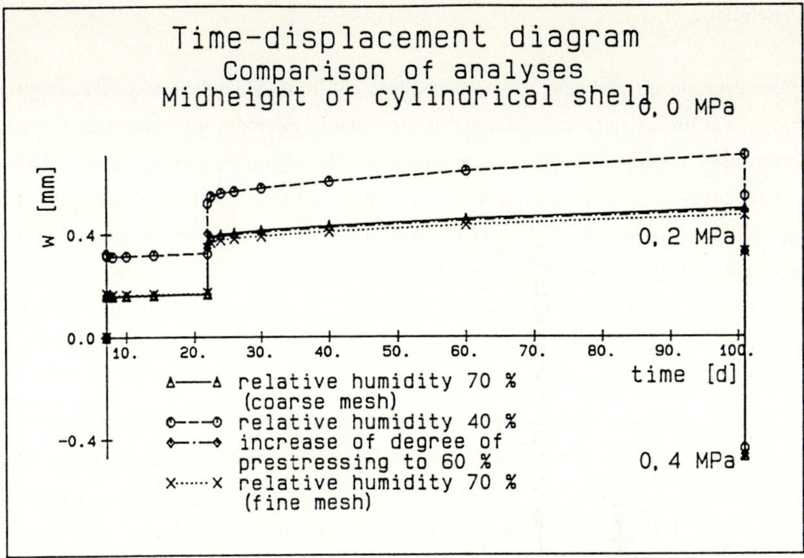

(a) Midheight of cylindrical shell

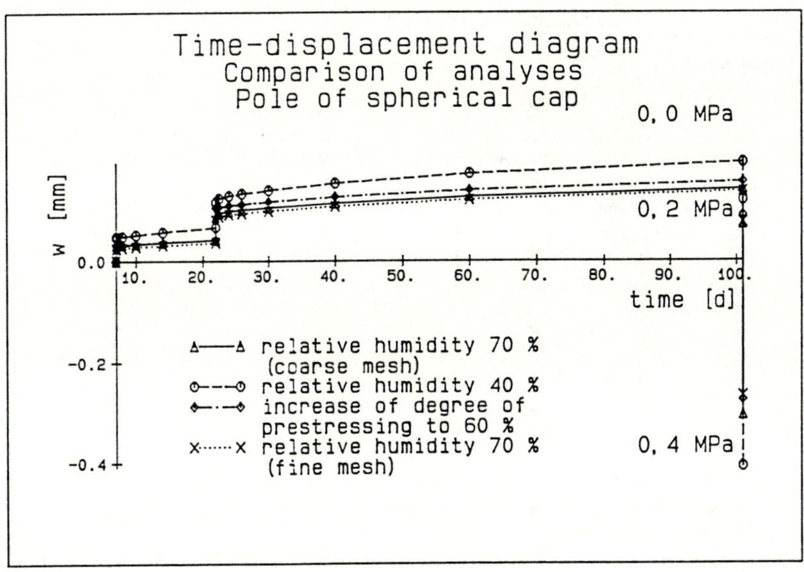

(b) Pole of spherical cap

Fig.18. Time-displacement paths for two points of the structure: a) midheight of the
cylindrical part of the shell; b) pole of the spherical cap

5. Conclusions

Consideration of nonlinear time-independent and time-dependent material properties of concrete, reinforcing steel and prestressing steel allows to model the deformational behavior of prestressed concrete shells more realistically. The algorithm developed for taking the long-time behavior of such structures into account is characterized by a high degree of flexibility and numerical stability. The analytic description of the geometry of the middle surface of the shell and of the tendon axes facilitates the formulation of an efficient algorithm for the automatic determination of the forces exerted by the tendons on the respective ducts and of changes of these forces within the framework of incremental-iterative finite element analysis.

The quality of the numerical results strongly depends on the quality of the material parameters. Because of the relatively large scatter of these parameters (especially of the ones required to model the long-time behavior of concrete), there are obvious limitations of deterministic finite element analyses of prestressed concrete shells considering long-time deformations. In spite of these limitations, such analyses represent a significant step in the direction of a realistic analytical description of the structural behavior of prestressed concrete shells.

List of References

[1]   Floegl, H.: Traglastermittlung dünner Stahlbetonschalen mittels der Methode der Finiten Elemente unter Berücksichtigung wirklichkeitsnahen Werkstoffverhaltens sowie geometrischer Nichtlinearität. Diss., TU-Wien, Wien 1981

[2]   Kang, Y.J.: Nonlinear geometric, material and time dependent analysis of reinforced and prestressed concrete frames. Struct.Eng.Struct.Mech. Rep. No. UC SESM 77-1, University of California at Berkeley, Berkeley 1977

[3]   Van Greunen, J.: Nonlinear geometric, material and time dependent analysis of reinforced and prestressed concrete slabs and panels. Struct.Mat.Res. Rep. No. UC SESM 79-3, University of California at Berkeley, Berkeley 1979

[4]   Magura, D.D.; Sozen, M.A.; Siess, C.P.: A study of stress relaxation in prestressing reinforcement. PCI J. 9, 1964

[5]   Kupfer, H.: Das Verhalten des Betons unter mehrachsiger Kurzzeitbelastung unter besonderer Berücksichtigung der zweiachsigen Beanspruchung. Deutscher Ausschuß für Stahlbeton, Vol. 229, Berlin: Wilhelm Ernst 1973

[6]   Walter, H.: Finite Elemente Berechnungen von Flächentragwerken aus Stahl- und Spannbeton unter Berücksichtigung von Langzeitverformungen und Zustand II. Diss., TU-Wien, Wien 1988

[7]   Walter, H.: Erweiterung eines zweidimensionalen Betonmodells für nichtproportionale Belastung und linear elastische Entlastung. ZAMM, 67 (1987) 386-387

[8]   Bazant, Z.P.; Wittmann, F.H. (eds.): Creep and shrinkage in concrete structures, London: John Wiley 1982

[9]   Chiorino, M.A.; Napoli, P.; Mola, F.; Koprna, M. (eds): Structural effects of time-dependent behaviour of concrete. CEM-Design Manual, Switzerland: Georgi Publ. Comp. 1984

[10]   American Concrete Institute Comm. 209, Subcomm. II: Prediction of creep, shrinkage and temperature effects in concrete structures in designing for effects of creep, shrinkage and temperature. Am.Concr.Inst.Spec.Publ. No. 27

[11]   Thomas, G.R.; Gallagher, R.H.: A triangular thin shell finite element: nonlinear analysis. NASA CR 2483, Washington, D.C. 1975

[12]   Koiter, W.T.: General equations of elastic stability for thin shells. Proc. of Symposium on the Theory of Shells to Honor Lloyd Hamilton Donnell, University of Houston, Houston 1967

[13]   Kabir, A.F.: Nonlinear analysis of reinforced concrete panels, slabs and shells for time-dependent effects. Rep. No. UC SESM 76-6, University of California at Berkeley, Berkeley 1976

[14]   Hofstetter, G.: Physikalisch und geometrisch nichtlineare Traglastanalysen von Spannbetonscheiben, -platten und -schalen mittels der Methode der Finiten Elemente. Diss., TU-Wien, Wien 1987

[15]   Vandepitte, D.; Rathe, J.; Weymeis, G.: Experimental investigation into the buckling and creep buckling of shallow spherical caps subjected to uniform radial pressure. In: Proc. of the IASS World Congress on Shell and Spatial Structures, Madrid 1979

[16]   McGregor, J.G.; Simmonds, S.H.; Rizkalla, S.H.: Test of a prestressed concrete secondary containment structure. Struct.Eng.Rep. No. 85, Dep. of Civ. Eng., University of Alberta, Edmonton 1980

S. N. Atluri, Atlanta, GA; **A. K. Amos,**
Washington, D.C. (Eds.)

Large Space Structures: Dynamics and Control

1988. IX, 363 pp. 166 figs. (Springer Series
in Computational Mechanics)
Hardcover DM 148,– ISBN 3-540-18900-9

From the contents: Continuum Modeling
of Large Lattice Structures: Status and
Projections. – Nonlinearities in the
Dynamics and Control of Space Struc-
tures: Some Issues for Computational
Mechanics. – Modal Cost Analysis for
Simple Continua. – Computational
Issues in Control-Structure Interaction
Analysis. – A Review of Modelling Tech-
niques for the Open and Closed-Loop
Dynamics of Large Space Structures. –
Adaptive Control of Large Space Struc-
tures: Uncertainty Estimation and Robust
Control Calibration. – An Integrated
Approach to the Minimum Weight and
Optimum Control Design of Space Struc-
tures.

Springer-Verlag Berlin
Heidelberg New York London
Paris Tokyo Hong Kong

Computational Mechanics

The purpose of this journal is to report original research of scholarly value and of reasonable permanence in those areas of computational mechanics which involve and enrich the rational application of mechanics, mathematics, and numerical methods in the practice of modern engineering. The scope of the research reported in this journal includes theoretical and computational methods, and their rational application, in:

- solid and structural mechanics, constitutive modeling, inelastic and finite deformation response, structural control;
- fluid mechanics and fluids engineering, compressible and incompressible flows, and aerodynamics;
- fracture mechanics and structural integrity;
- transport phenomena and heat transfer; and
- modern variational methods in mechanics in general.

Papers which emphasize the enhancement of the understanding of the underlying mechanics, physics, and engineering science in these areas through computational methods, in addition to those which deal with novel computational methods themselves, are in their own right particularly encouraged.

Editors-in-Chief: Satya N. Atluri, Atlanta; Genki Yagawa, Tokyo

The journal contains scholarly research papers, invited review papers, brief research notes, and letters of relevance to the practice of computational engineering mechanics.

Abstracted/Indexed in: Applied Mechanics Review, Current Contents

From the contents:

S. Vallippan, K. K. Ang:
η method of numerical integration

S. N. Atluri, R. Reissner:
On the formulation of variational theorems involving volume constraints

H. Le Dret:
Folded plates revisited

Y. H. Kim, S. W. Lee:
A small strain, moderately large deflection finite element beam model with cross-sectional warping effect

J. C. Simo, J. W. Ju:
On continuum damage-elasto-plasticity at finite strains